THE BLOOD GROUP ANTIGEN

FactsBook

Marion E. Reid
Christine Lomas-Francis
Immunohematology Laboratory
New York Blood Center, New York

ACADEMIC PRESS

An imprint of Elsevier

Amsterdam Boston Heidelberg London New York Oxford
Paris San Diego San Francisco Singapore Sydney Tokyo

Academic Press is an imprint of Elsevier
84 Theobald's Road, London WC1X 8RR, UK
Radarweg 29, PO Box 211, 1000 AE Amsterdam, The Netherlands
30 Corporate Drive, Suite 400, Burlington, MA 01803, USA
525 B Street, Suite 1900, San Diego, CA 92101-4495, USA

First edition 2004
Reprinted 2007

Notice
No responsibility is assumed by the publisher for any injury and/or damage to persons
or property as a matter of products liability, negligence or otherwise, or from any use
or operation of any methods, products, instructions or ideas contained in the material
herein. Because of rapid advances in the medical sciences, in particular, independent
verification of diagnoses and drug dosages should be made

British Library Cataloguing in Publication Data
A catalogue record for this book is available from the British Library

Library of Congress Catalog Number: 2003102995

ISBN: 978-0-12-586585-2

For information on all Elsevier publications
visit our website at books.elsevier.com

Transferred to Digital Printing in 2008

Working together to grow
libraries in developing countries

www.elsevier.com | www.bookaid.org | www.sabre.org

ELSEVIER   BOOK AID
            International   Sabre Foundation

THE
BLOOD GROUP
ANTIGEN
FactsBook

Other titles in the FactsBook series include:

Marie-Paule Lefranc and Gérard Lefranc
The Immunoglobulin FactsBook

Hans G. Drexler
The Leukemia-Lymphoma Cell Line FactsBook

Katherine A. Fitzgerald, Luke A.J. O'Neill, Andy J.H. Gearing, Robin E. Callard
The Cytokine FactsBook, 2nd edition

Marie-Paule Lefranc and Gérard Lefranc
The T Cell Receptor FactsBook

Vincent Laudet and Hinrich Gronemeyer
The Nuclear Receptor FactsBook

A. Neil Barclay, Marion H. Brown, S.K. Alex Law, Andrew J. McKnight,
Michael G. Tomlinson, P. Anton van der Merwe
The Leukocyte Antigen FactsBook, 2nd edition

Clare M. Isacke and Michael A. Horton
The Adhesion Molecule FactsBook, 2nd edition

Kris Vaddi, Margaret Keller and Robert Newton
The Chemokine FactsBook

Edward C. Conley
The Ion Channel FactsBook
I: Extracellular Ligand Gated Channels

Edward C. Conley
The Ion Channel FactsBook
II: Intracellular Ligand Gated Channels

Edward C. Conley and William J. Brammar
The Ion Channel FactsBook
IV: Voltage Gated Channels

Steven G.E. Marsh, Peter Parham and Linda Barber
The HLA FactsBook

See http://www.elsevier-international.com/catalogue/titlesearch.cfm for more
details

Contents

Preface   x

Abbreviations   xi

Section I THE INTRODUCTORY CHAPTERS

Chapter 1
Introduction   3

Chapter 2
Organization of the Data   7

Section II THE BLOOD GROUP SYSTEMS AND ANTIGENS

ABO Blood Group System	19
A antigen	23
B antigen	25
A_1 antigen	26
MNS Blood Group System	29
M antigen	34
N antigen	36
S antigen	38
s antigen	40
U antigen	41
He antigen	43
Mi^a antigen	46
M^c antigen	48
Vw antigen	49
Mur antigen	51
M^g antigen	53
Vr antigen	55
M^e antigen	57
Mt^a antigen	58
St^a antigen	59
Ri^a antigen	62
Cl^a antigen	63
Ny^a antigen	64
Hut antigen	65
Hil antigen	67
M^v antigen	69
Far antigen	71
s^D antigen	72

Mit antigen	73
Dantu antigen	75
Hop antigen	76
Nob antigen	78
En^a antigen	80
ENKT antigen	81
'N' antigen	83
Or antigen	84
DANE antigen	85
TSEN antigen	87
MINY antigen	89
MUT antigen	91
SAT antigen	92
ERIK antigen	94
Os^a antigen	96
ENEP antigen	97
ENEH antigen	98
HAG antigen	100
ENAV antigen	101
MARS antigen	102
P Blood Group System	105
P1 antigen	106
Rh Blood Group System	109
D antigen	121
C antigen	131
E antigen	133
c antigen	135

Contents

e antigen	137
f antigen	139
Ce antigen	140
C^W antigen	141
C^X antigen	143
V antigen	144
E^W antigen	146
G antigen	147
Hr_0 antigen	148
Hr antigen	150
hr^S antigen	151
VS antigen	152
C^G antigen	154
CE antigen	154
D^W antigen	155
Rh26 (c-like) antigen	157
cE antigen	158
hr^H antigen	159
Rh29 antigen	160
Go^a antigen	161
hr^B antigen	162
Rh32 antigen	164
Rh33 antigen	166
Hr^B antigen	167
Rh35 antigen	168
Be^a antigen	170
Evans antigen	171
Rh39 antigen	172
Tar antigen	173
Rh41 antigen	175
Rh42 antigen	175
Crawford antigen	176
Nou antigen	177
Riv antigen	178
Sec antigen	179
Dav antigen	180
JAL antigen	181
STEM antigen	182
FPTT antigen	183
MAR antigen	185
BARC antigen	187
JAHK antigen	188
DAK antigen	190
LOCR antigen	191
Lutheran Blood Group System	193
Lu^a antigen	197
Lu^b antigen	199
Lu3 antigen	200
Lu4 antigen	202
Lu5 antigen	203
Lu6 antigen	204
Lu7 antigen	206
Lu8 antigen	207
Lu9 antigen	209
Lu11 antigen	210
Lu12 antigen	212
Lu13 antigen	213
Lu14 antigen	214
Lu16 antigen	216
Lu17 antigen	217
Au^a antigen	218
Au^b antigen	220
Lu20 antigen	221
Lu21 antigen	223
Kell Blood Group System	225
K antigen	231
k antigen	233
Kp^a antigen	234
Kp^b antigen	236
Ku antigen	238
Js^a antigen	239
Js^b antigen	241
Ul^a antigen	242
K11 antigen	244
K12 antigen	245
K13 antigen	246
K14 antigen	248
K16 antigen	249
K17 antigen	249
K18 antigen	251
K19 antigen	252
Km antigen	253
Kp^c antigen	255
K22 antigen	256
K23 antigen	258
K24 antigen	259
VLAN antigen	260
TOU antigen	261
RAZ antigen	263
Lewis Blood Group System	265
Le^a antigen	269
Le^b antigen	270
Le^{ab} antigen	272
Le^{bH} antigen	273
ALe^b antigen	274
BLe^b antigen	276

Duffy Blood Group System	278
Fy^a antigen	281
Fy^b antigen	283
Fy3 antigen	284
Fy4 antigen	286
Fy5 antigen	286
Fy6 antigen	288
Kidd Blood Group System	290
Jk^a antigen	293
Jk^b antigen	294
Jk3 antigen	296
Diego Blood Group System	298
Di^a antigen	302
Di^b antigen	303
Wr^a antigen	305
Wr^b antigen	306
Wd^a antigen	308
Rb^a antigen	310
WARR antigen	311
ELO antigen	312
Wu antigen	314
Bp^a antigen	315
Mo^a antigen	316
Hg^a antigen	318
Vg^a antigen	319
Sw^a antigen	321
BOW antigen	322
NFLD antigen	323
Jn^a antigen	325
KREP antigen	326
Tr^a antigen	328
Fr^a antigen	329
SW1 antigen	330
Yt Blood Group System	332
Yt^a antigen	334
Yt^b antigen	336
Xg Blood Group System	338
Xg^a antigen	341
CD99 antigen	343
Scianna Blood Group System	344
Sc1 antigen	347
Sc2 antigen	348
Sc3 antigen	349
Sc4 antigen	351
Dombrock Blood Group System	353
Do^a antigen	356
Do^b antigen	357
Gy^a antigen	359
Hy antigen	360
Jo^a antigen	362
Colton Blood Group System	364
Co^a antigen	367
Co^b antigen	368
Co3 antigen	369
Landsteiner–Wiener Blood Group System	372
LW^a antigen	375
LW^{ab} antigen	377
LW^b antigen	378
Chido/Rodgers Blood Group System	381
Ch1 antigen	384
Ch2 antigen	385
Ch3 antigen	386
Ch4 antigen	386
Ch5 antigen	387
Ch6 antigen	387
WH antigen	388
Rg1 antigen	389
Rg2 antigen	390
Hh Blood Group System	391
H antigen	395
Xk Blood Group System	398
Kx antigen	401
Gerbich Blood Group System	403
Ge2 antigen	406
Ge3 antigen	408
Ge4 antigen	410
Wb antigen	412
Ls^a antigen	413
An^a antigen	415
Dh^a antigen	417
Cromer Blood Group System	419
Cr^a antigen	422
Tc^a antigen	424

Tc^b antigen 425
Tc^c antigen 427
Dr^a antigen 428
Es^a antigen 430
IFC antigen 431
WES^a antigen 433
WES^b antigen 434
UMC antigen 436
GUTI antigen 437

Knops Blood Group System 439
Kn^a antigen 443
Kn^b antigen 445
McC^a antigen 446
Sl^a antigen 448
Yk^a antigen 449
McC^b antigen 451
Vil antigen 452
Sl3 antigen 453

Indian Blood Group System 455
In^a antigen 458
In^b antigen 459

OK Blood Group System 461
Ok^a antigen 463

RAPH Blood Group System 465
MER2 antigen 465

JMH Blood Group System 467
JMH antigen 469

I Blood Group System 471
I antigen 473

Globoside Blood Group System 475
P antigen 477

GIL Blood Group System 479
GIL antigen 481

Cost Blood Group Collection 483
Cs^a antigen 483
Cs^b antigen 485

Ii Blood Group Collection 486
i antigen 487

Er Blood Group Collection 489
Er^a antigen 489
Er^b antigen 490

Globoside Blood Group
Collection 492
P^k antigen 495
LKE antigen 497

Unnamed Blood Group
Collection 499
Le^c antigen 499
Le^d antigen 500

The 700 Series of Low Incidence
Antigens 501

The 901 Series of High Incidence
Antigens 503
Vel antigen 503
Lan antigen 505
At^a antigen 506
Jr^a antigen 507
Emm antigen 508
AnWj antigen 509
Sd^a antigen 511
Duclos antigen 513
PEL antigen 514
ABTI antigen 515
MAM antigen 516

Section III OTHER USEFUL FACTS

Antigen-based Facts   521
Autoantibody-based and Drug Facts   530
Clinically Useful Information   535
Blood Group System and Protein-based Facts   541
Lectins and Polyagglutination Information   545
Gene-based Facts   548
Useful Definitions   551

Index   552

Preface

We thank Gail Coghlan, Laura Cooling, George Garratty, Carole Green, Peter Howell, Marilyn Moulds, Cyril Levene, Martin Olsson, Ragnhild Øyen, Joyce Poole, Jill Storry, Yoshihiko Tani, Marilyn Telen and Mikoto Uchikawa for providing information. We appreciate the help from Tessa Picknett and Mukesh V.S. Last but not least, we are indebted to Robert Ratner for his skill and patience in preparing the manuscript and figures. Without his help, this project would not have been possible.

In compiling the entries for this text, we were again impressed at the rapid pace with which new information became available. We encourage comments from readers on any errors, omissions and improvements so that they may be incorporated into the next edition. Please write to the authors, at the Immunohematology Laboratory, New York Blood Center, 310 East 67th Street, New York, NY 10021, USA (E-mail: mreid@nybloodcenter.org; clomas-francis@nybloodcenter.org).

Abbreviations

AChE	Acetylcholinesterase
AET	2-Aminoethylisothiouronium bromide
AIDS	Acquired immune deficiency syndrome
AIHA	Autoimmune hemolytic anemia
ART	ADP-ribosyltransferase
BCAM	B-cell adhesion molecule
bp	Base pair
CD	Cluster differentiation
CDA	Congenital dyserythropoetic anemia
CDG	Congenital disorder of glycosylation
cDNA	Complementary DNA
Cer	Ceramide
CHAD	Cold hemagglutinin disease
CHIP	Channel-forming integral protein
Chloroquine	Chloroquine diphosphate
CHO	Carbohydrate moiety
CGD	Chronic granulomatous disease
COOH	Carboxyl terminus
CR1	Complement receptor 1
CSF	Cerebral spinal fluid
CTH	Ceramide trihexoside
DAF	Decay accelerating factor
DAT	Direct antiglobulin test
DNA	Deoxyribosyl nucleic acid
DTT	Dithiothreitol
ER-Golgi	Endoplasmic reticulum-Golgi apparatus
Fuc	L-fucose
Gal	D-galactose
GalNAc	N-acetyl-D-galactosamine
Glc	Glucose
GlcNAc	N-acetyl-D-glucosamine
GP	Glycophorin
GPI	Glycosylphosphatidylinositol-linked
GYP	Glycophorin gene
HDN	Hemolytic disease of the newborn
HLA	Human leukocyte antigen
HEMPAS	Hereditary erythoblastic multinuclearity with positive acidified serum test
HUS	Hemolytic uremic syndrome
IAT	Indirect antiglobulin test
Ig	Immunoglobulin
ISBT	International Society of Blood Transfusion
ITP	Immune thrombocytopenia
kbp	Kilo base pair

LAD	Leukocyte adhesion deficiency
LISS	Low ionic strength solution
MAIEA	Monoclonal antibody immobilization of erythrocyte antigens
2-ME	2-Mercaptoethanol
MDS	Myeloidysplastic syndromes
M_r	Apparent relative molecular mass
NeuAc	N-acetyl neuraminic acid
NH_2	Amino terminus
NT	Not tested
PCH	Paroxysmal cold hemoglobinuria
PEG	Polyethylene glycol
PNH	Paroxysmal nocturnal hemoglobinuria
R	Remainder of carbohydrate chain
RBC	Red blood cell
RT	Room temperature
SCD	Sickle cell disease
SCR	Short consensus repeat
SDS-PAGE	Sodium dodecyl sulfate-polyacrylamide gel electrophoresis
Se	Secretor
SGP	Sialoglycoprotein
SLE	Systemic lupus erythematosus
SMP1	Small membrane protein 1
SSEA	Stage specific embryonic antigen
ter	Terminus of chromosome

THE
INTRODUCTORY
CHAPTERS

1 Introduction

AIMS OF THIS FACTSBOOK

The purpose of this FactsBook is to provide key information relating to the erythrocyte membrane components carrying blood group antigens, the genes encoding them, the molecular basis of the antigens and phenotypes, their characteristics, and the clinical significance of blood group antibodies. Only key references are given to allow the interested reader to obtain more details. The book is designed to be a convenient, easy-to-use reference for those involved in the field of transfusion medicine as well as medical technologists, students, physicians and researchers interested in erythrocyte blood group antigens.

This FactsBook contains information about the blood group antigens that have been numbered by the Committee on Terminology for Red Cell Surface Antigens of the International Society of Blood Transfusion (ISBT)[1-5]. The 29 blood group systems and the antigens within each system are listed by their commonly used name and are arranged in the same order as described by the ISBT Committee. See Table 1 for an overview of the blood group systems. Those antigens not in a blood group system are accommodated in Collections (200 series), in the 700 series of low incidence antigens or in the 901 series of high incidence antigens. (For the latest ISBT terminology, see http://www:iccbba.com/page25.htm.)

Table 1. *Blood group systems with gene name and location*[1-5]

System name	ISBT symbol	ISBT number	Number of antigens	Gene name	Chromosome location	CD number
ABO	ABO	001	4	*ABO*	9q34.1–q34.2	
MNS	MNS	002	40	*GYPA, GYPB, GYPE**	4q28.2–q31.1	CD235
P	P1	003	1	*P1*	22q11.2–qter	
Rh	RH	004	48	*RHD, RHCE*	1p36.13–p34.3	CD240
Lutheran	LU	005	18	*LU*	19q13.2	CD239
Kell	KEL	006	24	*KEL*	7q33	CD238
Lewis	LE	007	3	*FUT3**	19p13.3**	
Duffy	FY	008	6	*FY*	1q22–q23	CD234
Kidd	JK	009	3	*HUT11**	18q11–q12	
Diego	DI	010	21	*SLC4A1 (AE1)**	17q12–q21	CD233
Yt	YT	011	2	*ACHE**	7q22	
Xg	XG	012	2	*XG*	Xp22.32	CD99
Scianna	SC	013	4	*SC*	1p34	
Dombrock	DO	014	5	*DO (ART4)*	12p13.2–p12.1	
Colton	CO	015	3	*AQP1**	7p14	
Landsteiner-Wiener	LW	016	3	*LW*	19p13.3**	CD242
Chido/Rodgers	CH/RG	017	9	*C4A, C4B**	6p21.3	
Hh	H	018	1	*FUT1**	19q13.3**	CD173
Kx	XK	019	1	*XK*	Xp21.1	
Gerbich	GE	020	7	*GYPC*	2q14–q21	CD236
Cromer	CROM	021	11	*DAF**	1q32	CD55
Knops	KN	022	5	*CR1* (CD35)*	1q32	CD35
Indian	IN	023	2	*CD44**	11p13	CD44

System name	ISBT symbol	ISBT number	Number of antigens	Gene name	Chromosome location	CD number
OK	OK	024	1	*OK*	19p13.3**	CD147
RAPH	MER2	025	1	*MER2*	11p15.5	
JMH	JMH	026	1	*JMH (CD108)*	15q22.3–q23	CD108
I	I	027	1	*IGnT (GCNT2)*	6p24	
Globoside	GLOB	028	1	*βGalT3*	3q25	
GIL	GIL	029	1	*AQP3*	9p13	

* The Gene Mapping name. If genetic information is obtained by blood group typing, the gene name used should be the italic form of the blood group system ISBT symbol. For example, *ACHE* would be written *YT1* or *YT2*.
** Not at same locus.

SELECTION OF ENTRIES

Blood group antigens are surface markers on the outside of the red blood cell (RBC) membrane. They are proteins and carbohydrates attached to lipid or protein. A model for the types of membrane components carrying blood group antigens is shown in Fig. 1. A blood group antigen is defined serologically by an antibody and

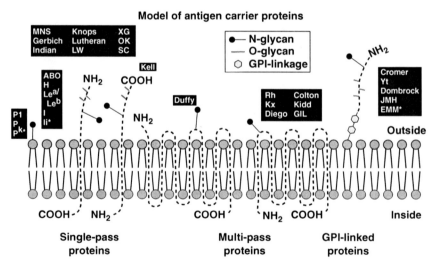

Figure 1 *Model of RBC membrane components that carry blood group antigens.*
* = *Blood group collections or high incidence antigens. Not shown are the Ch/Rg and RAPH blood group systems. Ch/Rg antigens are carried on C4d which is absorbed onto RBC membrane proteins and carbohydrates. The type of membrane component carrying RAPH blood group antigens is unknown.*

in order to be assigned a number by the ISBT Committee the antigen must be shown to be inherited. Antigens associated with forms of polyagglutination have not been numbered by the ISBT, however, in Section III we have included a table summarizing the characteristics of T, Tn, Tk and Cad.

TERMINOLOGY

The nomenclature used for erythrocyte blood group antigens is inconsistent. While several antigens were named after the proband whose red cells carried the antigen or who made the first known antibody, others were assigned an alphabetical or a numerical notation. Even within the same blood group system, antigens have been named using different schemes and this has resulted in a cumbersome terminology for describing phenotypes. The ISBT Committee established a system of upper case letters and numbers to represent blood group systems and blood group antigens in a format that will allow both infinite expansion and computer-based storage. These symbols and numbers are designed for use in computer databases (no lower case letters) and are short (for column headings). A comprehensive review of terminology and its recommended usage can be found in Garratty et al.[6]

Throughout this book, the systems and antigens are named by the traditional name but the ISBT symbol, the ISBT number and other names that have been used in the literature are also given. In this edition, we have included a brief description of how the blood group systems and antigens were named.

The following are examples of how to write antigens, antibodies, phenotypes and genotypes.

List of antigens: M, N, P1, K, Kp^b, K11, Fy^a, Fy^b, Lu3

List of antibodies: Anti-M, anti-K, anti-Fy^a, anti-Jk3 or Anti-M, -K, -Fy^a, -Jk3 or Antibodies directed against M, K, Fy^a and Jk3 antigens

Phenotype: $D + C - E - c + e +$; $M + N - S - s +$, Vw +; K+, k−, K11− (or K:−1,2,−11); Fy(a + b+); Jk(a + b−)

	Traditional	ISBT	International System for Gene Nomenclature (ISGN)
Antigen	Fy^a	FY1, 008001* or 8.1	Fy^a
Phenotype	Fy(a + b−)	FY:1,−2	Fy(a + b−)
Gene	Fy^a	FY1	FY*A
	FY	FY0	FY
Genotype	$Fy^a Fy^a$	FY1/1	FY*A/FY*B
	$Fy^a Fy$	FY1/0	FY*A/FY

* Throughout this FactsBook we have separated the three digits referring to the system from the three digits referring to the antigen with a period (e.g. 008.001) in order to make reading easier

References

[1] Daniels, G.L. et al. (1995) Vox Sang. 69, 265–279.
[2] Daniels, G.L. et al. (1996) Vox Sang. 71, 246–248.
[3] Daniels, G.L. et al. (1999) Vox Sang. 77, 52–57.
[4] Daniels, G.L. et al. (2001) Vox Sang. 80, 193–197.
[5] Daniels, G.L. et al. (2003) Vox Sang. 84, 244–247.
[6] Garratty, G. et al. (2000) Transfusion 40, 477–489.

2 Organization of the Data

Section II of this FactsBook is organized into four main parts: (i) the blood group systems; (ii) blood group collections; (iii) the 700 series of low incidence antigens; and (iv) the 901 series of high incidence antigens. Within each system are facts for individual antigens, listed in ISBT numerical order. The format for the facts sheets displaying the data about the systems and about the antigens is explained below. Section III consists of facts in lists and tables where the information encompasses more than one antigen or blood group system. Other related information is also included.

BLOOD GROUP SYSTEMS

Information, under the following headings, relating to each blood group system includes facts that apply to the system in general and to the membrane component on which the blood group antigens are carried. The total number of antigens in the system is indicated. Table 1 summarizes the blood group systems and antigens currently recognized by the ISBT Committee. If a number assigned to an antigen becomes inappropriate, the number becomes obsolete and is not reused (...).

Terminology

The commonly used name for the blood group system is used at the top of each page. The ISBT symbol, ISBT number[1-5] and other names that have been associated with the system will be given. In this edition we have included a brief description of how the system was named. We have also given CD numbers, which were assigned to certain blood group systems during the 7th Human Leucocyte Differentiation Antigens (HDLA) Workshop[6].

Expression

This section relates to the component on which the blood group system is carried. Several blood group antigens occur naturally in soluble form in body fluids (e.g. saliva, urine, plasma). If a soluble form of the carbohydrate antigen or carrier protein is available, it will be indicated in this section. Soluble forms of an antigen can be used for inhibition tests to confirm or eliminate antibody activity. If the soluble form of the antigen is to be used for inhibition studies, the substance must be obtained from a person who inherited the antigen of interest. An ideal negative dilution control for this test would be to obtain the particular fluid from a person who did not inherit the polymorphism of interest[7]. A soluble form of many antigens can be produced through recombinant technology.

Some of the components carrying blood group antigens have been detected on other blood cells and tissues by use of various methods including testing with polyclonal antibodies and monoclonal antibodies or Northern blot analysis. However, it cannot be assumed that detection of the carrier molecule equates with the expression of the red cell antigen(s).

Gene

The chromosomal location for the genes encoding proteins associated with blood group systems is taken from original papers and reviews of the chromosome

Table 2.1. *Blood group antigens assigned to each system*

Antigen number

System	001	002	003	004	005	006	007	008	009	010	011	012	013	014	015	016	017	018
001 ABO	A	B	A,B	A1	–													
002 MNS	M	N	S	s	U	He	Mi^a	M^c	Vw	Mur	M^g	Vr	M^e	Mt^a	St^a	Ri^a	Cl^a	Ny^a
003 P1	P1	–																
004 RH	D	C	E	c	e	f	Ce	C^w	C^x	V	E^w	G	–	–	–	–	Hr_0	Hr
005 LU	Lu^a	Lu^b	Lu3	Lu4	Lu5	Lu6	Lu7	Lu8	Lu9	–	Lu11	Lu12	Lu13	Lu14	–	Lu16	Lu17	Au^a
006 KEL	K	k	Kp^a	Kp^b	Ku	Js^a	Js^b	–	–	Ul^a	K11	K12	K13	K14	–	K16	K17	K18
007 LE	Le^a	Le^b	Le^{ab}	Le^{bh}	ALe^a	BLe^b												
008 FY	Fy^a	Fy^b	Fy3	Fy4	Fy5	Fy6												
009 JK	Jk^a	Jk^b	Jk3															
010 DI	Di^a	Di^b	Wr^a	Wr^b	Wd^a	Rb^a	WARR	Elo	Wu	Bp^a	Mo^a	Hg^a	Vg^a	Sw^a	BOW	NFLD	Jn^a	KREP
011 YT	Yt^a	Yt^b																
012 XG	Xg^a	CD99																
013 SC	Sc1	Sc2	Sc3	Rd														
014 DO	Do^a	Do^b	Gy^a	Hy	Jo^a													
015 CO	Co^a	Co^b	Co3															
016 LW	–			–	LW^a	LW^{ab}	LW^b											
017 CH/RG	Ch1	Ch2	Ch3	Ch4	Ch5	Ch6	WH				Rg1	Rg2						
018 H	H																	
019 XK	Kx																	
020 GE	–	Ge2	Ge3	Ge4	Wb	Ls^a	An^a	Dh^a										
021 CROM	Cr^a	Tc^a	Tc^b	Tc^c	Dr^a	Es^a	IFC	WES^a	WES^b	UMC	GUTI							
022 KN	Kn^a	Kn^b	McC^a	Sl^a	Yk^a	McC^b	Vil	Sl3	RAZ									
023 IN	In^a	In^b																
024 OK	Ok^a																	
025 RAPH	MER2																	
026 JMH	JMH																	
027 I	I																	
028 GLOB	P																	
029 GIL	GIL																	

System	019	020	021	022	023	024	025	026	027	028	029	030	031	032	033	034	035	036
002 MNS	Hut	Hil	M^v	Far	s^D	Mit	Dantu	Hop	Nob	En^a	ENKT	'N'	Or	DANE	TSEN	MINY	MUT	SAT
004 RH	hr^S	VS	C^G	CE	D^W	–	–	c-like	cE	hr^H	Rh29	Go^a	hr^B	Rh32	Rh33	Hr^B	Rh35	Be^a
005 LU	Au^b	Lu20	Lu21															
006 KEL	K19	Km	Kp^c	K22	K23	K24	VLAN	TOU	RAZ									
010 DI	(Tr^a)	Fr^a	SW1															

System	037	038	039	040	041	042	043	044	045	046	047	048	049	050	051	052	053	054	055
002 MNS	ERIK	Os^a	ENEP	ENEH	HAG	ENAV	MARS												
004 RH	Evans	–	Rh39	Tar	Rh41	Rh42	Crawford	Nou	Riv	Sec	Dav	JAL	STEM	FPTT	MAR	BARC	JAHK	DAK	LOCR

assignments[8] and how blood groups were cloned[9]. The chromosome number, arm [short p (upper on diagrams); long q (lower on diagrams)] and band number are given. The name of the gene used is that recommended by the ISBT with alternative names in parenthesis. If the presence of a blood group antigen is detected by serological means, the gene is named by the corresponding ISBT system symbol followed by the antigen number, in italics to indicate the specific allele, for example, FY1. The organization of the gene in terms of number of exons, kilobase pairs of gDNA, and a map is provided. The product name (and alternative names) is given.

Database accession numbers

Key GenBank accession numbers are given. Where polymorphic or variant proteins have been recorded, more than one accession number may be given. Access was obtained through the following worldwide web address: http://www.ncbi.nlm.nih.gov. This gives access to GenBank, enter accession number provided. The Blood Group Antigen Gene Mutation Database (http://www.bioc.aecom.ye.edu.bgmut/ index.htm) is a useful website, which gives information about genes and alleles relevant to blood groups and hyperlinks to other websites.

Amino acid sequence

The amino acid sequences are shown in the single or three letter code (Table 2). The predicted transmembrane sequence for single pass membrane proteins is underlined. In the literature, the numbering of amino acids and nucleotides is

Table 2. *Three-letter and single-letter amino acid codes[10]*

Amino acid	Three-letter code	Single-letter code	Properties	Molecular weight
Alanine	Ala	A	Nonpolar	89
Arginine	Arg	R	Polar, positively charged	174
Asparagine	Asn	N	Polar, uncharged	132
Aspartic acid	Asp	D	Polar, negatively charged	133
Cysteine	Cys	C	Polar, uncharged	121
Glutamine	Gln	Q	Polar, uncharged	146
Glutamic acid	Glu	E	Polar, negatively charged	147
Glycine	Gly	G	Polar, uncharged	75
Histidine	His	H	Polar, positively charged	155
Isoleucine	Ile	I	Nonpolar	131
Leucine	Leu	L	Nonpolar	131
Lysine	Lys	K	Polar, positively charged	146
Methionine	Met	M	Nonpolar	149
Phenylalanine	Phe	F	Nonpolar	165
Proline	Pro	P	Nonpolar	115
Serine	Ser	S	Polar, uncharged	105
Threonine	Thr	T	Polar, uncharged	119
Tryptophan	Trp	W	Nonpolar	204
Tyrosine	Tyr	Y	Polar, uncharged	181
Valine	Val	V	Nonpolar	117

inconsistent and potentially confusing. Therefore, we have included a sentence under each sequence stating which amino acid has been counted as number 1 and the number of amino acids that are believed to be cleaved once the protein is inserted into the membrane.

Carrier molecule

Molecules carrying blood group antigens are either carbohydrate chains, single-pass membrane proteins, multi-pass membrane proteins or glycosylphoshatidylinositol (GPI)-linked proteins. Single pass proteins can be oriented with the N-terminus outside (type I) or inside (type II) the membrane.

Carbohydrate antigens are depicted by the critical immunodominant sugars and linkages. Proteins carrying blood group antigens are depicted by a stick diagram within a gray band that represents the RBC lipid bilayer. The inside (cytoplasmic surface) of the membrane is always to the bottom of the page and the outside (exofacial surface) is to the top of the page. The predicted topology of the protein in the membrane is shown in the models. The predicted orientation of the amino-terminus and the carboxyl-terminus is indicated as are the total number of amino acids. O-glycans are depicted by an open circle (○) and N-glycans by a closed circle on a line (lollipop). GPI linkage will be depicted by the zig-zag symbol. On the RBC membrane, the presence of a third fatty acid chain on GPI-linked proteins makes the protein harder to cleave by phospholipases. For background reading about membrane proteins, the interested reader is referred to Alberts et al.[10].

The approximate locations of all antigens within the blood group that are a consequence of a single amino acid change are shown on this diagram. While one protein molecule can carry numerous high prevalence antigens, it is unlikely to carry more than one antigen of low prevalence.

Certain characteristics will be given:

M_r *(SDS-PAGE)*: The relative molecular mass (M_r) of a protein as determined by SDS-PAGE. The M_r of a protein (and in particular a glycoprotein) usually differs from the actual molecular weight and the molecular weight calculated from the amino acid sequence deduced from the nucleotide sequence.

Glycosylation: Potential N-linked glycosylation sites (Asn-X-Ser/Thr where X is any amino acid except Pro) are indicated. O-linked glycosylation occurs at Ser and Thr residues. Not all Ser and Thr are glycosylated.

Cysteine residues: The total number present is indicated.

Copies per RBC: The number of copies in the RBC membrane of the protein carrying a blood group antigen is indicated. Human polyclonal antibodies to a specific antigen and monoclonal antibodies to the protein have been used as intact immunoglobulin molecules and as Fab fragments to ascertain copy number. Depending on the technology used, these numbers can vary dramatically in different publications. The figures given for numbers of copies of a protein per RBC are only a guide and the interested investigator is encouraged to perform a thorough literature search.

Molecular basis of antigens

This edition includes a table summarizing antigens associated with a single point mutation.

Function

Function of the carrier protein is given if it is known; or the predicted function, based on homology with other proteins of known function, is given.

Disease association

This entry includes diseases caused by an absence of the protein carrying the blood group antigens and disease susceptibilities associated with an absence, an altered form, or a reduced number of copies/RBC of the protein. For more detail refer to *Transfusion Medicine Reviews*, Volume 14, No. 4, 2000[6].

Phenotypes

The incidence of phenotypes associated with the blood group system, the null phenotype, and any unusual phenotypes is given. In general, the figures given are for Caucasian populations (northern European) because that is the best studied group. Information was usually obtained from original publications, *Blood Groups in Man*[11], and the AABB *Technical Manual*[7].

Tables with useful facts pertaining to antigens or phenotypes in one system are given here.

Molecular basis of phenotypes

This edition includes a table summarizing the molecular basis associated with phenotypes.

Comments

Any fact or interesting information relevant to the blood group system or the carrier protein that does not fit elsewhere is placed here.

References

It is incompatible with the format of this book to provide a comprehensive list of references. However, when appropriate, key references for reviews or recent papers have been selected as a source of further relevant references. Certain reference books have been used throughout and rather than list them on each set of sheets they are listed below. These texts[1, 6, 7, 11–16] are a good source of references.

BLOOD GROUP ANTIGENS

Terminology

The traditional name for the blood group antigen is given at the top of the page. The ISBT symbol, ISBT number (parenthetically), and other names that have been associated with the antigen are also given. In this edition, we have given a brief history about the antigen, including how it was named. If anyone remembers fun facts about the antigens, we encourage them to let us know, so that we can include the information in the next edition.

Occurrence

The prevalence of an antigen is given for Caucasians and notable ethnic differences are given. Where no ethnicity is given, the figures refer to all populations tested. In general, the information was obtained from *Blood Groups in Man*[11], *Distribution of the Human Blood Groups and Other Polymorphisms*[17] and the *AABB Technical Manual*[7]. Antigen and phenotype prevalence are often obtained by averaging several series of tests and are given as a guide.

Antithetical antigen

If an antigen is polymorphic, the antithetical partner is indicated.

Expression

Entries here relate to serologically detectable antigens, however, in most instances the information also will apply to the carrier molecule.

Molecular basis associated with antigen

The name and position of specific amino acid(s) associated with the antigen is indicated. For the amino acid associated with the allele encoding a blood group antigen, it will be necessary to refer to the pages for the *antithetical antigen*. For those antigens that do not have defined antithetical partners, the amino acid associated with the wild type protein, or with the absence of a high prevalence antigen, will be given.

Where appropriate, the nucleotide and position of the base pair (bp) change are noted. Also noted are cases where the nucleotide substitution introduces or ablates a restriction enzyme site. For antigens on hybrid molecules or that are more complex than a single amino acid change, a stick diagram is given.

Effect of enzymes and chemicals on intact RBCs

The options for entries are: resistant, sensitive, weakened, variable. In those instances when reactions between antibody and antigen are markedly enhanced, a (↑) symbol is used: enhancement may not have been studied for all antigens. If no information is available we have used 'presumed' and extrapolated our interpretation based on the behavior of other antigens in the same system. The information given is to be used only as a guide because with all chemical treatment of RBCs, the effect varies depending on the exact conditions of treatment, purity of reagents and the age (condition) of the RBCs. It should be noted that the effect of enzymes on an isolated protein may not be the same when the protein is within the milieu of the RBC membrane.

If an antigen is sensitive to treatment of RBCs with 200 mM dithiothreitol (DTT), an additional entry will be made for the effect of 50 mM DTT. Other thiol-containing reagents, which include 2-mercaptoethanol (2-ME) and 2-aminoethylisothiouronium bromide (AET) would be expected to give similar results to those indicated for DTT treatment. The commonly used reagent, ZZAP[18], is not listed because its effect is simply a combination of DTT and papain.

The effect of acid treatment on antigens is included for those who wish to type RBCs after *in vivo* bound immunoglobulin has been removed by EDTA/acid/glycine method[19]. RBCs treated this way do not express antigens in the Kell blood group system, the Er collection or Bg antigens.

Most information for this section was obtained from original papers and from refs[14, 18, 20, and 21].

In vitro characteristics of alloantibody

An alloantibody can be made by a person who lacks the corresponding antigen. The immunoglobulin class of a blood group antibody is usually IgG and/or IgM. Blood bank techniques do not routinely include methods to detect IgA antibodies. IgD and IgE have not been described as blood group specific antibodies. In general, naturally occurring antibodies are IgM and react best by direct agglutination tests while immune antibodies are IgG and react best by the indirect antiglobulin test (IAT). Readers who are interested in information about the IgG subclass of blood group antibodies are referred to Garratty[22], p. 116.

The optimal technique for detection of an antibody to a given antigen is listed as room temperature (RT) or IAT. 'RT' means incubation at ambient temperature followed by centrifugation and examination for hemagglutination. 'IAT' represents the indirect antiglobulin test regardless of which enhancement medium (e.g. LISS, albumin, PEG) was used. 'Enzymes' means that the antibody agglutinates protease-treated (usually ficin or papain) RBCs, usually after incubation at 37 °C. Enzyme-treated RBCs also may be used by the IAT. If column technology is being used then RT indicates use of the neutral cassette and IAT indicates use of the antiglobulin cassette.

Complement binding is used to convey whether the alloantibody is known to bind complement during the *in vitro* interaction with its antigen. It is not intended to indicate the potential of an alloantibody to cause *in vivo* hemolysis of transfused antigen-positive blood.

Clinical significance of alloantibody

This section summarizes the type and degree of transfusion reaction(s) and the degree of clinical hemolytic disease of the newborn (HDN) that have been associated with the alloantibody in question. Many factors influence the clinical significance of a blood group antibody and the interested reader is referred to Mollison et al[15].

Under 'Transfusion reaction' the entries are: No/ + DAT/mild/moderate/severe; immediate/delayed/hemolytic and no data. 'Severe' usually means an immediate transfusion reaction and may be fatal. 'Delayed' transfusion reaction means a reduced RBC survival as indicated by hemoglobinemia, hemoglobinuria, or reduced RBC count or hematocrit. The options for HDN are No/+DAT but no clinical HDN/mild/moderate/severe and rare.

Autoantibody

If autoantibodies directed to the antigen in question have been described, they are indicated here[12, 22].

Comments

Any fact or interesting information relevant to the antigen and that does not fit elsewhere are placed here.

References

It is incompatible with the format of this book to provide a comprehensive list of references. However, appropriate key references for reviews or recent papers have been selected as a source of further relevant references. References given on the system page will not necessarily be repeated on each antigen page. Where no reference is given, refer to the system page. Certain textbooks have been used throughout and rather than list them on each antigen page they are listed below. These textbooks are a good source of references[6, 11, 13–16, 23].

BLOOD GROUP COLLECTIONS

Information relating to the collections of blood group antigens recognized by the ISBT Committee (COST, Ii, Er, Globoside and Unnamed) are given on separate pages. Antigens within each collection have serological, biochemical or genetic relationship but do not fulfil the criteria for System status.

The 700 series of low incidence antigens

Antigens in this series occur in less than 1% of most populations studied and are not known to belong to a blood group system.

The 901 series of high incidence antigens

Antigens in this series occur in more than 90% of the population and are not known to belong to a blood group system.

References
1   Daniels, G.L. et al. (1995) Vox Sang. 69, 265–279.
2   Daniels, G.L. et al. (1996) Vox Sang. 71, 246–248.
3   Daniels, G.L. et al. (1999) Vox Sang. 77, 52–57.

4   Daniels, G.L. et al. (2001) Vox Sang. 80, 193–197.
5   Daniels, G.L. et al. (2003) Vox Sang. 84, 244–247.
6   Schenkel-Brunner, H. (2000) Human Blood Groups: Chemical and Biochemical Basis of Antigen Specificity, 2nd Edition, Springer-Verlag Wien, New York.
7   Brecher, M.E. et al. (2002) Technical Manual, 14th Edition, American Association of Blood Banks, Bethesda, MD.
8   Reid, M.E. et al. (1998) Transf. Med. Rev. 12, 151–161.
9   Lögdberg, L. et al. (2002) Transf. Med. Rev. 16, 1–10.
10 Alberts, B. et al. (1994) Molecular Biology of the Cell, 3rd Edition, Garland Publishing Co., New York.
11 Race, R.R. and Sanger, R. (1975) Blood Groups in Man, 6th Edition, Blackwell Scientific, Oxford, UK.
12 Daniels, G. (2002) Human Blood Groups, 2nd Edition, Blackwell Science, Oxford.
13 Issitt, P.D. and Anstee, D.J. (1998) Applied Blood Group Serology, 4th Edition, Montgomery Scientific Publications, Durham, NC.
14 Daniels, G. (2002) Human Blood Groups, 2nd Edition, Blackwell Science Ltd, Oxford.
15 Mollison, P.L. et al. (1997) Blood Transfusion in Clinical Medicine, 10th Edition, Blackwell Science, Oxford, UK.
16 Cartron, J.-P. and Rouger, P. eds. (1995) Molecular Basis of Human Blood Group Antigens, Plenum Press, New York.
17 Mourant, A.E. et al. (1976) Distribution of the Human Blood Groups and Other Polymorphisms, 2nd Edition, Oxford University Press, London.
18 Branch, D.R. and Petz, L.D. (1982) Am. J. Clin. Pathol. 78, 161–167.
19 Byrne, P.C. (1991) Immunohematology 7, 46–47.
20 American Red Cross National Reference Laboratory Methods Manual Committee. (1993) Immunohematology Methods. American Red Cross National Reference Laboratory, Rockville, MD.
21 Judd, W.J. (1994) Methods in Immunohematology, 2nd Edition, Montgomery Scientific Publications, Durham, NC.
22 Garratty, G. (1989) In: Immune Destruction of Red Blood Cells (Nance, S.J. ed.) American Association of Blood Banks, Arlington, VA, pp. 109–169.
23 Garratty, G. ed. (1994) Immunobiology of Transfusion Medicine, Marcel Dekker New York.

THE
BLOOD GROUP
SYSTEMS AND
ANTIGENS

| Number of antigens | 4 (A,B not given) |

Terminology

ISBT symbol	ABO
ISBT number	001
History	In 1900, Landsteiner mixed sera and RBCs from his colleagues and observed agglutination. On the basis of the agglutination pattern, he named the first two blood group antigens A and B, using the first letters of the alphabet. RBCs not agglutinated by either sera were first called C but became known as "ohne A" and "ohne B" (*ohne* is German for "without") and finally O. In 1907, Jansky proposed using Roman numerals I, II, III, IV for O, A, B and AB respectively, and in 1910, Moss proposed using I, II, III and IV for AB, B, A and O, respectively. These numerical terminologies were used respectively in Europe and America until 1927 when Landsteiner suggested, in order to avoid confusion, to use throughout the world the symbols A, B, O and AB

Expression

Soluble form	Saliva and all body fluids except CSF (in secretors)
Other blood cells	Lymphocytes, platelets (adsorbed from plasma)
Tissues	On most epithelial cells (particularly glandular epithelia) and on endothelial cells. Broad tissue distribution (often termed "histo-blood group" antigens)

Gene[1,2]

Chromosome	9q34.1–q34.2
Name	*ABO*
Organization	Seven exons distributed over 19.5 kbp of gDNA
Product	3-α-N-Acetylgalactosaminyltransferase for A
	3-α-Galactosyltransferase for B
Gene map	

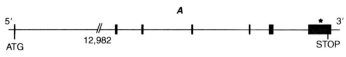

* Most variants are encoded by missense mutations in exon 7.

├────────┤ 1 kbp

Database accession numbers

AF134412-40
http://www.bioc.aecom.yu.edu/bgmut/index.htm

Amino acid sequence

A transferase (*ABO*A101*)

```
MAEVLRTLAG  KPKCHALRPM  ILFLIMLVLV  LFGYGVLSPR  SLMPGSLERG   50
FCMAVREPDH  LQRVSLPRMV  YPQPKVLTPC  RKDVLVVTPW  LAPIVWEGTF  100
NIDILNEQFR  LQNTTIGLTV  FAIKKYVAFL  KLFLETAEKH  FMVGHRVHYY  150
VFTDQPAAVP  RVTLGTGRQL  SVLEVRAYKR  WQDVSMRRME  MISDFCERRF  200
LSEVDYLVCV  DVDMEFRDHV  GVEILTPLFG  TLHPGFYGSS  REAFTYERRP  250
QSQAYIPKDE  GDFYYLGGFF  GGSVQEVQRL  TRACHQAMMV  DQANGIEAVW  300
HDESHLNKYL  LRHKPTKVLS  PEYLWDQQLL  GWPAVLRKLR  FTAVPKNHQA  350
VRNP                                                       354
```

B transferase (*ABO*B101*)

```
MAEVLRTLAG  KPKCHALRPM  ILFLIMLVLV  LFGYGVLSPR  SLMPGSLERG   50
FCMAVREPDH  LQRVSLPRMV  YPQPKVLTPC  RKDVLVVTPW  LAPIVWEGTF  100
NIDILNEQFR  LQNTTIGLTV  FAIKKYVAFL  KLFLETAEKH  FMVGHRVHYY  150
VFTDQPAAVP  RVTLGTGRQL  SVLEVGAYKR  WQDVSMRRME  MISDFCERRF  200
LSEVDYLVCV  DVDMEFRDHV  GVEILTPLFG  TLHPSFYGSS  REAFTYERRP  250
QSQAYIPKDE  GDFYYMGAFF  GGSVQEVQRL  TRACHQAMMV  DQANGIEAVW  300
HDESHLNKYL  LRHKPTKVLS  PEYLWDQQLL  GWPAVLRKLR  FTAVPKNHQA  350
VRNP                                                       354
```

Carrier molecule description

A and B antigens are not primary gene products.
Antigens are defined by immunodominant sugars (GalNAc for A; Gal for B) attached to 1 of 4 different types of oligosaccharide chains carried on glycosphingolipid and glycoprotein molecules. Expressed on N-glycans containing polylactosaminyl units carried predominantly on band 3, glucose transporter, RhAG, and CHIP-1.

Type 1   Galβ(1–3)GlcNAcβ(1–3)R; the predominant type in secretions, plasma and some tissues
Type 2   Galβ(1–4)GlcNAcβ(1–3)R; the predominant type in RBCs

The precursor of A and B antigens is the H antigen (H1)

Disease association

Expression of A and B antigens may be weakened as a result of chromosome 9 translocations and associated leukemia and of any disease inducing stress hemopoiesis, for example, thalassemia, Diamond Blackfan anemia. Stress hemopoiesis results in reduced branching of carbohydrate chains and thus less A, B, H and I antigens. Modifications in sugar chains are characteristic of cancer and erythroleukemia and therefore A and B antigens are important tumor markers. The acquired B antigen is a consequence of bacterial infection. ABO phenotypes are associated with susceptibility to numerous diseases.

Phenotypes (% occurrence)

	Caucasians	Blacks	Asian	Mexican
A_1	33	19	27	22
A_2	10	8	Rare	6
B	9	20	25	13
O	44	49	43	55
A_1B	3	3	5	4
A_2B	1	1	Rare	Rare

Null O is the amorph; O_h (Bombay).
Unusual Many subgroups of A and B.

Molecular basis associated with variant A transferases[3,4]

(*ABO*A101* taken as the reference allele sequence)

Phenotype	Nucleotide change	Amino acid change
A_1	467C>T	Pro156Leu
A_2	467C>T; 1059–1061delC	Pro156Leu; fs and 21 extra amino acids
A_2	1054C>T	Arg352Trp
A_2	1054C>G	Arg352Gly
A_2	526C>G; 703G>A; 829G>A	Arg176Gly; Gly235Ser; Val277Met
A_3	871G>A	Asp291Asn
A_x	646T>A	Phe216Ile
A_x	A or B–O^{lv} hybrid	Phe216Ile; Val277Met
A_{el}	798–804insG	fs
A_{el}	467C>T; 646T>A	Pro156Leu; Phe216Ile
A_w	407C>T; 467C>T; 1060delC	Thr136Met; Pro156Leu; Pro354fs
A_w	350C>G; 467C>T; 1060delC	Pro156Leu; Gly177Ala; Pro354fs
A_w	203G>C; 467C>T; 1060delC	Arg68Thr; Pro156Leu; Pro354fs
A_w	965A>G	Glu322Gly
A_w	502C>G	Arg168Gly

Silent mutations are not noted.
For more alleles and details, see http://www.bioc.aecom.yu.edu/bgmut/index.htm.

Molecular basis associated with variant B transferases[3]

(*ABO*B101* taken as the reference allele sequence)

Phenotype	Nucleotide change	Amino acid change
B_3	1054C>T	Arg352Trp
B_x	871G>A	Asp291Asn
B_{el}	641T>G	Met214Arg
B_{el}	669G>T	Glu223Asp
B_w	873C>G	Asp291Glu
B_w	721C>T	Arg241Trp
B_w	548A>G	Asp183Gly
B_w	539G>A	Arg180His
B_w	1036A>G	Lys346Glu
B_w	1055G>A	Arg352Gln
B_w	863T>G	Met288Arg

Silent mutations are not noted.
For more alleles and details, see http://www.bioc.aecom.yu.edu/bgmut/index.htm.

Molecular basis associated with the O phenotype[5]

(*ABO*A101* taken as the reference allele sequence)

Allele	Nucleotide change	Amino acid change
O_1	261delG 88 fs; codon 116Stop	
O_2	526C>G; 802G>A	Arg176Gly; Gly268Arg

Silent mutations are not noted
For more alleles and details, see http://www.bioc.aecom.yu.edu/bgmut/index.htm

Amino acid changes associated with hybrid transferases encoding both A and B

The relevant amino acids for A and B transferases are given for reference

	Amino acid number				
Phenotype	176	234	235	266	268
A	Arg	Pro	Gly	Leu	Gly
B	Gly	Pro	Ser	Met	Ala
Cis-AB	Arg	Pro	Gly	Leu	Ala
Cis-AB	Gly	Pro	Ser	Leu	Ala

Phenotype	Amino acid number				
	176	234	235	266	268
B(A)	Gly	Pro	Gly	Met	Ala
B(A)	Gly	Ala	Ser	Met	Ala

There are only four amino acid differences between the A and B transferases in the catalytic domain

Two residues, 266 and 268, are important for the transferase specificity

Comments

Due to the complexity of the membrane components that express A or B antigens, we recommend reading the special issue of *Transfusion Medicine*, volume 11; 2001, which is devoted to the ABO system.

Aberrant ABO results created by modern medical practices include: bone marrow transplants, *in vitro* fertilization, artificial insemination, surrogate motherhood.

References
[1] Yamamoto, F. et al. (1990) J. Biol. Chem. 265, 1146–1151.
[2] Yamamoto, F. et al. (1990) Nature 345, 229–235.
[3] Olsson, M.L. et al. (2001) Blood 98, 1585–1593.
[4] Seltsam, A. et al. (2002) Transfusion 42, 294–301.
[5] Olsson, M.L. and Chester, M.A. (2001) Transf. Med. 11, 295–313.

A ANTIGEN

Terminology

ISBT symbol (number)	ABO1 (001.001)
Other names and history	See system page

Occurrence

Caucasians	43%
Blacks	27%
Asians	28%
Mexicans	28%
South American Indians	0%

These numbers do not include group AB, which would increase the numbers (all except South American Indians) by approximately 4%

Expression

Cord RBCs   Weak
Altered   Weak in some variants; some diseases

Molecular basis associated with A antigen

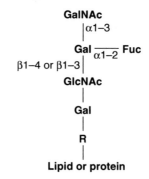

See ABO Blood Group System page for subgroups

Effect of enzymes/chemicals on A antigen on intact RBCs

Ficin/Papain   Resistant (↑↑)
Trypsin   Resistant (↑↑)
α-Chymotrypsin   Resistant (↑↑)
Pronase   Resistant (↑↑)
Sialidase   Resistant
DTT 200 mM   Resistant
Acid   Resistant

In vitro characteristics of alloanti-A

Immunoglobulin class   IgM; IgG
Optimal technique   RT or below
Neutralization   Saliva from A secretors
Complement binding   Yes; some hemolytic

Clinical significance of alloanti-A

Transfusion reaction   None to severe; immediate/delayed; intravascular/ extravascular
HDN   No to moderate (rarely severe)

Autoanti-A

Rare

Comments

Serum from group A individuals contains naturally occurring anti-B (see **ABO2**).

B ANTIGEN

Terminology

ISBT symbol (number)	ABO2 (001.002)
Other names and history	See system page

Occurrence

Caucasians	9%
Blacks	20%
Asians	27%
Mexicans	13%
South American Indians	0%

These numbers do not include group AB, which would increase the numbers (all except South American Indians) by approximately 4%

Expression

Cord RBCs	Weak
Altered	Weak in some variants; some diseases

Molecular basis associated with B antigen

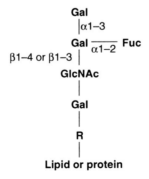

See ABO system page for subgroups

25

Effect of enzymes/chemicals on B antigen on intact RBCs

Ficin/papain	Resistant (↑↑)
Trypsin	Resistant (↑↑)
α-Chymotrypsin	Resistant (↑↑)
Pronase	Resistant (↑↑)
Sialidase	Resistant
DTT 200 mM	Resistant
Acid	Resistant

In vitro characteristics of alloanti-B

Immunoglobulin class	IgM; IgG
Optimal technique	RT or below
Neutralization	Saliva from B secretors
Complement binding	Yes; some hemolytic

Clinical significance of alloanti-B

Transfusion reaction	No to severe; immediate/delayed; intravascular/ extravascular
HDN	No to moderate

Autoanti-B

Rare

Comments

Serum from group B individuals contains naturally occurring anti-A (see **ABO1**).

A1 ANTIGEN

Terminology

ISBT symbol (number)	ABO4 (001.004)
History	Named in 1930 when the A antigen was subdivided by the reactivity or non-reactivity with an antibody (later called anti-A1) in the serum of some A people (later called the A_2 phenotype)

Occurrence

Caucasians	34%
Blacks	19%
Asians	27%

These numbers do not include group AB

Expression

Cord RBCs	Weak

Molecular basis associated with A1 antigen

Presence of the A type 3 epitope defines this antigen. A_1 and A_2 phenotypes are determined by qualitative differences between the A_1 and A_2 transferases (see ABO system page). In addition, the number of A antigens on A_1 RBCs is approximately five times more than on A_2 RBCs.

Other A subgroups (A_3, A_x, A_{el}, etc.) are phenotypes and do not possess specific antigens (see ABO system page).

Effect of enzymes/chemicals on A1 antigen on intact RBCs

Ficin/papain	Resistant (↑↑)
Trypsin	Resistant (↑↑)
α-Chymotrypsin	Resistant (↑↑)
Pronase	Resistant (↑↑)
Sialidase	Resistant
DTT 200 mM	Resistant
Acid	Resistant

In vitro characteristics of alloanti-A1

Immunoglobulin class	IgM more common than IgG
Optimal technique	RT or below
Neutralization	Saliva from A secretors
Complement binding	Rare

Clinical significance of alloanti-A1

Transfusion reaction	None to mild/delayed
HDN	No

Autoanti-A1

Rare

Comments

The transferase activity in serum from A_1 individuals is 5–10 times higher than that in A_2 (A1−) individuals.

A lectin with anti-A1 specificity can be prepared from *Dolichos biflorus* seeds. Anti-A1 is found in serum from 1–2% of A_2 and 25% of A_2B individuals and is a component of anti-A from group O and B people.

Number of antigens 43

Terminology

ISBT symbol	MNS
ISBT number	002
CD number	CD235A (GPA); CD235B (GPB)
Other name	MNSs
History	Discovered in 1927 by Landsteiner and Levine; named after the first three antigens identified: M, N and S.

Expression

Tissues   Renal endothelium and epithelium

Gene

Chromosome	4q28.2–q31.1
Name	*GYPA, GYPB*
Organization	*GYPA*: seven exons distributed over 60 kbp
	GYPB: five exons (1 pseudoexon) distributed over 58 kbp
Product	Glycophorin A (GPA; MIRL2; MN sialoglycoprotein; SGPα)
	Glycophorin B (GPB; Ss sialoglycoprotein; SGPδ)

A third gene (*GYPE*), which is adjacent to *GYPB*, may not encode a RBC membrane component, but participates in gene rearrangements resulting in variant alleles.

Gene map

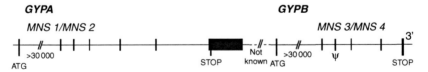

MNS 1/MNS 2 (*GYPA* 60C > T; 72G > A) encode M/N (Ser1Leu; Gly5Glu)
MNS 3/MNS 4 (*GYPB* 243T > C) encode S/s (Met29Thr)
ψ = Pseudo exon

GenBank accession numbers

GYPA: X51798; M60707
GYPB: JO2982; M60708

Amino acid sequence

Glycophorin A^M:

```
                                    MYGKIIFVL   LLSAIVSISA   -1
SSTTGVAMHT   STSSSVTKSY   ISSQTNDTHK   RDTYAATPRA   HEVSEISVRT   50
VYPPEEETGE   RVQLAHHFSE   PEITLIIFGV   MAGVIGTILL   ISYGIRRLIK  100
KSPSDVKPLP   SPDTDVPLSS   VEIENPETSD   Q                        131
```

Glycophorin B^S:

```
                                    MYGKIIFVL   LLSEIVSISA   -1
LSTTEVAMHT   STSSSVTKSY   ISSQTNGETG   QLVHRFTVPA   PVVIILIILC   50
VMAGIIGTIL   LISYSIRRLI   KA                                     72
```

Antigen mutations are numbered by counting Ser for GPA as 1 and Leu for GPB as 1.

Both *GYPA* and *GYPB* encode a leader sequence of 19 amino acid residues.

Carrier molecule[1,2]

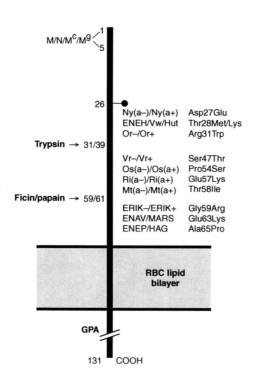

GPA and GPB are single-pass membrane sialoglycoproteins (type I). GPA is cleaved by trypsin at residues 31 and 39 on intact RBCs. GPB is cleaved by α-chymotrypsin at residue 32 on intact RBCs.

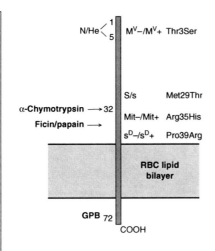

	GPA	GPB
M_r (SDS-PAGE)	43 000	25 000
CHO: N-glycan	1 site	0 site
CHO: O-glycan	15 sites	11 sites
Copies per RBC	800 000	200 000

Molecular basis of antigens involving single nucleotide mutations[3]

Antigen	Amino acid involved	Exon	Nt change
GPA			
ENEH/Vw/Hut	Thr28Met/Lys	3	140C>T>A
Vr	Ser47Tyr	3	197C>A
Mt^a	Thr58Ile	3	230C>T
Ri^a	Glu57Lys	3	226G>A
Ny^a	Asp27Glu	3	138T>A
Or	Arg31Trp	3	148C>T
ERIK	Gly59Arg	4	232G>A
Os^a	Pro54Ser	3	217C>T
ENEP/HAG	Ala65Pro	4	250G>C
ENAV/MARS	Glu63Lys	4	244C>A
GPB			
S/s	Met29Thr	4	143T>C
M^V	Thr3Ser	2	65C>G
s^D	Pro39Arg	4	173C>G
Mit	Arg35His	4	161G>A

Function

Receptor for complement, bacteria and viruses[4-6]. Chaperone for band 3 transport to RBC membrane. Major component contributing to the negatively charged RBC glycocalyx.

Disease association

Plasmodium falciparum invasion[7,8].

Phenotypes (% occurrence)

Phenotype	Caucasians	Blacks
M+N−S+s−	6	2
M+N−S+s+	14	7
M+N−S−s+	8	16
M+N+S+s−	4	2
M+N+S+s+	24	13
M+N+S−s+	22	33
M−N+S+s−	1	2
M−N+S+s+	6	5
M−N+S−s+	15	19
M+N−S−s−	0	0.4
M+N+S−s−	0	0.4
M−N+S−s−	0	0.7

Null: En(a−); U−; M^kM^k.
Unusual: Various hybrids[1,2,9].
          GPB is decreased in Rh_{null} and Rh_{mod} RBCs.

Glycophorin phenotypes and associated antigens

(Previously the Miltenberger subsystem)

ISBT MNS # →		Mi^a 8	Vw 9	Hut 19	Mur 10	MUT 35	Hil 20	TSEN 33	MINY 34	Hop 26	Nob 27	DANE 32
Mi.I	GP.Vw	+	+	0	0	0	0	0	0	0	0	0
Mi.II	GP.Hut	+	0	+	0	+	0	0	0	0	0	0
Mi.IIII	GP.Mur	+	0	0	+	+	+	0	+	0	0	0
Mi.IV	GP.Hop	+	0	0	+	+	0	+	+	+	0	0
Mi.V	GP.Hil	0	0	0	0	0	+	0	+	0	0	0
Mi.VI	GP.Bun	+	0	0	+	+	+	0	+	+	0	0
Mi.VII	GP.Nob	0	0	0	0	0	0	0	0	0	+	0
Mi.VIII	GP.Joh	0	0	0	0	0	0	NT	0	+	+	0
Mi.IX	GP.Dane	0	0	0	+	0	0	0	0	0	0	+
Mi.X	GP.HF	+	0	0	0	+	+	0	+	0	0	0
Mi.XI	GP.JL	0	0	0	0	0	0	+	+	0	0	0

Hybrid glycophorin molecules, phenotypes and associated low incidence antigens

Molecular basis	Glycophorin	Phenotype symbol	Associated novel antigens
GYP(A-B)	GP(A-B)	GP.Hil (Mi.V)	Hil, MINY
		GP.JL (Mi.XI)	TSEN, MINY
		GP.TK	SAT
GYP(B-A)	GP(B-A)	GP.Sch (M^r)	St^a
		GP.Dantu	Dantu
GYP(A-B-A)	GP(A-B-A)	GP.Mg	M^g
		GP.KI	Hil
GYP(B-A-B)	GP(B-A-B)	GP.Mur (Mi.III)	Mi^a, Mur, MUT, Hil, MINY
		GP.Bun (Mi.VI)	Mi^a, Mur, MUT, Hop, Hil, MINY
		GP.HF (Mi.X)	Mi^a, MUT, Hil, MINY
		GP.Hop (Mi.IV)	Mi^a, Mur, MUT, Hop, TSEN, MINY
	GP(A-B)	GP.He; (P2, GL)	He
GYP(B-A-ψB-A)	GP(A-A)	GP.Cal	He, St^a
GYP(A-ψB-A)	GP(A-B-A)	GP.Vw (Mi.I)	Mi^a, Vw
		GP.Hut (Mi.II)	Mi^a, Hut, MUT
		GP.Nob (Mi.VII)	Nob
		GP.Joh (Mi.VIII)	Nob, Hop
		GP.Dane (Mi.IX)	Mur, DANE
	GP(A-A)	GP.Zan (M^z)	St^a
GYPA 179G > A	GPA	GP.EBH	ERIK (from transcript 1)
	GP(A-A)	GP.EBH	St^a (from transcript 2)
GYP(A-ψE-A)	GP(A-A)	GP.Mar	St^a

Molecular basis of other phenotypes

Phenotype	Basis
M^{k}M^k	Deletion of *GYPA* (exon 2 to 7), *GYPB* (exon 1 to 5) and *GYPE* (exon 1)
En(a−)	Deletion of *GYPA* (exon 2 to 7), *GYPB* (exon 1)
S−s−U−	Deletion of *GYPB* (exon 2 to 4) and *GYPE* (exon 1)
S−s−U+var	See U (MNS5) and He (MNS6)
Wr(b−)	See Wr^b (DI4)

Comments

Linkage disequilibrium exists with M/N and S/s antigens.

MNS antigens not numbered by the ISBT Working Party include: Tm, Sj, M$_1$, Can, Sext and Hu. These antigens are associated with atypical glycosylation[10].

GPA and GPB are the major sialic acid containing structures of the RBC membrane. The majority of the sialic acid is on the O-glycans.

GPA-deficient RBCs have a weak expression of Ch and Rg antigens.

References
1. Huang, C.-H. and Blumenfeld, O.O. (1995) In: Molecular Basis of Human Blood Group Antigens (Cartron J.-P. and Rouger, P. eds) Plenum Press, New York, pp. 153–188.
2. Reid, M.E. (1994) Transf. Med. 4, 99–111.
3. Reid, M.E. and Storry, J.R. (2001) Immunohematology 17, 76–81.
4. Daniels, G. (2002) Human Blood Groups, 2nd Edition, Blackwell Science, Oxford.
5. Moulds, J.M. et al. (1996) Transfusion 36, 362–374.
6. Daniels, G. (1999) Blood Rev. 13, 14–35.
7. Hadley, T.J. et al. (1991) Transf. Med. Rev. 5, 108–113.
8. Miller, L.H. (1994) Proc. Natl. Acad. Sci. USA 91, 2415–2419.
9. Tippett, P. et al. (1992) Transf. Med. Rev. 6, 170–182.
10. Dahr, W. et al. (1991) Biol. Chem. Hoppe-Seyler 372, 573–584.

M ANTIGEN

Terminology

ISBT symbol (number)	MNS1 (002.001)
History	M, identified in 1927, was the first antigen of the MNS system; named after the second letter of "immune" because anti-M was the result of immunizing rabbits with human RBCs

Occurrence

Caucasians	78%
Blacks	74%

Antithetical antigen

N (MNS2)

Expression

Cord RBCs	Expressed
Altered	On some hybrid glycophorin molecules

Molecular basis associated with M antigen[1]

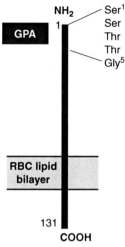

Nucleotide   C at bp 59, G at bp 71 and T at bp 72 in exon 2
Recognition of antigen by anti-M is usually dependent on O-glycans attached to amino acid residues 2, 3 and 4.

Effect of enzymes/chemicals on M antigen on intact RBCs

Ficin/papain	Sensitive
Trypsin	Sensitive
α-Chymotrypsin	Resistant
Pronase	Sensitive
Sialidase	Variable
DTT 200 mM	Resistant
Acid	Resistant

In vitro characteristics of alloanti-M

Immunoglobulin class	IgG (cold reactive; many agglutinating) and IgM
Optimal technique	4 °C; RT; rarely also reactive by IAT
Complement binding	No

Clinical significance of alloanti-M

Transfusion reactions	No (except in extremely rare cases)
HDN	No (except in extremely rare cases)

Autoanti-M

Rare; reactive at low temperatures.

Comments

Acidification of serum enhances the reactivity of some anti-M. Anti-M often react more strongly with M+N−RBCs than with M+N+RBCs (i.e. they show dosage).

Anti-M is more common in children than adults, and in patients with bacterial infections. It is not uncommon for pregnant M− women to produce anti-M but to give birth to an M− baby.

Reference
[1] Dahr, W. et al. (1977) Hum. Genet. 35, 335–343.

N ANTIGEN

Terminology

ISBT symbol (number)	MNS2 (002.002)
History	Identified shortly after the M antigen in 1927; named as the next letter after M and for the fifth letter of "immune" because anti-N was the result of immunizing rabbits with human RBCs

Occurrence

Caucasians	72%
Blacks	75%

Antithetical antigen

M (MNS1)

Expression

Cord RBCs	Expressed
Altered	On some hybrid glycophorin molecules

Molecular basis associated with N antigen[1]

Nucleotide T at bp 59, A at bp 71 and G at bp 72 in exon 2
Recognition of antigen by anti-N is often also dependent on O-glycans
attached to amino acid residues 2, 3 and 4.

Effect of enzymes/chemicals on N antigen on intact RBCs

	GPA	*GPB*
Ficin/papain	Sensitive	Sensitive
Trypsin	Sensitive	Resistant
α-Chymotrypsin	Resistant	Sensitive
Pronase	Sensitive	Sensitive
Sialidase	Variable	Variable
DTT 200 mM	Resistant	Resistant
Acid	Presumed resistant	Presumed resistant

In vitro characteristics of alloanti-N

Immunoglobulin class	IgM; IgG (some agglutinating)
Optimal technique	4 °C; RT; rarely also reactive by IAT
Complement binding	No

Clinical significance of alloanti-N

Transfusion reaction	No
HDN	No

Rare S−s−U− individuals make an antibody that reacts with N on GPA and GPB and may be clinically significant.

Autoanti-N

Rare. Found in patients on dialysis when equipment was sterilized with formaldehyde (anti-Nf).

Comments

The N antigen on GPB is denoted as 'N' (MNS30) to distinguish it from N on GPA. Anti-N typing reagents are formulated to detect N antigen on GPA but not on GPB. Monoclonal anti-N are more specific at alkaline pH.

Reference
[1] Dahr, W. et al. (1977) Hum. Genet. 35, 335–343.

S ANTIGEN

Terminology

ISBT symbol (number)	MNS3 (002.003)
History	S was named after the city of Sydney (Australia), where the first example of anti-S was identified in 1947

Occurrence

Caucasians	55%
Blacks	31%

Antithetical antigen

s (MNS4)

Expression

Cord RBCs	Expressed
Altered	On Rh_{null}, M^v+, Mit+ and TSEN+ RBCs

Molecular basis associated with S antigen[1]

Amino acid   Met 29 of GPB
Nucleotide   T at bp 143 in exon 4
In addition to Met 29, some anti-S also require Thr 25 (and/or the GalNAc attached to this residue), Glu 28, His 34 and Arg 35[2].

Effect of enzymes/chemicals on S antigen on intact RBCs

Ficin/papain   Variable
Trypsin   Resistant
α-Chymotrypsin   Sensitive
Pronase   Sensitive
Sialidase   Variable
DTT 200 mM   Resistant
Acid   Resistant

In vitro characteristics of alloanti-S

Immunoglobulin class   IgM less common than IgG
Optimal technique   RT; IAT
Complement binding   Some

Clinical significance of alloanti-S

Transfusion reaction   No to moderate (rare)
HDN   No to severe (rare)

Autoanti-S

Rare

Comments

There are approximately 1.5 times more copies of GPB in S+s− than in S−s+ RBCs. S+s+ RBCs have an intermediate amount of GPB[2].
S antigen is sensitive to trace amounts of chlorine[3,4].
Sera containing anti-S frequently contain antibodies to low incidence antigens.

References
[1] Dahr, W. et al. (1980) Hoppe-Seylers Z. Physiol. Chem. 361, 895–906.
[2] Dahr, W. (1986) In: Recent Advances in Blood Group Biochemistry (Vengelen-Tyler, V. and Judd, W.J. eds) American Association of Blood Banks, Arlington, VA, pp. 23–65.
[3] Rygiel, S.A. et al. (1985) Transfusion 25, 274–277.
[4] Long, A. et al. (2002) Immunohematology 18, 120–122.

s ANTIGEN

Terminology

ISBT symbol (number)	MNS4 (002.004)
History	Anti-s identified in 1951; reacted with an antigen antithetical to S

Occurrence

Caucasians	89%
Blacks	93%

Antithetical antigen

S (MNS3)

Expression

Cord RBCs	Expressed
Altered	Dantu+, Mit+, M^v+, s^D+, GP.Mur, GP.Hil and some Rh_{null}

Molecular basis associated with s antigen[1]

Amino acid	Thr 29
Nucleotide	C at bp 143 in exon 4

In addition to Thr 29, some anti-s also require Thr 25 (and/or GalNAc attached to this residue), Glu 28, His 34 and Arg 35[2].

Effect of enzymes/chemicals on s antigen on intact RBCs

Ficin/papain	Variable
Trypsin	Resistant
α-Chymotrypsin	Sensitive
Pronase	Sensitive
Sialidase	Variable
DTT 200 mM	Resistant
Acid	Resistant

In vitro characteristics of alloanti-s

Immunoglobulin class	IgG; IgM
Optimal technique	IAT (often after incubation at RT or 4°C)
Complement binding	Rare

Clinical significance of alloanti-s

Transfusion reaction   No to mild (rare)
HDN   No to severe (rare)

Comment

A pH of 6.0 enhances the reactivity of some anti-s.

References

[1] Dahr, W. et al. (1980) Hoppe-Seylers Z. Physiol. Chem. 361, 895–906.
[2] Dahr W. (1986) In: Recent Advances in Blood Group Biochemistry (Vengelen-Tyler, V. and Judd, W.J. eds) American Association of Blood Banks, Arlington, VA, pp. 23–65.

U ANTIGEN

Terminology

ISBT symbol (number)   MNS5 (002.005)
History   Described in 1953; named "U" from "the almost universal distribution" of the antigen

Occurrence

Caucasians   99.9%
Blacks   99%

Expression

Cord RBCs   Expressed
Altered   GPB variants and on regulator type of Rh_{null} and on Rh_{mod} RBCs

Molecular basis associated with U antigen[1]

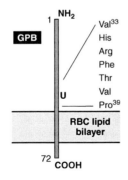

Expression of U may require an interaction with another membrane protein, possibly the Rh associated glycoprotein (RhAG)[2,3].

The U-negative phenotype is associated with an absence of GPB or with altered forms of GPB [see He (MNS 6)][4].

Effect of enzymes/chemicals on U antigen on intact RBCs

Ficin/papain	Resistant
Trypsin	Resistant
α-Chymotrypsin	Resistant
Pronase	Resistant
Sialidase	Resistant
DTT 200 mM	Resistant
Acid	Resistant

In vitro characteristics of alloanti-U

Immunoglobulin class	IgG
Optimal technique	IAT
Complement binding	No

Clinical significance of alloanti-U

Transfusion reaction	Mild to severe
HDN	Mild to severe (one report of fetus requiring interuterine transfusion)[5]

Autoanti-U

Yes

Comments

U– RBCs (except Dantu+ and some Rh_{null}/Rh_{mod} RBCs) are S–s–. Of S–s– RBCs, approximately 16% are U+, albeit weakly (U+^{var}), and are encoded by a hybrid glycophorin gene, of these approximately 23% are He+[4]. Antibodies that detect the altered protein should be more correctly called anti-U/GPB[4].

References

1 Dahr, W. and Moulds, J.J. (1987) Biol. Chem. Hoppe-Seyler 368, 659–667.
2 Mallinson, G. et al. (1990) Transfusion 30, 222–225.
3 Ballas, S.K. et al. (1986) Biochim. Biophys. Acta 884, 337–343.
4 Reid, M.E. et al. (1996) Transfusion 36, 719–724.
5 Win, N. et al. (1996) Transf. Med. 6 (Suppl. 2), 39 (abstract).

He ANTIGEN

Terminology

ISBT symbol (number)	MNS6 (002.006)
Other names	Henshaw
History	Named for the first He+ proband, Mr. Henshaw; the original anti-He, present in a rabbit anti-M serum was identified in 1951

Occurrence

Only found in Blacks: African Americans 3%; up to 7% in Natal.

Antithetical antigen

"N" (MNS30)

Expression

Cord RBCs	Presumed expressed
Altered	On S−s− GPB variants[1]

Molecular basis associated with He antigen[1–3]

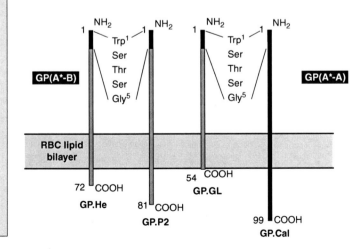

Variant glycophorin

GP.He	GPA(1–4*)-GPA^M(5–26)-GPB(27–72)
GP.He(P2)	GPA(1–4*)-GPA^M(5–26)-GPB^S(27–39)-GPB(40–81*)
GP.He(GL)	GPA(1–4*)-GPA^M(5–26)-GPB(27–59)
GP.He(Cal)	GPA(1–4*)-GPA^M(5–99)

* An altered sequence.

Contribution by parent glycophorin

GP.He	GPA(1–4*)-GPA^M(5–26)-GPB(27–72)
GP.He(P2)	GPA(1–4*)-GPA^M(5–26)-GPB^S(27–39 then new sequence 40–81)
GP.He(GL)	GPA(1–4*)-GPA^M(5–26)-GPB(40–72)
GP.He(Cal)	GPA(1–4*)-GPA^M(5–26)-GPA(59–131)

* An altered sequence.

Gene arrangement and mechanism[2]

GP.He	*GYP(B-A-B)*	Gene conversion
GP.He(P2)	*GYP(B-A-B)*	Gene conversion with a G>T mutation of intron 5, which causes altered splicing and chain elongation with a novel transmembrane amino acid sequence. GP.He(P2) is hard to detect in the RBC membrane; the S−s− RBCs are He+^W due to expression of low levels of GP.He.
GP.He(GL)	*GYP(B-A-B)*	Gene conversion. There are 4 transcripts: t1 is GP.He; t2 has a T>G mutation in the acceptor splice site (Intron 3 at nt −6) leading to skipping of exon 4 [GP.He(GL)]; t3 has a partial deletion of exon 5 due to a C>G mutation in exon 5, a frame shift and a premature stop codon; t4 has the T>G in intron 3 and deletion of exon 4 and the C>G in exon 5, which results in a partial deletion of exon 5. Products of t3 and t4 have not been demonstrated in the RBC membrane[4].
GP.Cal	*GYP(B-A-ψB-A)*	Gene conversion and splice site interaction. The *GYPA* recombination site is in exon 2 so the mature protein, after cleavage of the leader peptide, is GP(A-A). The *GYPB* also contributes the pseudoexon, which is out-spliced. There are 2 He-specific transcripts: t1 has a junction of exon 2 to exon 4 and generates the amino acid sequence associated with the St^a antigen [GP.He(Cal)]; t2 has a junction of exon 2 to exon 5, which is unlikely to be translated[5].
GP.He(NY)	*GYP(B-A-B)*	Gene conversion with partial deletion of exon 5 that alters the open reading frame, predicted to encode a protein of 43 amino acids, which has not been demonstrated in the RBC membrane.

The S−s− RBCs are He+^{W} due to expression of low levels of GP.He[6].

Phenotype with antigen strength

	Antigens expressed		
Glycophorin	He	S/s	U
GP.He	Strong	S or s	Strong
GP.He(P2)	Wk/mod*	No	Wk/mod
GP.He(GL)	Strong**	No	No
GP.He(Cal)	Wk/mod	No	No
GP.He(NY)	Wk/mod*	No	Wk/mod

* The RBCs express He due to low levels of GP.He
** The RBCs also express GP.He

Effect of enzymes/chemicals on He antigen on intact RBCs

Ficin/papain	Sensitive
Trypsin	Resistant
α-Chymotrypsin	Variable
Pronase	Sensitive
Sialidase	Variable
DTT 200 mM	Resistant
Acid	Presumed resistant

In vitro characteristics of alloanti-He

Immunoglobulin class	IgM; IgG
Optimal technique	RT; IAT
Complement binding	No

Clinical significance of alloanti-He

No data are available. Human anti-He is rare.

Comments

Approximately 23% of S−s−RBCs have the He antigen[7] and approximately a half have an altered *GYPB* [see U (**MNS 5**)]
GPB carrying He does not express 'N'.

References
1 Dahr, W. et al. (1984) Eur. J. Biochem. 141, 51–55.
2 Huang, C.-H. and Blumenfeld, O.O. (1995) In: Molecular Basis of Human Blood Group Antigens (Cartron, J.-P. and Rouger, P. eds) Plenum Press, New York, 153–188.
3 Huang, C.-H. et al. (1994) J. Biol. Chem. 269, 10804–10812.
4 Huang, C.-H. et al. (1997) Blood 90, 391–397.
5 Huang, C.-H. et al. (1994) Blood 83, 3369–3376.
6 Storry, J.R. and Reid, M.E. (2000) Transfusion 40 (Suppl.), 13S (abstract).
7 Reid, M.E. et al. (1996) Transfusion 36, 719–724.

Mi^a ANTIGEN

Terminology

ISBT symbol (number)	MNS7 (002.007)
Other names	Miltenberger
History	In 1951, the serum of Mrs. Miltenberger revealed a 'new' low prevalence antigen, named Mi^a. When other related antigens and antisera were found they formed a subsystem named Miltenberger. These antigens are in the MNS blood group system and a terminology based on glycophorin (e.g. GP.Vw, GP.Hop, etc.) is now commonly used[1].

Occurrence

Most populations less than 0.01%; Chinese and SE Asians up to 15%.

Expression

Cord RBCs	Expressed

Molecular basis associated with Miᵃ antigen[2]

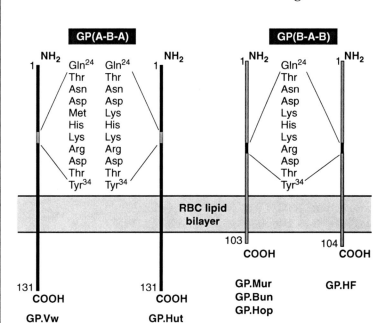

Anti-Miᵃ recognizes the amino acid sequence QTND M/K HKRDTY[34] but does not require residues C-terminal of the tyrosine at position 34[2].

Effect of enzymes/chemicals on Miᵃ antigen on intact RBCs

	GP.Vw; GP.Hut	GP.Mur; GP.Hop	GP.Bun; GP.HF
Ficin/papain	Sensitive	Sensitive/weakened	Sensitive/weakened
Trypsin	Sensitive	Resistant	Resistant
α-Chymotrypsin	Resistant	Sensitive	Sensitive/weakened
Pronase	Sensitive	Presumed sensitive	Presumed sensitive
Sialidase	Sensitive	Variable	Variable
DTT 200 mM	Resistant	Resistant	Resistant
Acid	Resistant	Resistant	Resistant

In vitro characteristics and clinical significance of alloanti-Miᵃ

Transfusion reaction	Rare
HDN	Mild to severe

Comments

Anti-Mi^a is often present in serum containing anti-Vw. Production of mono clonal anti-Mi^a (GAMA 210 and CBC-172) showed that anti-Mi^a exists as a single specificity and that Mi^a is a discrete antigen[3].

Due to the relatively high prevalence of some Mi(a+) phenotypes (particularly GP.Mur (Mi.III); up to 15% in parts of Taiwan)[4,5] in Chinese and SE Asian populations, it is recommended to include GP.Mur phenotype RBCs in antibody investigations.

References

1 Tippett, P. et al. (1992) Transf. Med. Rev. 6, 170–182.
2 Dahr, W. (1992) Vox Sang. 62, 129–135.
3 Chen, V. et al. (2001) Vox Sang. 80, 230–233.
4 Mak, K.H. et al. (1994) Transfusion 34, 238–241.
5 Broadberry, R.E. and Lin, M. (1994) Transfusion 34, 349–352.

M^c ANTIGEN

Terminology

ISBT symbol (number)	MNS8 (002.008)
History	M^c was described in 1953. The antigen appeared to be intermediate between M and N, and as such, was analogous to the situation described in 1948 for the Rh antigen c^v; hence the name M^c was used.

Occurrence

Less than 0.01%; all probands are of European origin.

Molecular basis associated with M^c antigen[1]

O-glycosylation of residues 2, 3 and 4 is normal.

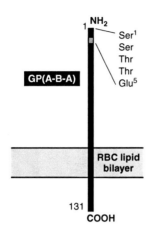

Variant glycophorin GPA^M(1–4)-GPB(5)-GPA(6–131)
Gene arrangement *GYP(A-B-A)*
Mechanism Gene conversion

Comments

No alloanti-M^c has been described. M^c is defined by the reaction of defined anti-M and anti-N: a majority of anti-M and a minority of anti-N react with M^c+ RBCs.

Reference

[1] Huang, C.-H. and Blumenfeld, O.O. (1995) In: Molecular Basis of Human Blood Group Antigens (Cartron, J.-P. and Rouger, P. eds) Plenum Press, New York, pp. 153–188.

Vw ANTIGEN

Terminology

ISBT symbol (number) MNS9 (002.009)
Other names Gr; Verweyst; Mi.I
History Identified in 1954; named for Mr. Verweyst; anti-Vw caused positive DAT on the RBCs of a Verweyst baby

Occurrence

About 0.057% in White populations; up to 1.43% in S.E. Switzerland.

Antithetical antigens

Hut (MNS19); ENEH (MNS40).

Expression

Cord RBCs   Expressed

Molecular basis associated with Vw antigen[1,2]

Amino acid	Met 28 of GPA
Nucleotide	T at bp 140 in exon 3
Variant glycophorin	GPA(1–27)-GPB(28)-GPA(29–131)
Gene arrangement	*GYP(A-ψB-A)*
Mechanism	Gene conversion with untemplated mutation or a single nucleotide substitution in *GYPA*.

The N-glycosylation consensus sequence is changed so that Asn 26 is not N-glycosylated, which results in a decreased M_r of about 3000.

Effect of enzymes/chemicals on Vw antigen on intact RBCs

Ficin/papain	Sensitive
Trypsin	Sensitive
α-Chymotrypsin	Resistant
Pronase	Sensitive
Sialidase	Resistant
DTT 200 mM	Resistant
Acid	Resistant

In vitro characteristics of alloanti-Vw

Immunoglobulin class	IgM; IgG
Optimal technique	RT; IAT
Complement binding	No

Clinical significance of alloanti-Vw

Transfusion reaction	No to severe
HDN	Mild to severe

Comments

The altered GPA carrying Vw usually carries N (MNS2).

One Vw+ homozygote person has been described who made anti-En[a] TS (anti-ENEH)[3].

Anti-Vw is found in 1% of sera and is a frequent component of multi-specific sera.

Anti-Vw is commonly found in sera of AIHA patients.

References

1 Dahr, W. (1992) Vox Sang. 62, 129–135.
2 Huang, C.-H. and Blumenfeld, O.O. (1995) In: Molecular Basis of Human Blood Group Antigens (Cartron, J.-P. and Rouger, P. eds) Plenum Press, New York, 153–188.
3 Spruell, P. et al. (1993) Transfusion 33, 848–851.

Mur ANTIGEN

Terminology

ISBT symbol (number)	MNS10 (002.010)
Other names	Murrell; Mu
History	Identified in 1961 as the cause of HDN in the Murrell family

Occurrence

Less than 0.1% in most populations; 6% in Chinese; 7% in Taiwanese; 9% in Thais.

Expression

Cord RBCs	Expressed

Molecular basis associated with Mur antigen[1,2]

Mur antigen is expressed when a sequence of amino acids ([34]YPAHTANE[41]) is encoded by the pseudoexon 3 of *GYPB*.

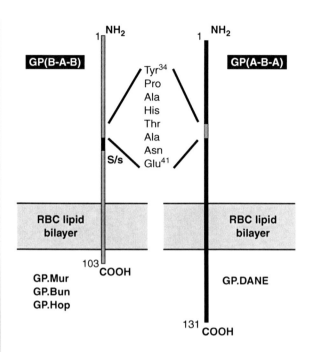

Variant glycophorin:

GP.Mur (Mi.III)	GPB(1–26)-GPψB(27–48)-GPA(49–57)-GPB^s(58–103)
GP.Bun (Mi.VI)	GPB(1–26)-GPψB(27–50)-GPA(51–57)-GPB^s(58–103)
GP.Hop (Mi.IV)	GPB(1–26)-GPψB(27–50)-GPA(51–57)-GPB^S(58–103)
GP.DANE (Mi.IX)	GPA(1–34)-GPψB(35–40)-GPA(41–131)

Contribution by parent glycophorin:

GP.Mur	GPB(1–26)-GPψB-GPA(49–57)-GPB(27–72)
GP.Bun, GP.Hop	GPB(1–26)-GPψB-GPA(51–57)-GPB(27–72)
GP.DANE	GPA(1–34)-GPψB-GPA(41–131)

Gene arrangement and mechanism:

GP.Mur, GP.Bun, GP.Hop	*GYP(B-A-B)* Gene conversion and splice site reactivation
GP.DANE	*GYP(A-ψB-A)* Gene conversion with untemplated mutation

Effect of enzymes/chemicals on Mur antigen on intact RBCs

Ficin/papain	Sensitive
Trypsin	Resistant
α-Chymotrypsin	Sensitive (resistant on GP.DANE)
Pronase	Sensitive

Sialidase	Presumed resistant
DTT 200 mM	Resistant
Acid	Resistant

In vitro characteristics of alloanti-Mur

Immunoglobulin class	IgM less common than IgG
Optimal technique	RT; IAT
Complement binding	No

Clinical significance of alloanti-Mur

Transfusion reaction	No to severe
HDN	No to severe

Comments

Anti-Mur occurs as a single specificity and is a common separable component of "anti-Mi^a" sera. Sera with inseparable anti-Mur and anti-Hut are now considered to contain an additional specificity, anti-MUT (anti-MNS35).

Anti-Mur is common in S.E. Asian and Oriental populations (0.2, 0.28, and 0.06% of patients in Thailand, Taiwan, and Hong Kong, respectively).

Mg+ (MNS11) RBCs reacted with serum from Mrs. Murrell but not with other anti-Mur[4].

References

[1] Huang, C.-H. and Blumenfeld, O.O. (1995) In: Molecular Basis of Human Blood Group Antigens (Cartron, J.-P. and Rouger, P. eds) Plenum Press, New York, pp. 153–188.
[2] Storry, J.R. et al. (2000) Transfusion 40, 560–565.
[3] Johe, K.K. et al. (1991) Blood 78, 2456–2461.
[4] Green, C. et al. (1994) Vox Sang. 66, 237–241.

M^g ANTIGEN

Terminology

ISBT symbol (number)	MNS11 (002.011)
Other names	Gilfeather
History	Identified in 1958; RBCs of a patient, Mr. Gilfeather, reacted with the serum of a donor

Occurrence

Less than 0.01%. In Switzerland and Sicily the incidence is 0.15%. One M^gM^g homozygote person has been described.

Expression

Cord RBCs   Expressed

Molecular basis associated with M^g antigen[1]

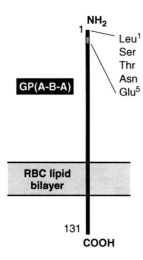

GP.M^g variant glycophorin   GPA^N(1–4)-GPB(5)-GPA(6–131)
Gene arrangement $GYP(A^N\text{-}B\text{-}A)$
Mechanism   Gene conversion with possible untemplated mutation.

The O-glycans attached to residues 2 and 3 are altered. There is no O-glycan attached to residue 4[2]. This causes a reduction in sialic acid content and decreased electrophoretic mobility.

Effect of enzymes/chemicals on M^g antigen on intact RBCs

Ficin/papain   Sensitive
Trypsin   Sensitive
α-Chymotrypsin   Resistant
Pronase   Sensitive

Sialidase	Resistant (mostly)
DTT 200 mM	Resistant
Acid	Resistant

In vitro characteristics of alloanti-M^g

Immunoglobulin class	IgM more common than IgG
Optimal technique	RT; 37 °C; IAT
Complement binding	No

Clinical significance of alloanti-M^g

No data available.

Comments

Human and rabbit anti-M and anti-N do not detect the M^g antigen. Some monoclonal anti-M react with M- M^g+ RBCs. Two of six anti-M^g reacted with a variant M^g+ RBC sample that had a higher level of glycosylation than other M^g+ samples.

M^g can be aligned with s and S, the latter alignment may be indicative of a Sicilian background. M^g+ RBCs carry DANE (MNS32) and were agglutinated by anti-Mur from Mrs. Murrell but not with other anti-Mur[3].

Anti-M^g is present in 1–2% of sera.

References
[1] Huang, C.-H. and Blumenfeld, O.O. (1995) In: Molecular Basis of Human Blood Group Antigens (Cartron, J.-P. and Rouger, P. eds) Plenum Press, New York, pp. 153–188.
[2] Dahr, W. et al. (1981) Hoppe-Seylers Z. Physiol. Chem. 362, 81–85.
[3] Green, C. et al. (1994) Vox Sang. 66, 237–241.

Vr ANTIGEN

Terminology

ISBT symbol (number)	MNS12 (002.012)
Other names	Verdegaal
History	Identified in 1958; named for the family in which the antigen and antibody were found

Occurrence

Only found in a few Dutch families.

Expression

Cord RBCs   Expressed

Molecular basis associated with Vr antigen[1]

Amino acid   Tyr 47 of GPA
Nucleotide   A at bp 197 in exon 3
Vr– form (wild type) has Ser 47 and C at position 197.

Effect of enzymes/chemicals on Vr antigen on intact RBCs[1]

Ficin/papain   Sensitive
Trypsin   Resistant
α-Chymotrypsin   Sensitive
Pronase   Sensitive
Sialidase   Resistant
DTT 200 mM   Resistant
Acid   Resistant

In vitro characteristics of alloanti-Vr

Immunoglobulin class   IgM and IgG
Optimal technique   RT; IAT
Complement binding   No

Clinical significance of alloanti-Vr

Transfusion reaction   No data
HDN   The original maker of anti-Vr had three
                                    Vr+ children; none had HDN

Comments

Inherited with Ms[2].
Anti-Vr has been found in multi-specific sera.

The Ser47Tyr substitution in GPA is predicted to introduce a novel α-chymotrypsin cleavage site, which would explain the (unexpected) sensitivity of the Vr antigen to α-chymotrypsin. Vr is located to the carboxyl side of the major trypsin cleavage site on GPA thus making it resistant to trypsin treatment.

References
1 Storry, J.R. et al. (2000) Vox Sang. 78, 52–56.
2 van der Hart, M. et al. (1958) Vox Sang. 3, 261–265.

Me ANTIGEN

Terminology

ISBT symbol (number)	MNS13 (002.013)
History	Anti-M^e identified in 1961; named M^e because epitope expressed on M+ RBCs and on He+ RBCs regardless of MN type

Molecular basis associated with M^e antigen[1]

M^e antigen is expressed when glycine occupies residue 5 of either GPA (M antigen [MNS1]) or GPB (He antigen [MNS6]).

Comments

Some anti-M (anti-M^e) have a component that reacts with glycine at residue 5 of GPA or GPB^{He}. The characteristics of these antibodies are the same as for anti-M (see MNS1). M^e on GPB^{He} is resistant to trypsin treatment and sensitive to α-chymotrypsin treatment.

Reference
1 Dahr, W. (1986) In: Recent Advances in Blood Group Biochemistry (Vengelen-Tyler, V. and Judd, W.J. eds) American Association of Blood Banks, Arlington, VA, pp. 23–65.

Mtª ANTIGEN

Terminology

ISBT symbol (number)	MNS14 (002.014)
Other names	Martin
History	Reported in 1962; named for the first antigen positive donor

Occurrence

Present in 0.24% of White Americans, 0.35% of Swiss, 0.1% of Black Americans and 1% of Thais.

Expression

Cord RBCs	Expressed

Molecular basis associated with Mtª antigen[1]

Amino acid	Ile 58 of GPA
Nucleotide	T at bp 230 in exon 3
Restriction enzyme	Ablates a *Msp* I site

Mt(a–) form (wild type) has Thr 58 and C at bp 230

Effect of enzymes/chemicals on Mtª antigen on intact RBCs

Ficin/papain	Variable
Trypsin	Resistant
α-Chymotrypsin	Resistant
Pronase	Sensitive
Sialidase	Resistant
DTT 200 mM	Resistant
Acid	Resistant
Chloroquine	Sensitive

In vitro characteristics of alloanti-Mtª

Immunoglobulin class	IgM; IgG
Optimal technique	RT; IAT
Complement binding	No

Clinical significance of alloanti-Mt^a

Transfusion reaction	No data
HDN	No to severe[2]

Comments

Inherited with Ns[3,4].

The variable susceptibility of Mt^a to ficin and papain treatment may reflect slight differences in the epitope recognized by certain anti-Mt^a or may result from the proximity of residue 58 to Arg 61, one of the two proteolytic sites on GPA.

Anti-Mt^a is found as single specificity and occasionally in multi-specific sera.

References

1 Storry, J.R. et al. (2000) Vox Sang. 78, 52–56.
2 Cheung, C.C. et al. (2002) Immunohematology 18, 37–39.
3 Swanson, J. and Matson, G.A. (1962) Vox Sang. 7, 585–590.
4 Konugres, A.A. et al. (1965) Vox Sang. 10, 206–207.

St^a ANTIGEN

Terminology

ISBT symbol (number)	MNS15 (002.015)
Other names	Stones
History	Antigen named in 1962 after the first producer of the antibody

Occurrence

Less than 0.1% in Caucasians, 2% in Asians and 6% in Japanese.

Expression

Cord RBCs	Presumed expressed

Molecular basis associated with St[a] antigen[1,2]

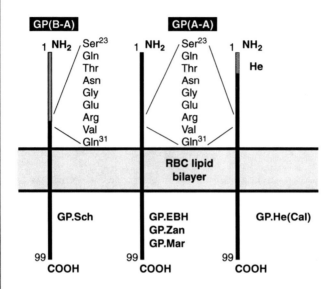

The St[a] antigen arises when amino acid at residue 26 of GPB or GPA joins to GPA at residue 59.

Variant glycophorin:

GP.Sch (M[r])	GPB(1–26)-GPA(27–99)
GP.Zan† (M[z]), GP.EBH t2	GPA(1–26)-GPA(27–99)
GP.Mar	GPA(1–26)-GPA(27–99)
GP.He(Cal)	GPA(1–5*)-GPA(6–99)

* Altered sequence

Contribution by parent glycophorin:

GP.Sch	GPB(1–26)-GPA(59–131)
GP.Zan†, GP.EBH t2	GPA(1–26)-GPA(59–131)
GP.Mar	GPA(1–26)-GPA(59–131)
GP.He(Cal)	GPA(1–5*)-GPA(6–26)-GPA(59–131)

* Altered sequence

Gene arrangement and mechanism:

GP.Sch	*GYP(B-A)*	Single crossover
GP.Zan†	*GYP(A-ψB-A)*	Gene conversion and splice site interaction
GP.EBH	*GYPA*	G>A mutation in exon III alters splice site causing deletion of exon 3 in one transcript (t2) whose product expresses St[a] but not ERIK [see ERIK (MNS37)]
GP.Mar	*GYP(A-ψE-A)*	Gene conversion and splice site interaction
GP.He(Cal)	*GYP(B-A-ψB-A)*	Gene conversion and splice site interaction

Effect of enzymes/chemicals on St^a antigen on intact RBCs

Ficin/papain	Variable
Trypsin	Resistant
α-Chymotrypsin	Resistant
Pronase	Sensitive
Sialidase	Presumed resistant
DTT 200 mM	Resistant
Acid	Resistant

In vitro characteristics of alloanti-St^a

Immunoglobulin class	IgM; IgG
Optimal technique	RT; IAT
Complement binding	No

Clinical significance of alloanti-St^a

No data are available.

Comments

GYP.Sch is the reciprocal gene rearrangement product of *GYP.Hil* (see Hil antigen [MNS20]) and *GYP.JL* (see TSEN antigen [MNS33]).

The shortened product from transcript 2 (t2) of the *GYP.EBH* gene, which lacks amino acids encoded by exon 3, expresses St^a but not ERIK antigen. The full length product (GP.EBH) expresses ERIK [MNS37] but not St^{a}[1]. (See MNS system page.)

GP.Zan and GP.Mar lack amino acids encoded by exon 3 and each has a trypsin-resistant M antigen.

One St(a+) homozygote person has been described.

Anti-St^a is a rare specificity and occurs in multispecific sera (especially anti-S). Anti-St^a is notorious for deteriorating *in vitro*.

References

[1] Huang, C.-H. and Blumenfeld, O.O. (1995) In: Molecular Basis of Human Blood Group Antigens (Cartron, J.-P. and Rouger, P. eds) Plenum Press, New York, pp. 153–188.

[2] Huang, C.-H., et al. (1994). Blood 84 (Suppl. 1), 238a (abstract).

Ri[a] ANTIGEN

Terminology

ISBT symbol (number)	MNS16 (002.016)
Other name	Ridley
History	Identified in 1962; named for the original Ri(a+) person

Occurrence

Only found and studied in one large family[1].

Expression

Cord RBCs	Presumed expressed

Molecular basis associated with Ri[a] antigen[2]

Amino acid	Lys 57 of GPA
Nucleotide	A at bp 226 in exon 3

Ri(a-) form (wild type) has Glu 57 and G at bp 226

Effect of enzymes/chemicals on Ri[a] antigen on intact RBCs

Ficin/papain	Partially sensitive
Trypsin	Sensitive
α-Chymotrypsin	Resistant
Pronase	Resistant
Sialidase	Not known
DTT 200 mM	Resistant
Acid	Resistant

In vitro characteristics of alloanti-Ri[a]

Immunoglobulin class	IgM (12 of 13 anti-Ri[a] were IgM, one was IgG)
Optimal technique	RT; IAT
Complement binding	Some

Clinical significance of alloanti-Ri[a]

No data are available.

Comments

The Glu57Lys substitution in GPA is predicted to introduce a novel trypsin cleavage site. Anti-Ri[a], likely to be naturally-occurring, was found in sera containing multiple antibodies to low incidence antigens[3]. Anti-S [see MNS3] often contain anti-Ri[a].
Inherited with MS[3].

References
[1] Cleghorn, T.E. (1962) Nature 195, 297–298.
[2] Storry, J.R. and Reid, M.E. (2001) Immunohematology 17, 76–81.
[3] Contreras, M. et al. (1984) Vox Sang. 46, 360–365.

Cl[a] ANTIGEN

Terminology

ISBT symbol (number)	MNS17 (002.017)
Other name	Caldwell
History	Identified in 1963; antibody found in an anti-B typing serum; named for the antigen positive person

Occurrence

Only found in one Scottish and one Irish family.

Expression

Cord RBCs	Presumed expressed

Effect of enzymes/chemicals on Cl[a] antigen on intact RBCs

Ficin/papain	Sensitive
Trypsin	Sensitive
α-Chymotrypsin	Resistant
Pronase	Sensitive
Sialidase	Not known
DTT 200 mM	Resistant
Acid	Presumed resistant

In vitro characteristics of alloanti-Cl^a

Immunoglobulin class	IgM
Optimal technique	RT
Complement binding	No

Clinical significance of alloanti-Cl^a

No data are available.

Comments

Anti-Cl^a was found in 24 of 5000 British blood donor sera. Inherited with Ms[1].

Reference

[1] Wallace, J. and Izatt, M.M. (1963) Nature 200, 689–690.

Ny^a ANTIGEN

Terminology

ISBT symbol (number)	MNS18 (002.018)
Other name	Nyberg
History	Identified in 1964; named for Mr. Nyberg, the first Ny(a+) person

Occurrence

Found in 0.2% of Norwegians, in one Swiss family and an American of non-Scandinavian descent.

Expression

Cord RBCs	Expressed

Molecular basis associated with Ny^a antigen[1]

Amino acid	Glu 27 of GPA
Nucleotide	A at bp 138 in exon 3

Ny(a–) form (wild type) has Asp 27 and T at bp 138.

Effect of enzymes/chemicals on Ny^a antigen on intact RBCs

Ficin/papain	Sensitive
Trypsin	Sensitive
α-Chymotrypsin	Resistant
Pronase	Sensitive
Sialidase	Not known
DTT 200 mM	Resistant
Acid	Resistant

In vitro characteristics of alloanti-Ny^a

Immunoglobulin class	IgM
Optimal technique	RT
Complement binding	No

Clinical significance of alloanti-Ny^a

No data are available.

Comments

Inherited with Ns[2,3].
Anti-Ny^a appears to be naturally-occurring, found in about 0.1% of sera studied, and has been produced in rabbits.

References

[1] Daniels, G.L. et al. (2000) Transfusion 40, 555–559.
[2] Örjasaeter, H. et al. (1964) Vox Sang. 9, 673–683.
[3] Kornstad, L. et al. (1971) Am. J. Hum. Genet. 23, 612–613.

Hut ANTIGEN

Terminology

ISBT symbol (number)	MNS19 (002.019)
Other name	Mi.II
History	Anti-Hut, reported in 1962, was redefined in 1982; first identified in 1958 as the cause of HDN in the Hutchinson family; considered to be anti-Mi^a at that time but later shown to be different

Occurrence

Less than 0.01%.

Antithetical antigen

Vw (MNS9); ENEH (MNS40)

Expression

Cord RBCs   Expressed

Molecular basis associated with Hut antigen[1,2]

Amino acid   Lys 28 of GPA
Nucleotide   A at bp140 in exon 3
Variant glycophorin   GPA(1–27)-GPB(28)-GPA(29–131)
Gene arrangement   GYP(A-ψB-A)
Mechanism   Gene conversion or a single nucleotide
                              substitution in GYPA

The N-glycosylation consensus sequence is changed so that Asn 26 is not N-glycosylated, which results in a decreased M_r of about 3000.

Effect of enzymes/chemicals on Hut antigen on intact RBCs

Ficin/papain   Sensitive
Trypsin   Sensitive
α-Chymotrypsin   Resistant
Pronase   Sensitive
Sialidase   Resistant
DTT 200 mM   Resistant
Acid   Resistant

In vitro characteristics of alloanti-Hut

Immunoglobulin class   IgM more common than IgG
Optimal technique   RT; IAT
Complement binding   No

Clinical significance of alloanti-Hut

Transfusion reaction	No data
HDN	No to moderate

Comments

Hut has been aligned with *MS, Ns* and *Ms* in decreasing order of frequency but not with *NS*.

The specificity originally called anti-Hut is now called anti-MUT (see MNS35) since Hut+, Mur+ RBCs are reactive. Anti-Hut reacts with Hut+ only.

References
1 Dahr, W. (1992) Vox Sang. 62, 129–135.
2 Huang, C.-H. and Blumenfeld, O.O. (1995) In: Molecular Basis of Human Blood Group Antigens (Cartron, J.-P. and Rouger, P. eds) Plenum Press, New York, 153–188.

Hil ANTIGEN

Terminology

ISBT symbol (number)	MNS20 (002.020)
Other name	Hill
History	Antibody identified in 1963 as the cause of HDN in the Hill family; in 1966 named Hil

Occurrence

Most populations less than 0.01%; Chinese 6%. One GP.Hil homozygote has been described.

Expression

Cord RBCs	Expressed

67

Molecular basis associated with Hil antigen[1-4]

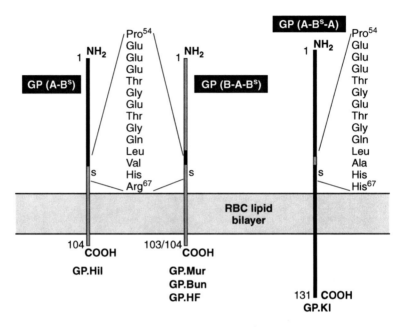

A PCR-RFLP method detects Hil+ phenotypes[5].

Variant glycophorin:

GP.Hil (Mi.V)	GPA(1–58)-GPBˢ(59–104)
GP.Mur (Mi.III)	GPB(1–26)-GPψB(27–48)-GPA(49–57)-GPBˢ(58–103)
GP.Bun (Mi.VI)	GPB(1–26)-GPψB(27–50)-GPA(51–57)-GPBˢ(58–103)
GP.HF (Mi.X)	GPB(1–26)-GPψB(27–34)-GPA(35–58)-GPBˢ(59–104)
GP.KI	GPA(1–60)-GPB(61–62)-GPA(63–131)

Contribution by parent glycophorin:

GP.Hil	GPA(1–58)-GPB(27–72)
GP.Mur	GPB(1–26)-GPψB-GPA(49–57)-GPB(27–72)
GP.Bun	GPB(1–26)-GPψB-GPA(51–57)-GPB(27–72)
GP.HF	GPB(1–26)-GPψB-GPA(35–58)-GPB(27–72)
GP.KI	GPA(1–60)-GPB(29–30)-GPA(63–131)

Gene arrangement and mechanism:

GP.Hil	GYP(A-B) Single crossover
GP.Mur	
GP.Bun	GYP(B-A-B) Gene conversion with splice site
GP.HF	reactivation
GP.KI	GYP(A-B-A) Gene rearrangement

Effect of enzymes/chemicals on Hil antigen on intact RBCs

Ficin/papain	Sensitive
Trypsin	Resistant
α-Chymotrypsin	Sensitive

Pronase	Sensitive
Sialidase	Resistant
DTT 200 mM	Resistant
Acid	Resistant

In vitro characteristics of alloanti-Hil

Immunoglobulin class	IgM and IgG
Optimal technique	RT; IAT
Complement binding	No

Clinical significance of alloanti-Hil

Transfusion reaction	No data
HDN	No to moderate

Comments

Reciprocal product to *GYP.Hil* is *GYP.Sch* (see St^a antigen).
Hil+ RBCs, except those carrying GP.KI, are also MINY+.

References
[1] Huang, C.-H. and Blumenfeld, O.O. (1995) In: Molecular Basis of Human Blood Group Antigens (Cartron, J.-P. and Rouger P. eds) Plenum Press, New York, pp. 153–188.
[2] Poole, J. (2000) Blood Rev. 14, 31–43.
[3] Dahr, W. (1992) Vox Sang. 62, 129–135.
[4] Poole, J. et al. (1998). Transfusion 38 (Suppl.), 103S (abstract).
[5] Shih, M.C. et al. (2000) Transfusion 40, 54–61.

M^v ANTIGEN

Terminology

ISBT symbol (number)	MNS21 (002.021)
Other name	Armstrong
History	Found in 1961 when a serum containing anti-N agglutinated RBCs from 1 in 400 M+ N- whites; described in detail in 1966; the 'v' of M^v is for the 'variant' antigen detected

Occurrence

Less than 0.01%.

Expression

Cord RBCs   Expressed

Molecular basis associated with M^v antigen[1]

Amino acid   Ser 3 of GPB
Nucleotide   G at bp 65 in exon 2
M^v- form (wild type) has Thr 3 and C at bp 65.

Effect of enzymes/chemicals on M^v antigen on intact RBCs

Ficin/papain   Sensitive
Trypsin   Resistant
α-Chymotrypsin   Sensitive
Pronase   Sensitive
Sialidase   Sensitive
DTT 200 mM   Resistant
Acid   Resistant

In vitro characteristics of alloanti-M^v

Immunoglobulin class   IgG and IgM
Optimal technique   IAT
Complement binding   No

Clinical significance of alloanti-M^v

Transfusion reaction   No data
HDN   No to moderate

Comments

M^v+ RBCs have a decreased level of GPB and a weak expression of s (MNS4)
    and may have a slight weakening of S when M^v is associated with MS[1,2].
Inherited with *Ms* in 14 families and *MS* in two families.
GPB carrying M^v does not express 'N'.

References
[1] Storry, J.R. et al. (2001) Transfusion 41, 269–275.
[2] Dahr, W. and Longster, G. (1984) Blut 49, 299–306.

Far ANTIGEN

Terminology

ISBT symbol (number)	MNS22 (002.022)
Other names	Kam; Kamhuber
History	The Kam antigen, reported in 1966, and the Far antigen, reported in 1968, were shown to be the same in 1977. The name Far was chosen. Anti-'Kam' caused a severe transfusion reaction in a multiply transfused hemophiliac. Probably immunized following transfusion with blood of the same donor 11 years previously!!

Occurrence

Found in only two families.

Expression

Cord RBCs	Expressed

Effect of enzymes/chemicals on Far antigen on intact RBCs

Ficin/papain	Resistant
Trypsin	Resistant
α-Chymotrypsin	Not known
Pronase	Not known
Sialidase	Not known
DTT 200 mM	Resistant
Acid	Presumed resistant

In vitro characteristics of alloanti-Far

Immunoglobulin class	IgG
Optimal technique	IAT
Complement binding	No

71

Clinical significance of alloanti-Far

Transfusion reaction	Severe in one
HDN	Severe in one

Comments

Travels with Ns[1] and MS[2].
Only two examples of anti-Far have been reported.

References
1. Cregut, R. et al. (1974) Vox Sang. 26, 194–198.
2. Speiser, P. et al. (1966) Vox Sang. 11, 113–115.

s^D ANTIGEN

Terminology

ISBT symbol (number)	MNS23 (002.023)
Other name	Dreyer
History	Named in 1981; 's' was used because the s antigen is expressed weakly and 'D' from the family name

Occurrence

Found only in one white South African family.

Expression

Cord RBCs	Expressed

Molecular basis associated with s^D antigen[1]

Amino acid	Arg 39 of GPB
Nucleotide	G at bp 173 in exon 4

Wild type (s^D−form) has Pro 39 and C at position 173.

Effect of enzymes/chemicals on s^D antigen on intact RBCs

Ficin/papain	Partially sensitive
Trypsin	Resistant
α-Chymotrypsin	Resistant
Pronase	Not known
Sialidase	Presumed resistant
DTT 200 mM	Presumed resistant
Acid	Presumed resistant

In vitro characteristics of alloanti-s^D

Immunoglobulin class	IgG
Optimal technique	IAT
Complement binding	No

Clinical significance of alloanti-s^D

Transfusion reaction	No data
HDN	No to Severe

Comments

S + s + s^D + RBCs have a weakened expression of the s antigen (MNS4)[2].
Inherited with Ms[2].
The Pro39Arg introduces a novel papain cleavage site. However, the close proximity of the antigen to the lipid bilayer may make the site relatively inaccessible.

References
[1] Storry, J.R. et al. (2001) Transfusion 41, 269–275.
[2] Shapiro, M. and Le Roux, M.E. (1981) Transfusion 21, 614 (abstract).

Mit ANTIGEN

Terminology

ISBT symbol (number)	MNS24 (002.024)
Other name	Mitchell
History	Named in 1980 after the family where the antigen (father's RBCs) and antibody (mother's serum) were found

Occurrence

Found in 0.1% of Western Europeans.

Expression

Cord RBCs   Expressed

Molecular basis associated with Mit antigen[1]

Amino acid   His 35 of GPB
Nucleotide   A at bp 161 in exon 4
Mit− form (wild type) has Arg 35 and G at bp 161.

Effect of enzymes/chemicals on Mit antigen on intact RBCs

Ficin   Resistant
Papain   Partially sensitive
Trypsin   Resistant
α-Chymotrypsin   Resistant
Pronase   Weakened
Sialidase   Variable
DTT 200 mM   Resistant
Acid   Resistant

In vitro characteristics of alloanti-Mit

Immunoglobulin class   IgG
Optimal technique   IAT
Complement binding   No

Clinical significance of alloanti-Mit

Transfusion reaction   No data
HDN   Positive DAT; no clinical HDN

Comments

Mit+ RBCs have weakened expression of S antigen[2,3] or s antigen[1].
Mit is usually associated with *MS* and rarely with *NS* or *Ms*.

References
[1]   Storry, J.R. et al. (2001) Transfusion 41, 269–275.
[2]   Skradski, K.J. et al. (1983) Transfusion 23, 409 (abstract).
[3]   Eichhorn, M. et al. (1981) Transfusion 21, 614 (abstract).

Dantu ANTIGEN

Terminology

ISBT symbol (number)   MNS25 (002.025)
History   Named in 1984 after the first proband

Occurrence

Found in 0.5% of Blacks.

Expression

Cord RBCs   Expressed

Molecular basis associated with Dantu antigen[1,2]

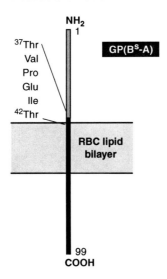

Variant glycophorin   GPB^s(1–39)-GPA(40–99)
Contribution by parent glycophorin   GPB(1–39)-GPA(70–131)
Gene arrangement   *GYP(B-A)*
Mechanism   Single crossover

The MD type is associated with a chromosome carrying *GYPA*, *GYP(B-A)* and *GYPB* genes. The NE type is associated with a chromosome carrying *GYPA*, *GYP(B-A)* and a duplicated *GYP(B-A)*. The Ph type, which had a higher ratio of GP(B-A) molecules to GPA than the NE type[3], was not studied at the molecular level but may be associated with a chromosome carrying *GYPA*, *GYP(B-A)*.

Effect of enzymes and chemicals on Dantu antigen on intact RBCs

Ficin/papain	Resistant
Trypsin	Resistant
α-Chymotrypsin	Resistant
Pronase	Resistant
Sialidase	Presumed resistant
DTT 200 mM	Resistant
Acid	Resistant

In vitro characteristics of alloanti-Dantu

Immunoglobulin class	IgM and IgG
Optimal technique	RT; IAT
Complement binding	No

Clinical significance of alloanti-Dantu

Transfusion reaction	No data
HDN	Positive DAT; no clinical HDN

Comments

Dantu positive RBCs (NE type) have a weak expression of s and are U negative.

The reciprocal product of *GYP.Dantu* is *GYP.TK* (see SAT antigen).

References

1 Huang, C.-H. and Blumenfeld, O.O. (1995) In: Molecular Basis of Human Blood Group Antigens (Cartron, J.-P. and Rouger, P. eds) Plenum Press, New York, pp. 153–188.
2 Blumenfeld, O.O. et al. (1987) J. Biol. Chem. 262, 11864–11870.
3 Merry, A.H. et al. (1986) Biochem. J. 233, 93–98.

Hop ANTIGEN

Terminology

ISBT symbol (number)	MNS26 (002.026)
History	Reported in 1977 and named after the first donor whose RBCs expressed the antigen

Occurrence

Most populations less than 0.01%; Thais 0.68%.

Expression

Cord RBCs   Presumed expressed

Molecular basis associated with Hop antigen[1-3]

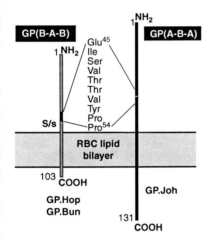

Variant glycophorin:
GP.Hop (Mi.IV)   GPB(1–26)-GPψB(27–50)-GPA(51–57)-GPB^S(58–103)
GP.Bun (Mi.VI)   GPB(1–26)-GPψB(27–50)-GPA(51–57)-GPB^s(58–103)
GP.Joh (Mi.VIII)   GPA(1–48)-GPψB(49)-GPA(50–131)
Contribution by parent glycophorin:
GP.Hop, GP.Bun   GPB(1–26)-GPψB-GPA(51–57)-GPB(27–72)
GP.Joh   GPA(1–48)-GPψB-GPA(50–131)
Gene arrangement and mechanism:
GP.Hop, GP.Bun   *GYP(B-A-B)* Gene conversion with splice site reactivation
GP.Joh   *GYP(A-B-A)* Gene conversion

Effect of enzymes/chemicals on Hop antigen on intact RBCs

	GP.Hop (Mi.IV)	GP.Bun (Mi.VI)	GP.Joh (Mi.VIII)
Ficin/papain	Sensitve	Sensitive	Sensitive
Trypsin	Resistant	Resistant	Sensitive
α-Chymotrypsin	Sensitive	Variable	Resistant
Pronase	Sensitive	Sensitive	Sensitive
Sialidase	Variable	Variable	Variable
DTT 200 mM	Resistant	Resistant	Resistant
Acid	Resistant	Resistant	Resistant

In vitro characteristics of alloanti-Hop

Immunoglobulin class	IgG
Optimal technique	IAT
Complement binding	No

Clinical significance of alloanti-Hop

No data are available.

Comments

Antigen defined by Anek serum (predominantly anti-Hop, weak anti-Nob). Sera which contain anti-Hop may also contain anti-Nob (see MNS27).

References

[1] Huang, C.-H. and Blumenfeld, O.O. (1995) In: Molecular Basis of Human Blood Group Antigens (Cartron, J.-P. and Rouger, P. eds) Plenum Press, New York, pp. 153–188.

[2] Storry, J.R. et al. (2000) Transfusion 40, 560–565.

[3] Dahr, W. (1992) Vox Sang. 62, 129–135.

Nob ANTIGEN

Terminology

ISBT symbol (number)	MNS27 (002.027)
History	The antigen is defined by the Lane serum and was named Nob after the person whose RBCs carried the antigen

Occurrence

Less than 0.01%.

Antithetical antigen

ENKT (MNS29)

Expression

Cord RBCs	Presumed expressed

Molecular basis associated with Nob antigen[1,2]

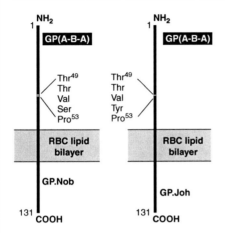

Variant glycophorin:
GP.Nob(Mi.VII)   GPA(1–48)-GPψB(49–52)-GPA(53–131)
GP.Joh(Mi.VIII)   GPA(1–48)-GPψB(49)-GPA(50–131)
Contribution by parent glycophorin:
GP.Nob   GPA(1–48)-GPψB-GPA(53–131)
GP.Joh   GPA(1–48)-GPψB-GPA(50–131)
Gene arrangement and mechanism:
GP.Nob, GP.Joh   GYP(A-B-A) Gene conversion

Effect of enzymes/chemicals on Nob antigen on intact RBCs

Ficin/papain   Sensitive
Trypsin   Resistant
α-Chymotrypsin   Resistant
Pronase   Sensitive
Sialidase   Variable
DTT 200 mM   Resistant
Acid   Resistant

In vitro characteristics of alloanti-Nob

Immunoglobulin class   IgM less common than IgG
Optimal technique   RT; IAT
Complement binding   No

Clinical significance of alloanti-Nob

Transfusion reaction   Mild in one case
HDN   No data

Comments

The Raddon serum is predominantly anti-Nob with a weak anti-Hop. Sera which contain anti-Nob may also contain anti-Hop (see **MNS26**).

References
[1] Huang, C.-H. and Blumenfeld O.O. (1995) In: Molecular Basis of Human Blood Group Antigens (Cartron J.-P. and Rouger P., eds) Plenum Press, New York, pp. 153–188.
[2] Dahr, W. (1992) Vox Sang. 62, 129–135.

En^a ANTIGEN

Terminology

ISBT symbol (number)	MNS28 (002.028)
History	Named in 1965 when it was recognized that the antigen was carried on an important component of the *en*velope of the RBC. Joined the MNS system in 1985

Occurrence

All populations	100%

Expression

Cord RBCs	Expressed

Molecular basis associated with En^a antigen[1]

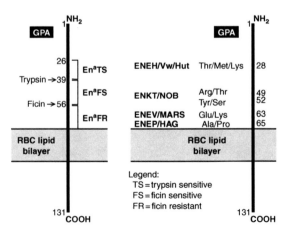

Effect of enzymes/chemicals on En^a antigen on intact RBCs

Ficin/papain	See figure
Trypsin	See figure
α-Chymotrypsin	Resistant
Pronase	Most are sensitive
Sialidase	Variable
DTT 200 mM	Resistant
Acid	Resistant

In vitro characteristics of alloanti-En^a

Immunoglobulin class	IgM and IgG
Optimal technique	RT; IAT
Complement binding	Rare

Clinical significance of alloanti-En^a

Transfusion reaction	No to severe
HDN	No to severe

Autoantibody

Yes (anti-En^{a}TS, anti-En^{a}FS and anti-En^{a}FR).

Comments

RBCs that lack GPA lack all En^a antigens, type as Wr(b-) and have reduced
  levels of sialic acid (40% of normal).

Reference
[1] Issitt, P.D. et al. (1981) Transfusion 21, 473–474.

ENKT ANTIGEN

Terminology

ISBT symbol (number)	MNS29 (002.029)
Other name	En^{a}FS, En^{a}KT
History	Reported as En^{a}FS in 1985. In 1988, it was named 'EN' because it is a high prevalence antigen on GPA and 'KT' for the initials of the first antigen-negative proband

Occurrence

All populations   100%

Antithetical antigen

Nob (MNS27)

Expression

Cord RBCs   Presumed expressed

Molecular basis associated with ENKT antigen[1]

Arg 49 and Tyr 52 of GPA

Effect of enzymes/chemicals on ENKT antigen on intact RBCs

Ficin/papain   Sensitive
Trypsin   Resistant
α-Chymotrypsin   Resistant
Pronase   Sensitive
Sialidase   Presumed resistant
DTT 200 mM   Resistant
Acid   Resistant

In vitro characteristics of alloanti-ENKT

Immunoglobulin class   IgG
Optimal technique   IAT

Clinical significance of alloanti-ENKT

No data are available.

Reference

[1] Dahr, W. (1992) Vox Sang. 62, 129–135.

'N' ANTIGEN

Terminology

ISBT symbol (number)   MNS30 (002.030)
Other name   GPB^N
History   Named when it was realized that the N-terminal amino acid sequence of GPB was the same as GPA carrying the N antigen. Quotation marks were used to distinguish GPB^N from GPA^N. Allocated an MNS number in 1985 by the ISBT

Occurrence

Present in all cells except those deficient in GPB or RBCs with GPB expressing He or M^v antigen.

Antithetical antigen

He (MNS6)

Expression

Cord RBCs   Expressed

Molecular basis associated with 'N' antigen[1]

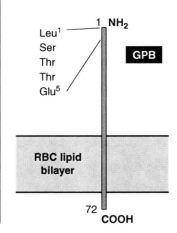

Effect of enzymes/chemicals on 'N' antigen on intact RBCs

Ficin/papain	Sensitive
Trypsin	Resistant
α-Chymotrypsin	Sensitive
Pronase	Sensitive
Sialidase	Variable
DTT 200 mM	Resistant
Acid	Resistant

Comment

See N antigen (MNS2). No anti-'N' exists.

Reference

[1] Blanchard, D. et al. (1987) J. Biol. Chem. 262, 5808–5811.

Or ANTIGEN

Terminology

ISBT symbol (number)	MNS31 (002.031)
Other names	Orriss; Or[a]
History	Named in 1987 after the family in which the antigen was first found

Occurrence

Found in two Japanese, one Australian, one African American and one Jamaican.

Expression

Cord RBCs	Expressed

Molecular basis associated with Or antigen[1,2]

Amino acid	Trp 31 of GPA
Nucleotide	T at bp 148 in exon 3

Or− form (wild type) Arg 31 and C at bp 148.

Effect of enzymes/chemicals on Or antigen on intact RBCs

Ficin/papain	Sensitive
Trypsin	Variable
α-Chymotrypsin	Resistant
Pronase	Sensitive
Sialidase	Sensitive
DTT 200 mM	Resistant
Acid	Resistant

In vitro characteristics of alloanti-Or

Immunoglobulin class	IgM more common than IgG
Optimal technique	RT
Complement binding	No

Clinical significance of alloanti-Or

Transfusion reactions	No data
HDN	Moderate[2]

Comments

The M (MNS1) antigen on Or+ RBCs is more resistant to trypsin treatment than normal M, presumably due to the close proximity of the mutation to the major trypsin cleavage site[3].

References

[1] Tsuneyama, H. et al. (1998) Vox Sang. 74 (Suppl. 1), 1446 (abstract).
[2] Reid, M.E. et al. (2000) Vox Sang. 79, 180–182.
[3] Bacon, J.M. et al. (1987) Vox Sang. 52, 330–334.

DANE ANTIGEN

Terminology

ISBT symbol (number)	MNS32 (002.032)
History	Named in 1991 after it was found in four Danish families

Occurrence

Most populations less than 0.01%; Danes 0.43%.

Expression

Cord RBCs   Presumed expressed

Molecular basis associated with DANE antigen[1]

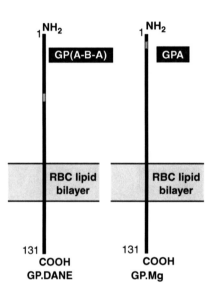

Variant glycophorin GP.DANE (Mi.IX)	GPA(1–34)-GPψB(35–40)-GPA(41–131)
Contribution by parent glycophorins	GPA(1–34)-GPψB-GPA(41–131)
Gene arrangement and mechanism	*GYP(A-B-A)* Gene conversion with untemplated mutation of Ile 46 of GPA to Asn 45 of GP.DANE

Effect of enzymes/chemicals on DANE antigen on intact RBCs

Ficin/papain	Sensitive
Trypsin	Sensitive
α-Chymotrypsin	Resistant
Pronase	Sensitive
Sialidase	Presumed resistant
DTT 200 mM	Resistant
Acid	Resistant

In vitro characteristics of alloanti-DANE

Immunoglobulin class	IgG
Optimal technique	IAT
Complement binding	No

Clinical significance of alloanti-DANE

Clinical significance is unknown since only one example, in an untransfused male, has been described.

Comments

GP.DANE has trypsin-resistant M (MNS1) and Mur (MNS10) antigens but does not express other low prevalence MNS antigens[2]. Mg+ RBCs are DANE+, maybe due to the presence of Asn 45 in the hybrid[3].

DANE is inherited with *MS*.

References

[1] Huang, C.-H. et al. (1992) Blood 80, 2379–2387.
[2] Skov, F. et al. (1991) Vox Sang. 61, 130–136.
[3] Green, C. et al. (1994) Vox Sang. 66, 237–241.

TSEN ANTIGEN

Terminology

ISBT symbol (number)	MNS33 (002.033)
History	Named in 1992 after the last name of the producer of the first antibody

Occurrence

Less than 0.01%.

Expression

Cord RBCs	Expressed

Molecular basis associated with TSEN antigen[1,2]

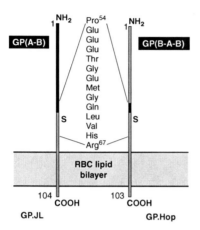

Variant glycophorin:

GP.JL (Mi.XI)	GPA(1–58)-GPB^S(59–104)
GP.Hop (Mi.IV)	GPB(1–26)-GPψB(27–50)-GPA(51–57)-GPB^S(58–103)

Contribution by parent glycophorin:

GP.JL	GPA(1–58)-GPB(27–72)
GP.Hop	GPB(1–26)-GPψB-GPA(51–57)-GPB(27–72)

Gene arrangement and mechanism:

GP.JL	*GYP(A-B)* Single crossover
GP.Hop	*GYP(B-A-B)* Gene conversion with splice site reactivation

Effect of enzymes/chemicals on TSEN antigen on intact RBCs

Ficin/papain	Sensitive
Trypsin	Resistant
α-Chymotrypsin	Sensitive
Pronase	Sensitive
Sialidase	Resistant
DTT 200 mM	Resistant
Acid	Resistant

In vitro characteristics of alloanti-TSEN

Immunoglobulin class	IgM and IgG
Optimal technique	RT; IAT
Complement binding	No

Clinical significance of alloanti-TSEN

Transfusion reactions	No
HDN	No

Comments

Reciprocal product of *GYP.JL* is *GYP.Sch* (see St^a antigen [MNS15]).
TSEN+ RBCs are also MINY+ (MNS34).
Several examples of anti-TSEN have been described[3].
Some anti-S do not agglutinate S + s+ TSEN+ RBCs.

References

1 Huang, C.-H. and Blumenfeld, O.O. (1995) In: Molecular Basis of Human Blood Group Antigens (Cartron, J.-P. and Rouger, P. eds) Plenum Press, New York, pp. 153–188.
2 Reid, M.E. et al. (1992) Vox Sang. 63, 122–128.
3 Storry, J.R. et al. (2000) Vox Sang. 79, 175–179.

MINY ANTIGEN

Terminology

ISBT symbol (number)	MNS34 (002.034)
History	Named in 1992 after the only producer of the antibody

Occurrence

Less than 0.01% in most populations; 6% in Chinese.

Expression

Cord RBCs	Presumed expressed

Molecular basis associated with MINY antigen[1]

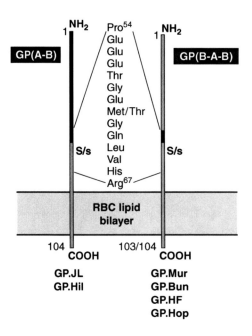

GP(A-B)   Pro⁵⁴   GP(B-A-B)

GP.JL   GP.Mur
GP.Hil   GP.Bun
           GP.HF
           GP.Hop

For details of variant glycophorin, contribution by parent glycophorin, gene arrangement and mechanism, see Hil (MNS20) and TSEN (MNS33).

Effect of enzymes/chemicals on MINY antigen on intact RBCs

Ficin/papain	Sensitive
Trypsin	Resistant
α-Chymotrypsin	Sensitive
Pronase	Sensitive
Sialidase	Resistant
DTT 200 mM	Resistant
Acid	Resistant

In vitro characteristics of alloanti-MINY

Immunoglobulin class	IgM
Optimal technique	RT

Clinical significance of alloanti-MINY

No data

Comments

All Hil+ (MNS20) and TSEN+ (MNS33) RBCs are MINY positive except
when the Hil antigen is carried on GP.KI (MNS20)[2].

References

[1] Reid, M.E. et al. (1992) Vox Sang. 63, 129–132.
[2] Poole, J. et al. (1998) Transfusion 38 (Suppl.), 103S (abstract)

MUT ANTIGEN

Terminology

ISBT symbol (number)	MNS35 (002.035)
History	The specificity originally called anti-Hut was renamed anti-MUT in 1984 because both <u>Mu</u>r+ and <u>Hut</u>+ RBCs are reactive

Occurrence

Less than 0.01% in most populations; 6% in Chinese.

Expression

Cord RBCs	Presumed expressed

Molecular basis associated with MUT antigen[1]

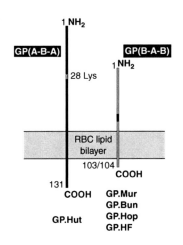

For details of variant glycophorin, contributions by parent glycophorin, gene arrangement and mechanism, see Hut (MNS19), Hil (MNS20) and Hop (MNS26).

Effect of enzymes and chemicals on MUT antigen on intact RBCs

	GP(A-B-A)	GP(B-A-B)
Ficin/papain	Sensitive	Sensitive
Trypsin	Sensitive	Resistant
α-Chymotrypsin	Resistant	Sensitive
Pronase	Sensitive	Sensitive
Sialidase	Presumed resistant	Presumed resistant
DTT 200 mM	Resistant	Resistant
Acid	Resistant	Resistant

In vitro characteristics of alloanti-MUT

Immunoglobulin class	IgM and IgG
Optimal technique	RT; IAT
Complement binding	No

Clinical significance of alloanti-MUT

No data are available.

Comments

Anti-MUT often in serum with (and is separable from) anti-Hut (see MNS19).

Reference
[1] Huang, C.-H. and Blumenfeld, O.O. (1995) In: Molecular Basis of Human Blood Group Antigens (Cartron, J.-P. and Rouger, P. eds) Plenum Press, New York, pp. 153–188.

SAT ANTIGEN

Terminology

ISBT symbol (number)	MNS36 (002.036)
History	Reported in 1991 and named after the first proband whose RBCs carried the antigen; joined the MNS system in 1994

Occurrence

Less than 0.01%.

Expression

Cord RBCs   Presumed expressed

Molecular basis associated with SAT antigen[1,2]

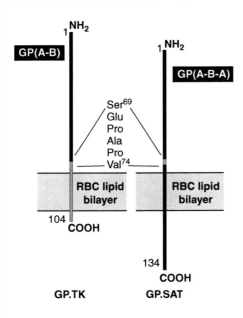

Variant glycophorin:
GP.TK   GPA(1–71)-GPB(72–104)
GP.SAT   GPA(1–71)-GPB(72–74)-GPA(75–134)
Contribution by parent glycophorin:
GP.TK   GPA(1–71)-GPB(40–72)
GP.SAT   GPA(1–71)-GPB(40–42)-GPA(72–131)
Gene arrangement and mechanism:
GP.TK   *GYP(A-B)* single crossover
GP.SAT   *GYP(A-B-A)* gene conversion

Effect of enzymes/chemicals on SAT antigen on intact RBCs

Ficin/papain   Sensitive
Trypsin   Sensitive (GP.TK); resistant (GP.SAT)

α-Chymotrypsin	Sensitive
Pronase	Sensitive
Sialidase	GP.SAT resistant
DTT 200 mM	Resistant
Acid	Resistant

In vitro characteristics of alloanti-SAT

Immunoglobulin class	IgG
Optimal technique	IAT
Complement binding	No

Clinical significance of alloanti-SAT

No data are available.

Comments

The reciprocal product of *GYP.TK* is GP.Dantu (see MNS25).

References
[1] Huang, C.-H. et al. (1995) Blood 85, 2222–2227.
[2] Uchikawa, M. et al. (1994) Vox Sang. 67 (Suppl. 2), 116 (abstract).

ERIK ANTIGEN

Terminology

ISBT symbol (number)	MNS37 (002.037)
History	Named in 1993 after the proband whose St(a+) RBCs had another low prevalence antigen

Occurrence

Less than 0.01%.

Expression

Cord RBCs	Presumed expressed

Molecular basis associated with ERIK antigen[1,2]

Amino acid   Arg 59 of GPA
Nucleotide   A at bp 232 in exon 4
ERIK− form (wild type) has Gly 59 and G at position 232. ERIK has been associated with a *GYP(A-E-A)*, which encodes a variant of GPA carrying St^a.

Effect of enzymes/chemicals on ERIK antigen on intact RBCs

Ficin/papain	Variable
Papain	Sensitive
Trypsin	Partially sensitive
α-Chymotrypsin	Resistant
Pronase	Sensitive
Sialidase	Resistant
DTT 200 mM	Resistant
Acid	Resistant

In vitro characteristics of alloanti-ERIK

Immunoglobulin class	IgG
Optimal technique	IAT
Complement binding	No

Clinical significance of alloanti-ERIK

Transfusion reaction	No data
HDN	Positive DAT

Comments

Alternative splicing of *GYP.EBH* gives rise to a variant glycophorin GP.EBH(t2) expressing the St^a antigen (see MNS15). Thus, in ERIK+ RBCs, ERIK and St^a antigens are carried on different glycophorin molecules. (See table on MNS system page.)
The Gly59Arg mutation introduces a trypsin cleavage site.

References
[1] Huang, C.-H. et al. (1993) J. Biol. Chem. 268, 25902–25908.
[2] Huang, C.-H. et al. (1994). Blood 84 (Suppl. 1), 238a (abstract).

Os^a ANTIGEN

Terminology

ISBT symbol (number)	MNS38 (002.038)
Other names	700.033
History	Named in 1983 after Osaka, the town where the antibody and antigen were first found; joined the MNS system in 1994

Occurrence

Only studied in one Japanese family[1].

Expression

Cord RBCs	Presumed expressed

Molecular basis associated with Os^a antigen[2]

Amino acid	Ser 54 of GPA
Nucleotide	T at bp 217 in exon 3

Os(a−) (wild type) has Pro 54 and C bp 217.

Effect of enzymes/chemicals on Os^a antigen on intact RBCs

Ficin/papain	Sensitive
Trypsin	Resistant
α-Chymotrypsin	Resistant
Pronase	Sensitive
Sialidase	Resistant
DTT 200 mM	Resistant
Acid	Resistant

In vitro characteristics of alloanti-Os^a

Immunoglobulin class	IgG
Optimal technique	IAT
Complement binding	No

Clinical significance of alloanti-Os[a]

No data are available.

Comments

Anti-Os[a] found in several sera containing antibodies to multiple low prevalence antigens.

References
1. Seno, T. et al. (1983) Vox Sang. 45, 60–61.
2. Daniels, G.L. et al. (2000) Transfusion 40, 555–559.

ENEP ANTIGEN

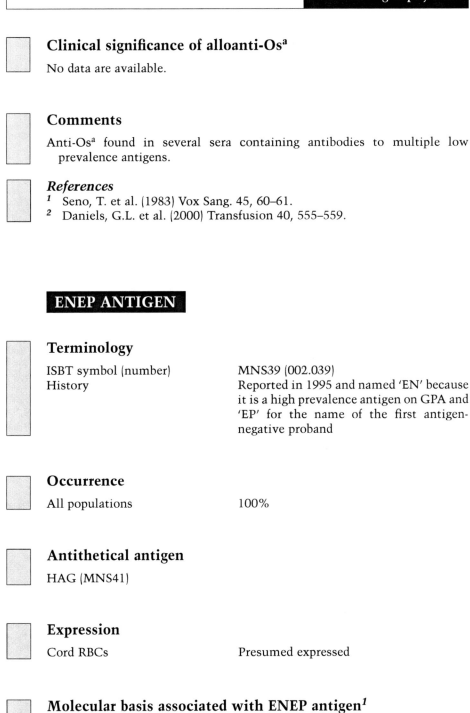

Terminology

ISBT symbol (number)	MNS39 (002.039)
History	Reported in 1995 and named 'EN' because it is a high prevalence antigen on GPA and 'EP' for the name of the first antigen-negative proband

Occurrence

All populations	100%

Antithetical antigen

HAG (MNS41)

Expression

Cord RBCs	Presumed expressed

Molecular basis associated with ENEP antigen[1]

Amino acid	Ala 65 of GPA
Nucleotide	G at bp 250 in exon 4

Effect of enzymes/chemicals on ENEP antigen on intact RBCs

Ficin/Papain	Ficin resistant; papain sensitive
Trypsin	Resistant
α-Chymotrypsin	Resistant
Pronase	Presumed sensitive
Sialidase	Presumed resistant
DTT 200 mM	Presumed resistant
Acid	Presumed resistant

In vitro characteristics of alloanti-ENEP

Immunoglobulin class	IgG
Optimal technique	IAT
Complement binding	No

Clinical significance of alloanti-ENEP

No data are available.

Comments

Anti-ENEP (anti-En^{a}FR) was made by a person homozygous for *GYP.HAG*. RBCs lacking ENEP (HAG+; MNS41) have an altered expression of Wr^b (DI4) antigen[1].

Reference
1   Poole, J. et al. (1999) Transf. Med. 9, 167–174.

ENEH ANTIGEN

Terminology

ISBT symbol (number)	MNS40 (002.040)
History	Named 'EN' because it is a high prevalence antigen on GPA and 'EH' from the initials of the first antigen-negative proband

Occurrence

All populations	100%

Antithetical antigen

Vw (**MNS9**); Hut (**MNS19**)

Expression

Cord RBCs   Expressed

Molecular basis associated with ENEH antigen[1]

Amino acid   Thr 28
Nucleotide   C at bp 140 in exon 3

Effect of enzymes/chemicals on ENEH antigen on intact RBCs

Ficin/papain   Sensitive
Trypsin   Sensitive
α-Chymotrypsin   Resistant
Pronase   Sensitive
Sialidase   Resistant
DTT 200 mM   Presumed resistant
Acid   Resistant

In vitro characteristics of alloanti-ENEH

Immunoglobulin class   IgM and IgG (only one example of anti-
                              ENEH described)[2]
Optimal technique   RT; IAT
Complement binding   No

Clinical significance of alloanti-ENEH

Transfusion reactions   No data are available
HDN   The anti-ENEH (anti-En^aTS) did not cause
                              HDN[2]

References
[1] Huang, C.-H. et al. (1992) Blood 80, 257–263.
[2] Spruell, P. et al. (1993) Transfusion 33, 848–851.

HAG ANTIGEN

Terminology

ISBT symbol (number)	MNS41 (002.041)
History	Reported in 1995 and named after the transfused man whose serum contained an antibody to a high prevalence antigen (ENEP) and whose RBCs had a double dose of this low prevalence antigen

Occurrence

Two probands, both Israeli.

Antithetical antigen

ENEP (MNS39)

Expression

Cord RBCs	Presumed expressed

Molecular basis associated with HAG antigen[1]

Amino acid	Pro 65 of GPA
Nucleotide	C at bp 250 in exon 4

Effect of enzymes/chemicals on HAG antigen on intact RBCs

Ficin/papain	Resistant
Trypsin	Resistant
α-Chymotrypsin	Resistant
Pronase	Presumed resistant
Sialidase	Presumed resistant
DTT 200 mM	Presumed resistant
Acid	Resistant

In vitro characteristics of alloanti-HAG

Immunoglobulin class	IgG
Optimal technique	IAT
Complement binding	No

Clinical significance of alloanti-HAG

No data are available.

Comments

RBCs with a double dose expression of HAG (ENEP−; [MNS39]) have an altered expression of Wr[b] (DI4)[1].

Reference
[1] Poole, J. et al. (1999) Transf. Med. 9, 167–174.

ENAV ANTIGEN

Terminology

ISBT symbol (number)	MNS42 (002.042)
Other names	Avis
History	Reported in 1996; named 'EN' because it is a high prevalence antigen on GPA and 'AV' after the name of the proband whose serum contained the antibody

Occurrence

All populations	100%

Antithetical antigen

MARS (MNS43)

Expression

Cord RBCs	Expressed

101

Molecular basis associated with ENAV antigen[1]

Amino acid	Glu 63 of GPA
Nucleotide	C at bp 244 in exon 4

Effect of enzymes/chemicals on ENAV antigen on intact RBCs

Ficin/papain	Resistant
Trypsin	Resistant
α-Chymotrypsin	Resistant
Pronase	Presumed resistant
Sialidase	Resistant
DTT 200 mM	Resistant
Acid	Resistant

In vitro characteristics of alloanti-ENAV

Immunoglobulin class	IgG
Optimal technique	IAT
Complement binding	No

Clinical significance of alloanti-ENAV

No data are available.

Comments

ENAV– RBCs have a weak expression of Wr^b (see DI4).

Reference
[1] Jarolim, P. et al. (1997) Transfusion 37 (Suppl.), 90S (abstract).

MARS ANTIGEN

Terminology

ISBT symbol (number)	MNS43 (002.043)
History	Reported in 1996 and named after the Native American proband (Marsden)

whose serum contained antibodies to several low prevalence antigens and reacted with ENAV− RBCs (MNS42)

Occurrence

Most populations: 0%; 15% of Choctaw tribe of Native Americans

Antithetical antigen

ENAV (MNS42)

Expression

Cord RBCs   Presumed expressed

Molecular basis associated with MARS antigen[1]

Amino acid   Lys 63 of GPA
Nucleotide   A at bp 244 in exon 4

Effect of enzymes/chemicals on MARS antigen on intact RBCs

Ficin/Papain   Resistant
Trypsin   Resistant
α-Chymotrypsin   Resistant
Pronase   Presumed resistant
Sialidase   Resistant
DTT 200 mM   Resistant
Acid   Resistant

In vitro characteristics of alloanti-MARS

Immunoglobulin class   IgG
Optimal technique   IAT
Complement binding   No

Clinical significance of alloanti-MARS

No data are available.

Comments

RBCs with a double dose of the MARS antigens have a weak expression of Wr^b (see **DI4**).

Reference
1 Jarolim, P. et al. (1997). Transfusion 37 (Suppl.), 90S (abstract).

Number of antigens   1

Terminology

ISBT symbol	P1
ISBT number	003
History	Discovered by Landsteiner and Levine in 1927 by immunizing rabbits; named P because this was the first letter after the already assigned M, N and O. The P system originally also contained P, P^k and LKE antigens; however, a different locus and biochemical pathway are involved in the production of these and they were moved in 1994 to the Globoside collection.

Expression

Soluble form	Pigeon egg white, hydatid cyst fluid, *Echinococcus* cyst fluid
Other blood cells	Lymphocytes, granulocytes, monocytes, platelets

Gene

Chromosome	22q11.2-qter
Name	*P1*
Product	The gene encoding this galactosyltransferase has not been cloned

Carrier molecule[1,2]

Paragloboside is the precursor for P1 antigen.

```
Gal
 | β1−4
GlcNAc
 | β1−3                                                Paragloboside (Lacto-N-
Gal                                                    neotetraosylceramide) (nLc4)
 | β1−4          Lactotrisylceramide (Lc3)
Glc
 | β1−1   Lactosylceramide (CDH)
Ceramide
```

Copies per RBC   500 000

Function

Receptor for *E. coli.*

Disease association

Can be weakened in carcinoma.

Phenotypes (% occurrence)

Phenotype	Caucasians	Blacks	Cambodians and Vietnamese
P_1	79	94	20
P_2	21	6	80

Null P_2 (P1–); p [See **GLOB** Collection (209)]

Comment

RBCs with either the P_1 or the P_2 phenotype express P and P^k antigens.

References

1 Bailly, P. and Bouhours, J.F. (1995) In: Molecular Basis of Human Blood Group Antigens (Cartron J.-P. and Rouger P., eds) Plenum Press, New York, pp. 300–329.
2 Spitalnik, P.F. and Spitalnik, S.L. (1995) Transf. Med. Rev. 9, 110–122.

P1 ANTIGEN

Terminology

ISBT symbol (number)	P1 (003.001)
Other names	P; P_1
History	Discovered in 1927; named P antigen because the letters M, N and O had been used; renamed P_1 and then P1

Occurrence

Caucasians	79%
Blacks	94%
Cambodian and Vietnamese	20%

Expression

Cord RBCs   Weaker than on RBCs from adults
Altered   There is considerable variation in the
                          strength of P1 expression on RBCs. This
                          variation is inherited

Molecular basis associated with P1 antigen[1]

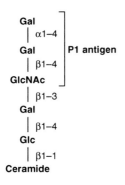

P1 antigen is derived by the addition of an α-galactosyl residue to paragloboside.

Effect of enzymes/chemicals on P1 antigen on intact RBCs

Ficin/papain   Resistant (↑↑)
Trypsin   Resistant (↑↑)
α-Chymotrypsin   Resistant (↑↑)
Pronase   Resistant (↑↑)
Sialidase   Resistant
DTT 200 mM   Resistant
Acid   Resistant

In vitro characteristics of alloanti-P1

Immunoglobulin class   IgM (IgG rare)
Optimal technique   RT (or lower)
Neutralization   Hydatid cyst fluid, pigeon egg white,
                         Echinococcus cyst fluid
Complement binding   Rare

Clinical significance of alloanti-P1

Transfusion reaction   No to moderate/delayed (rare)
HDN   No

Comments

The P1 determinant is widely distributed throughout nature. It has been detected in, for example, liver flukes, and pigeon egg white. The determinant is a receptor for a variety of microorganisms, including strains of *E. coli* and shiga toxin[2].

Anti-P1 is a naturally occurring antibody in many P1– individuals. Anti-P1 is frequently present in serum from patients with hydatid disease, liver fluke disease, acute hepatic fascioliasis.

References

[1] Bailly, P. and Bouhours, J.F. (1995) In: Molecular Basis of Human Blood Group Antigens (Cartron J.-P. and Rouger P. eds) Plenum Press, New York, pp. 300–329.

[2] Moulds, J.M. et al. (1996) Transfusion 36, 362–374.

Number of antigens   48

Polymorphic	D, C, E, c, e, f, Ce, G, hr^S, C^G, CE, c-like, cE, hr^B, Rh39, Rh41
Low prevalence	C^W, C^X, V, E^W, VS, D^W, hr^H, Go^a, Rh32, Rh33, Rh35, Be^a, Evans, Tar, Rh42, Crawford, Riv, JAL, STEM, FPTT, BARC, JAHK, DAK, LOCR
High prevalence	Hr$_0$, Hr, Rh29, Hr^B, Nou, Sec, Dav, MAR

Terminology

ISBT symbol	RH
ISBT number	004
CD number	CD240D (RhD); CD240CE (RhCcEe)
Other name	At times incorrectly called *Rhesus* (a genus of monkey)
History	Antibodies, made in 1940 by Landsteiner and Wiener, in rabbits in response to injected rhesus monkey (*Macacus rhesus*) RBCs, were thought to be the same specificity as the human antibody investigated in 1939 and the antigen detected by them was named Rh.

Expression

Cord RBCs	Expressed
Tissues	May be erythroid specific

Gene[1,2]

Chromosome	1p36.13–p34.3
Name	*RHD, RHCE*
Organization	*RHD* and *RHCE*, each with 10 exons, are distributed over 69 kbp of DNA in opposite orientation with their 3' ends facing each other. The genes are separated by a region of about 30 kbp of DNA that contains the small membrane protein 1 (SMP1) gene. The 3' and 5' ends of *RHD* are flanked by two 9 kbp homologous regions of DNA named the *Rhesus boxes*[3].
Product	RhD polypeptide (alternative names: Rh30; Rh30B; Rh30D; D$_{30}$)
	RhCE polypeptide (alternative names: Rh30; Rh30A; Rh30C)

Gene map

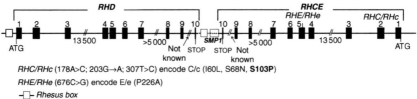

RHC/RHc (178A>C; 203G→A; 307T>C) encode C/c (I60L, S68N, **S103P**)

RHE/RHe (676C>G) encode E/e (P226A)

–□– Rhesus box

⊢⊣1 kbp

In subsequent diagrams representing *RH* exons, the information for *RHCE* is presented in the order of exon 1 to exon 10.

The opposite orientation of *RHD* and *RHCE* and a putative 'hairpin' formation allows homologous DNA segments to come into close proximity and most gene recombination occurs through gene conversion rather than unequal cross-over.

RHAG

A third homologous gene (*RHAG*), located on chromosome 6 at 6p11–p21.1, encodes the Rh-associated glycoprotein (RhAG; Rh50) and is essential for the expression of Rh antigens[4]. *RHAG* has 10 exons distributed over 32 kbp of DNA.

Database accession numbers

GenBank	*RHD*	X63094, X630976, U66341
	RHCE	X54534, M34015, U66340
	RHAG	X64594

http://www.bioc.aecom.yu.edu/bgmut/rh.htm

Amino acid sequence[5,6]

The full sequence is the RhCE (C and E) protein. Differences in the sequence for c, e and D proteins are shown.

RhCE and RhD

```
RhC:  MSSKYPRSVR  RCLPLCALTL  EAALILLFYF  FTHYDASLED  QKGLVASYQV  50
Rhc:              W
RhD:              W
RhC:  GQDLTVMAAI  GLGFLTSSFR  RHSWSSVAFN  LFMLALGVQW  AILLDGFLSQ  100
Rhc:          L           N
RhD:          I           S
RhC:  FPSGKVVITL  FSIRLATMSA  MSVLISAGAV  LGKVNLAQLV  VMVLVEVTAL  150
Rhc:  P
RhD:  S                     L           VD
```

```
RhC: GTLRMVISNI  FNTDYHMNLR  HFYVFAAYFG  LTVAWCLPKP  LPKGTEDNDQ 200
RhD: N                   MM  I                    S           E   K
RhE: RATIPSLSAM  LGALFLWMFW  PSVNSPLLRS  PIQRKNAMFN  TYYALAVSVV 250
Rhe:                                A
RhD: T                       F  A       E     V          V
RhC: TAISGSSLAH  PQRKISMTYV  HSAVLAGGVA  VGTSCHLIPS  PWLAMVLGLV 300
RhD:               G  K
RhC: AGLISIGGAK  CLPVCCNRVL  GIHHISVMHS  IFSLLGLLGE  ITYIVLLVLH 350
RhD:          V      Y  G       P S I GY  N             I      D
RhC: TVWNGNGMIG  FQVLLSIGEL  SLAIVIALTS  GLLTGLLLNL  KIWKAPHVAK 400
RhD:       GA                                                E
RhC: YFDDQVFWKF  PHLAVGF                                        417
RhD:
```

Amino acids are numbered by counting Met as 1.

RhAG

```
MRFTFPLMAI  VLEIAMIVLF  GLFVEYETDQ  TVLEQLNITK  PTDMGIFFEL   50
YPLFQDVHVM  IFVGFGFLMT  FLKKYGFSSV  GINLLVAALG  LQWGTIVQGI  100
LQSQGQKFNI  GIKNMINADF  SAATVLISFG  AVLGKTSPTQ  MLIMTILEIV  150
FFAHNEYLVS  EIFKASDIGA  SMTIHAFGAY  FGLAVAGILY  RSGLRKGHEN  200
EESAYYSDLF  AMIGTLFLWM  FWPSFNSAIA  EPGDKQCRAI  VDTYFSLAAC  250
VLTAFAFSSL  VEHRGKLNMV  HIQNATLAGG  VAVGTCADMA  IHPFGSMIIG  300
SIAGMVSVLG  YKFLTPLFTT  KLRIHDTCGV  HNLHGLPGVV  GGLAGIVAVA  350
MGASNTSMAM  QAAALGSSIG  TAVVGGLMTG  LILKLPLWGQ  PSDQNCYDDS  400
VYWKVPKTR                                                   409
```

Amino acids are numbered by counting Met as 1.

Carrier molecule

The assembly of the Rh proteins (RhD, RhCE) and the Rh-associated glyco-protein (RhAG) as a core complex in the RBC membrane appears to be essential for Rh antigen expression. The core complex is predicted to be a tetramer of two RhAG molecules and two RhD or RhCE molecules stabilized by N- and C-terminal domain associations.

RhD and RhCE

RhD and RhCE are multipass, acylated, palmitoylated, non-glycosylated proteins.

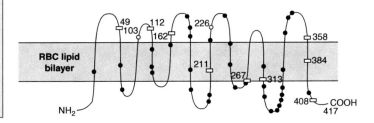

Black circles indicate the amino acid positions that differ between RhD and RhCE; white circles depict amino acids critical for C/c (Ser103Pro) or E/e (Ala226Pro) antigen expression. Segments of RhD and RhCE encoded by a particular exon are defined by numbered boxes, representing the start and finish of each exon.

M_r (SDS-PAGE)	30 000–32 000
Cysteine residues	4 in RhD; 6 in RhCE
Palmitoylation sites	2 in RhD: Cys 12, Cys 186
	3 in RhCE: Cys 12, Cys 186, Cys 311
Copies per RBC	100 000–200 000, RhD and RhCE combined

RhAG

The topology of RhAG closely resembles RhD and RhCE but the protein has a complex N-glycan on the first external loop and is not palmitoylated.

M_r (SDS-PAGE)	45 000 to 100 000 with a predominant band of 50 000
CHO: N-glycan	1 (plus 1 potential site)
Cysteine residues	5
Copies per RBC	100 000–200 000

Molecular basis of RhD antigens

See D Antigen pages. See website: http://www.uni-ulm.de/~wflegel/Rh/.

Molecular basis of RhCE antigens

Also see individual antigen pages for additional information.

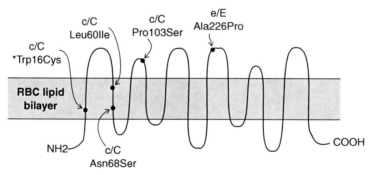

*Trp usually but not exclusively associated with c antigen
 Cys usually but not exclusively associated with C antigen

Seventy-four per cent of C−c+ black Americans with normal c have Cys16.

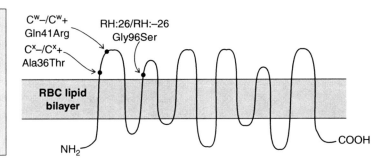

Function

The Rh membrane core complex interacts with band 3, GPA, GPB, LW and CD47 and is associated with the RBC membrane skeleton via ankyrin and protein 4.2. This complex maintains erythrocyte membrane integrity as demonstrated by the abnormal morphology and functioning of stomatocytic Rh_{null} RBCs[7].

The predicted structure of the Rh core proteins in the membrane indicates that they may function as ammonium transporters[8] or as channels for CO_2[9].

Rh and RhAG homologues are expressed in other tissues[4,10].

Disease association

Rh incompatibility is still the main cause of HDN.

Compensated hemolytic anemia occurs in some individuals with Rh_{null} or Rh_{mod} RBCs. Reduced expression of Rh antigens and Rh mosaicism can occur in leukemia, myeloid metaplasia, myelofibrosis, and polycythemia.

Rh and one form of hereditary spherocytosis are linked because both genes are on chromosome 1. Some Rh antigens are expressed weakly on South East Asian ovalocytes.

Phenotypes (% occurrence)

Haplotype	Caucasians	Blacks	Native Americans	Asians
DCe (R_1)	42	17	44	70
Ce (r′)	2	2	2	2
DcE (R_2)	14	11	34	21
cE (r″)	1	0	6	0
Dce (R_0)	4	44	2	3
ce (r)	37	26	6	3
DCE (R_z)	0	0	6	1
CE (r_y)	0	0	0	0

Phenotype (alternative)	Caucasians	Blacks	Asians	D antigen copy
D-positive				
R_1R_1 (R_1r')	18.5	2.0	51.8	14 500–19 300
R_2R_2 (R_2r'')	2.3	0.2	4.4	15 800–33 300
R_1r (R_1R_0; R_0r')	34.9	21.0	8.5	9900–14 600
R_2r (R_2R_0; R_0r'')	11.8	18.6	2.5	14 000–16 000
R_0r (R_0R_0)	2.1	45.8	0.3	12 000–20 000
R_zR_z (R_zr_y)	0.01	Rare	Rare	
R_1R_z (R_zr'; R_1r_y)	0.2	Rare	1.4	
R_2R_z (R_zr''; R_2r_y)	0.1	Rare	0.4	
R_1R_2 (R_1r''; R_2r'; R_zr; R_0Rz; R_0r_y)	13.3	4.0	30.0	23 000–36 000
D-negative				
r'r	0.8	Rare	0.1	
r'r'	Rare	Rare	0.1	
r''r	0.9	Rare	Rare	
r''r''	Rare	Rare	Rare	
rr	15.1	6.8	0.1	
r'r'' (r_yr)	0.05	Rare	Rare	
r'r_y; r''r_y; r_yr_y	Rare	Rare	Rare	
r'^Sr	0	1–2	0	

Null: Rh_{null}
Unusual: Rh_{mod}; many variants

Serological reactions of some unusual Rh complexes

Haplotype	Reactions with anti-							Low incidence antigens expressed
	D	C	c	E	e	f	G	
D – –	+↑	0	0	0	0	0	+	None
D••	+↑	0	0	0	0	0	+	Evans
D(C^W)–	+↑	0	0	0	0	0	+	C^W
D(c)–	+↑	0	(+)	0	0	(+)/0	+	None
D^{IVa}(C)–	+↑	(+)	0	0	0	(+)	+	Go^a, Rh33, Riv, FPTT
D^{Har}c(e) (R_0^{Har})	(+)/0	0	+	0	(+)	(+)	0	Rh33, FPTT
R_1^{Lisa}	+	(+)	0	0	(+)	0	+	Rh33 (weak), FPTT
D(C)(e) $\overline{\overline{R}}^N$	+/+↑	(+)	0	0	(+)	0	+	Rh32, DAK
D(C)(e)	+	(+)	0	0	(+)	0	+	Rh35
D(C)(e)	+↑	(+)	0	0	(+)	0	+	None
D(C)(e)	+	(+)	0	0	(+)	0	+	None
D(C)C^W(e)	+	+	0	0	+	0	+	C^W
D(C)C^X(e)	+	+	0	0	(+)	0	+	C^X
D(C)E (R_z)	+	(+)	0	+	0	0	+	None
D(C)(e)	+	(+)	0	0	(+)	0	+	JAL
D(c)(e)	+	0	(+)	0	(+)	0	+	JAL
Dce hr^S–	+	0	+	0	(+)	?	+	STEM, and/or VS
Dce hr^B–	+	0	+	0	(+)	?	+	STEM, and/or VS

Haplotype	Reactions with anti-							Low incidence antigens expressed
	D	C	c	E	e	f	G	
D(C)(e^S)	+	(+)	0	0	(+)	0	+	FPTT, VS (weak)
DC(e)	+	+	0	0	(+)	0	+	FPTT; DC(e) could not be distinguished from C(e)
(C)ce^S (r'^{S})	0	(+)	+	0	(+)	+	+	Rh42, VS
r^G	0	(+)	0	0	(+)	0	+	JAHK
r''^{G}	0	(+)/0	+	+	0	0	+	None
r^{Gs}	0	(+)	?	0	(+)	?	(+)	VS
C^{X}ce^S	0	0	+	0	(+)	0	0	C^X, VS
(c)(e)	0	0	(+)	0	(+)	(+)	0	Be^a
(c)(e)	0	0	(+)	0	(+)	(+)	0	LOCR
(C)E (r$_y$)	0	(+)	0	+	0	0	+	None
D(c)(E) or (c)(E)	?	0	(+)	(+)	0	0	?	Not known if D is encoded
DcE^W	+	0	+	(+)	0	0	+	E^W

↑ = elevated expression.

(+) = indicates qualitative or quantitative variation of antigen expression. Cells may either give weaker reactions with some sera or fail to react with some sera and could give normal strength reactions with others. These distinctions are often more obvious using single donor antisera.

? = Not known either because family studies were not informative or because testing was not done.

Other haplotypes have been reported but are no longer extant: r^t; r^L; r^m; r^{yn}; Dce^i.

Low prevalence antigens, associated haplotypes and molecular information[11-14]

Antigen	Associated phenotypes	Molecular information and comments
C^W	D(C)C^{W}e, D(C)C^{W}E, DC^W− (C)C^{W}E, C^{W}ce	Substitution in RhCE Gln41Arg
C^X	D(C)C^{X}e, (C)C^{X}e, C^{X}ce^S	Substitution in RhCE Ala36Thr
E^W	DcE^W	Not all anti-E react with E^W + RBCs
V	Dce^S, DCe^S, ce^S	Expression requires Leu245Val (VS+) and Gly336 in RhCE. Gly336Cys prevents V expression
VS	Dce^S, ce^S, (C)ce^S	Leu245Val in RhCE or hybrid *RHD-CE-D*
D^W	D^{Va}Ce, D^{Va}ce, D^{Va}cE	Part or whole *RHCE* exon 5 inserts into *RHD*
Go^a	D^{IVa}ce, D^{IVa}(C)−	Insert of *RHCE* into *RHD* in either exon 3 or exon 7
Rh32	D(C)(e), DBT	Association of *RHD* exon 4 with *RHCE* exon 5 in various hybrids

115

Antigen	Associated phenotypes	Molecular information and comments
Rh33	D^{Har}c(e), D^{IVa}(C)−, R$_0^{JOH}$, R$_1^{Lisa}$	For D^{Har}c(e): *RHD* exon 5 insert into *RHCE* The *CeVa* allele may encode R$_1^{Lisa}$
Be^a	(c)(e)	Associated with weak c, e and f expression
Evans	D.., D^{IVb}Ce, D^{IVb}cE	Hybrid gene, possible association with *RHCE* exon 7 adjoining *RHD* exon 6 in *RHD* background
Tar	D^{VII}Ce	RhD substitution: Leu110Pro
	D^c	RhD substitution: Ser103Pro and Leu110Pro. This RhD variant expresses weak c antigen
Rh42	(C)ce^S	Leu245Val in hybrid *RHD-CE-D*
Crawford	(C)ce^S	Most positives are (C)ce^S(r′^{S})
Riv	D^{IVa}(C)−	Haplotype also expresses Go^a, Rh33 and FPTT
JAL	D(C)(e), D(c)(e)	D(C)(e) in Whites; D(c)(e) in Blacks
STEM	Dce hr^S−, Dce hr^B−, and rare complexes	Possible marker for a variant e antigen
FPTT	D^{DFR}Ce, D^{DFR}cE, D^{Har}c(e), R$_1^{Lisa}$, D^{IVa}(C)− Rare complexes with depressed C and/or e	Hybrid genes: *RHCE* exon 4 joined to *RHD* exon 5 in various backgrounds
BARC	D^{VI}Ce (DVI type 2, 3, 4)	*RHCE* exon 6 joined to *RHD* exon 7 in *RHD* background
JAHK	r^G [(C)(e)G]	Hybrid *RHCE(1)-RHD(2)-RHCE(3–10)*
DAK	DIIIa, DOL, $\overline{R}^N$ other rare complexes	Not defined; for molecular basis, see D Antigen pages
LOCR	(c)(e)	Associated with weak c, e and f expression

Molecular basis of phenotypes

For D variants see D Antigen pages; for E variants see E antigen pages.

RHCE haplotypes and associated information[15–20]

	Exon 1 2 3 4 5 6 7 8 9 10 / substitutions	Associated Antigens Low	CE/other	Number of Probands	Ethnic Origin	Made anti-
D^HAR	226A	FPTT Rh33	c(e) G–	Many	C	D
r^G	226A	JAHK	(C)(e)(G)	Several	C	
R̄^N	226A	Rh32 DAK	(C)(e) Rh46–	Many	B	Rh46
R̄^N	T152N 226A	Rh32 DAK	(C)(e) Rh46–	Many	B	Rh46
ceBP (e^u)	Deletion 229R		e+/–	One	B	
ce	16C 226A		e+/–	Many/Few	B/C	
ce^S	L245V	V VS	e+/–	Many	B	
ce^S	16C 226A L245V	V VS	e+/–	Many	B	
(C)ce^S	16C L245V G336C	V– VS	(e) Rh34–	Many	B	Rh34, hr^B (D)
ceAR	16C 103P 226A 1306V M238V M267K L245V R263G	(V) VS– Frequently paired with *DAR*	e+/– Rh18– hr^S–	Many	B	Rh18 (D, C)
ce^S(340)	R114W L245V	(V/VS)	(hr^S)	Few	B	e, Ce
ce^S(697)	16C 233E L245V		e+/–	Few	B	
ce^S(748)	16C L245V V250M	(V/VS)	e+/–	Few	B	
ceMO	16C V223F 226A		e+/– hr^S–	Few	B	
ceEK	16C M238V M267K R263G		e+/– Rh18– hr^S–	Few	B	Rh18
ceBI	16C M238V A273V L378V		e+/– Rh18– hr^S–	Few	B	
cEMI	9nt 226P		E–e– hr^S–	One	B	
CeVA	16C 226A	Rh33 FPTT	(C)(e)	Several	C	
CeMA	16C R114W		(C)(e)	Several	C	
DCE(R_Z)	16W 60L 103S 68N		(C)	Many	All	

( ) denotes reduced antigen expression

+/– positive with some antibodies (could be weak), negative with other antibodies

B = Black   C = Caucasian   J = Japanese

Gray boxes show exons encoded by *RHCE*; black boxes show exons encoded by *RHD*; hatched boxes depict exons not expressed. Amino acid substitutions, rather than nucleotide substitutions are shown under the exons.

RHE variants[21,22]

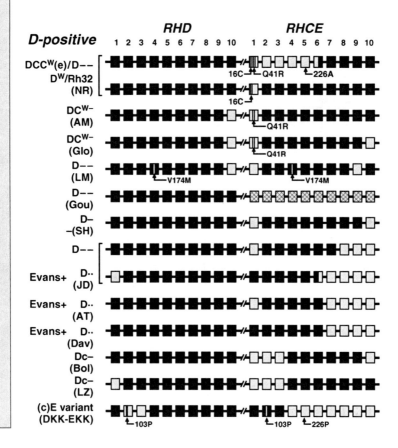

	1 2 3 4 5 6 7 8 9 10	Associated CE antigens	Number of probands	Ethnic origin
E type I	103P M167K 226P	E+/–	Several	C
E type II (EKK)	103P 226P	E+/– (c)	Few	C, J
E type III (EFM)	103P 226P M238V, Q233E	E+/– c+ (normal)	Few	J
E type IV	103P R201T 226P	E+/–	Several	C, J
EKH	103P R154T 226P	E+/– (c)		J

( ) denotes reduced antigen expression

Gray boxes show exons encoded by *RHCE*; black boxes show exons encoded by *RHD*. Amino acid substitutions, rather than nucleotide substitutions are shown under the exons.

Rearranged *RHD* and *RHCE*[4,7,17,22]

D-positive

RHD 1 2 3 4 5 6 7 8 9 10   RHCE 1 2 3 4 5 6 7 8 9 10

DCCᵂ(e)/D-- Dᵂ/Rh32	16C Q41R 226A
(NR)	16C
DCᵂ– (AM)	Q41R
DCᵂ– (Glo)	Q41R
D-- (LM)	V174M V174M
D-- (Gou)	
D– –(SH)	
D--	
Evans+ D·· (JD)	
Evans+ D·· (AT)	
Evans+ D·· (Dav)	
Dc– (Bol)	
Dc– (LZ)	
(c)E variant (DKK-EKK)	103P 103P 226P

118

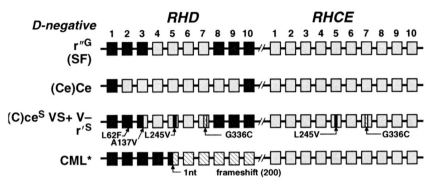

* *RHD* and *RHCE* identified in a D-positive patient, with chronic myeloid leukemia, who became D-negative

Molecular Basis of Rh$_{null}$ and Rh$_{mod}$ Phenotypes[4,23,24]

Changes in RHD/RHCE leading to amorph type of Rh$_{null}$	*Proband*
RHD del; *RHCe* frameshift Pro323; del2nt G398Stop	German (DR)
RHD del; *RHce*; intron 4, 5′ splice site mutation gt>tt	Spanish (DAA)
RHD del; *RHce* del TCTTC in exon 1, Leu31Stop	Japanese
Changes in RHAG leading to regulator type of Rh$_{null}$	
Gly279Glu and nt +1>G>A; intron 5′ splice site	Australian (YT)
G>A intron 6 acceptor site; exon 7 skipped	Spanish (AC)
G>T intron 6 acceptor site; exon 7 skipped	Japanese
G>A intron 7 donor site; exon 7 skipped	Japanese (TT)
G>A intron 1, first nt 5′ donor splice site; Pro52Stop	White American (AL)
Partial skipping of exon 9; Gly380Val	Japanese (WO)
CCTC>GA Tyr51, frameshift, Ile107Stop	SF, JL (S. African)
Changes in RHAG leading to Rh$_{mod}$ Phenotype	
3G>T in exon 1; Met1Ile	Jewish/Russian (SM)
236G>A in exon 2; Ser79Asn	White American (VL)
1195G>T in exon 8; Asp399Tyr	French (CB)
1183 del A (AAC>AC) in exon 9; frameshift; 52 additional amino acids	Japanese

Proteins altered on Rh$_{null}$ RBCs

Protein	Gene location	M_r	Copies per RBC	Comments
RhD/RhCE	1p36.13–34.3	30 000–32 000	10 000–30 000	Absent
RhAG	6p21.1–p11	45 000–100 000	100 000–200 000	Absent
CD47	3q13	47 000–52 000	10 000–50 000	Reduced
LW	19p13.3	37 000–47 000	3 000–5 000	Absent
GPB	4q28–q31	20 000–25 000	80 000–300 000	30% of normal
Duffy (Fy5)	1q22–q23	35 000–45 000	12 000–17 000	Fy5 antigen absent

Comparison of Rh$_{null}$ and Rh$_{mod}$ RBCs

Phenotype	Rh proteins/ antigens	RhAG	LW	CD47	GPB/ S, s, U antigens	Altered gene
Amorph Rh$_{null}$	Absent	Reduced (20%)	Absent	Reduced by 90%	Reduced by 50% S/s normal U weak	RHCE (RHD deleted)*
Regulator Rh$_{null}$	Absent	Absent	Absent	Reduced	Reduced by 70% S/s weak U absent	RHAG**
Rh$_{mod}$	Reduced (variable)	Absent or reduced (variable)	Absent or reduced	Reduced (variable)	Reduced (variable) S/s normal U normal /weak	RHAG**

* Express one Rh haplotype.
**Express both Rh haplotypes.

References
[1] Anstee, D.J. and Tanner, M.J. (1993) Baillieres Clin. Haematol. 6, 401–422.
[2] Cartron, J.-P. (1994) Blood Rev. 8, 199–212.
[3] Wagner, F.F. and Flegel, W.A. (2000) Blood 95, 3662–3668.
[4] Huang, C.-H. et al. (2000) Semin. Hematol. 37, 150–165.
[5] Arce, M.A. et al. (1993) Blood 82, 651–655.
[6] Chérif-Zahar, B. et al. (1990) Proc. Natl. Acad. Sci. USA 87, 6243–6247.
[7] Avent, N.D. and Reid, M.E. (2000) Blood 95, 375–387.
[8] Westhoff, C.M. et al. (2002) J. Biol. Chem. 277, 12499–12502.
[9] Soupene, E. et al. (2002) Proc. Natl. Acad. Sci. USA 99, 7769–7773.
[10] Huang, C.H. and Liu, P.Z. (2001) Blood Cells Mol. Dis. 27, 90–101.
[11] Noizat-Pirenne, F. et al. (2002) Transfusion 42, 627–633.
[12] Faas, B.H.W. et al. (2001) Transfusion 41, 1136–1142.
[13] Green, C. et al. (2002) Transfus. Med. 12, 55–61.
[14] Coghlan, G. et al. (1994) Transfusion 34, 492–495.
[15] Noizat-Pirenne, F. et al. (2002) Blood 100, 4223–4231.
[16] Noizat-Pirenne, F. et al. (2001) Br. J. Haematol. 113, 672–679.
[17] Westhoff, C.M. et al. (2000) Transfusion 40 (Suppl.), 7S (abstract).
[18] Hemker M.B., et al. (1999) Blood 94, 4337–4342.
[19] Noizat-Pirenne, F. et al. (1999) Transfusion 39 (Suppl.), 103S (abstract).
[20] Daniels, G.L. et al. (1998) Transfusion 38, 951–958.
[21] Noizat-Pirenne, F. et al. (1998) Br. J. Haematol. 103, 429–436.
[22] Kashiwase, K. et al. (2001) Transfusion 41, 1408–1412.
[23] Cartron, J.P. (1999) Baillieres Best. Pract. Res. Clin. Haematol. 12, 655–689.
[24] Kamesaki, T. et al. (2002) Transfusion 42, 383–384.

D ANTIGEN

Terminology

ISBT symbol (number)	Rh1 (004.001)
Other names	Rh_0
History	The original 'Rh' antigen stimulated a transfusion reaction, which was investigated by Levine and Stetson in 1939. The reactions of this antibody paralleled those of the anti-'Rh' reported by Landsteiner and Wiener in 1940 but stimulated in animals. Some years later, upon recognition that the human and the animal anti-'Rh' did not react with the same antigen, the accumulation of publications about the clinically important human anti-'Rh' made a name change undesirable. Ultimately however, the antigen name switched to D and the system took the Rh name.

Occurrence

Caucasians	85%
Blacks	92%
Asians	99%
Native Americans	99%

Expression

Cord RBCs	Expressed
Altered	Partial and weak D phenotypes; exalted on D deletion phenotypes; appears exalted on GPA-deficient RBCs because of reduction of sialic acid

Number of D antigen sites per RBC:

Common D phenotypes	10 000–33 000
Weak D phenotypes	200–10 000
Exalted D phenotypes	75 000–200 000

Effect of enzymes/chemicals on D antigen on intact RBCs

Ficin/papain	Resistant (↑↑)
Trypsin	Resistant (↑)

α-Chymotrypsin   Resistant (↑)
Pronase   Resistant (↑↑)
Sialidase   Resistant (↑)
DTT 200 mM   Resistant
Acid   Resistant

In vitro characteristics of alloanti-D

Immunoglobulin class   Most IgG, some IgM (IgA rare)
Optimal technique   IAT; enzymes
Complement binding   Very rarely

Clinical significance of alloanti-D

Transfusion reaction   Mild to severe/immediate or delayed
HDN   Mild to severe

Autoanti-D

Yes: may appear as mimicking alloantibody

Molecular and phenotypic information

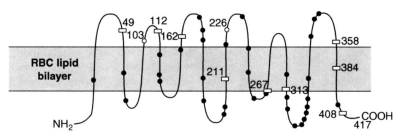

Depending on the Rh phenotype, RhD differs from RhCE by 32–35 amino acids (black circle).

Substitution of Ser103Pro in RhD results in the D+ G− phenotype.

The requirements for D antigen expression on RBCs are not fully understood. The D antigen is unusual in that it is not derived from an amino acid polymorphism but from the presence of the entire RhD. The expression of D antigen varies qualitatively and quantitatively.

Qualitative variation

The D antigen is a mosaic of different epitopes. People with RBCs lacking one or more of these epitopes (referred to as having partial expression of D antigen) can make alloanti-D directed at the missing D epitopes.

D categories

Cells	Anti-D from						
	II	IIIa	IIIc	IVa	IVb	Va	VI
II	0	+	+	+	V	+	+
IIIa	+	0	0	+	+	+	+
IIIb	+	0	0	+	+	V	+
IIIc	+	0	0	+	+	+	+
IVa	0	V	0	0	0	+	+
IVb	0	V	V	0	0	0	+/0
Va	+	0	0	+	+	0	+/0
VI	0	0	0	+/0	+/0	0	0
VII	+	+/0	+/0	+	+	+/0	+
DFR	+	0	+/0	+	+	0	0
DBT	0	NT	V	0	0	+/0	+/0

+ = positive; +/0 = positive with some sera and negative with other sera; 0 = negative; V = variable strength of positive reaction and some sera negative; NT = not tested.

Partial D phenotypes were classified into seven D categories based on the interaction of the RBCs and sera of the D category members (see table). Low prevalence marker antigens aided in their identification. Other partial D were added later [e.g. DFR, DBT, DHAR (R_0^{Har})].

Monoclonal anti-D revealed different reaction patterns and each reaction pattern recognizes a different epitope (epD) of the D mosaic. Seven reaction patterns were initially recognized and these were expanded to nine patterns (see table of nine epitope model) with an awareness that more epitopes would be identified.

Recognition of new partial D phenotypes and use of hundreds of monoclonal anti-D has sub-split the nine epitopes. The nine epitope model, which was directly related to the original D categories, was expanded to accommodate the new reaction patterns (see table of 30 epitope model). Sub-splits of the patterns by reactions observed with new unique partial D are being denoted by a dot followed by a second Arabic number, for example, the sub-split of epD1 was defined by reactions with DFR cells: anti-epD1.1 are positive and anti-epD1.2 are negative with DFR cells. New reaction patterns defined with monoclonal anti-D have been assigned numbers above 9 (see table).

These patterns were, and all future splits will be, defined through multi-center ISBT workshops for a standardized and logical approach.

Epitope profiles of partial D antigens: the nine epitope model[1]

| | Reactions with monoclonal anti-D, anti- | | | | | | | |
	epD1	epD2	epD3	epD4	epD5	epD6/7	epD8	epD9
DII	+	+/0	+	0	+	+	+	0
DIIIa	+	+	+	+	+	+	+	+
DIIIb	+	+	+	+	+	+	+	+
DIIIc	+	+	+	+	+	+	+	+
DIVa	0	0	0	+	+	+	+	0
DIVb	0	0	0	0	+	+	+	0
DVa	0	+	+	+	0	+	+	+
DVI	0	0	+	+	0	0	0	+
DVII	+	+	+	+	+	+	0	+
DFR	+/0	+/0	+	+	+/0	+/0	0	+
DBT	0	0	0	0	0	+/0	+	0
R_0^{Har}	0	0	0	0	+/0	+/0	0	0

+ = positive; +/0 = positive with some anti-D, negative with other anti-D; 0 = negative.

Nomenclature of partial D recommended by ISBT Committee for Terminology for Red Cell Surface Antigens[2]

D category phenotypes will retain the current numbering system, but the Roman numeral will not be a superscript, for example, D^{Va} will become DVa.

Subtypes of D categories will be denoted by the Arabic numerals, for example, DVa type 1.

Other (and new) partial D will be denoted by upper case letters, for example, DBT, DHAR (for R_0^{Har}).

Overall weak expression of D will be referred to as weak D (see later)[2].

Molecular basis of partial D phenotypes[3–16]

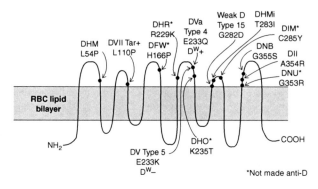

Serologically similar phenotypes have different molecular backgrounds.

Associated low prevalence antigens and other relevant serological findings are given next to the exon diagram.

Epitope profiles of partial D antigens: the expanded 30 epitope model[2]

Anti-epD	Partial D																		
	DII	DIII	DIVa	DIVb	DVa1	DVa2	DVa3	DVa4	DVa5	DVI	DVII	DFR	DBT	DHAR	DHMi	DNB	DAR	DNU	DOL
1.1	+	+	0	0	0	0	0	0	0	0	+	+	0	0	+	+	v	v	v
1.2	+	+	0	0	0	0	0	0	0	0	+	0	0	0	+	+	v	v	+
2.1	+	+	0	+	+	+	+	+	0	0	+	+	0	0	+	+	+	+	+
2.2	+	+	0	+	+	+	+	+	0	0	+	0	0	0	0	+	0	+	+
3.1	0	+	0	+	+	+	+	+	+	+	+	+	0	0	+	+	+	+	+
4.1	+	+	0	+	+	+	+	+	0	0	+	+	0	0	+	+	+	+	+
5.1	+	+	+	0	0	0	0	0	0	0	+	+	0	0	+	+	+	+	+
5.2	+	+	+	+	+	+	+	+	+	0	+	+	0	+	+	+	+	0	0
5.3	+	+	+	0	0	0	0	0	0	0	+	+	0	0	+	+	+	+	v
5.4	+	+	+	+	+	+	+	+	+	0	+	+	0	+	+	+	+	+	+
5.5	+	+	+	0	0	0	0	0	0	0	+	0	0	0	+	+	+	+	+
6.1	+	+	+	+	+	+	+	+	+	0	+	0	+	+	+	+	+	+	+
6.2	+	+	+	+	+	+	+	+	+	0	+	0	+	0	+	+	+	+	+
6.3	+	+	+	+	+	+	+	+	+	0	+	0	0	0	+	+	+	+	+
6.4	+	+	+	+	+	+	+	+	+	0	+	+	+	+	+	+	+	+	+
6.5	+	+	+	+	+	+	+	+	+	0	+	+	+	0	+	+	+	+	v
6.6	+	+	+	+	+	+	+	+	+	0	+	+	0	+	v	v	+	+	+
6.7	+	+	+	+	+	+	+	+	+	0	+	0	0	0	>	+	+	+	+
6.8	+	+	+	+	+	+	+	+	+	0	+	0	0	0	v	+	+	+	+
8.1	+	+	+	+	+	+	+	+	+	0	0	0	0	0	>	+	+	+	+
8.2	+	+	0	0	0	0	0	0	0	0	0	0	+	0	v	0	0	0	+
8.3	0	+	0	0	+	+	+	+	+	0	0	0	+	0	+	+	+	+	+
9.1	+	+	+	+	+	+	+	+	+	0	0	0	0	0	0	+	+	+	+
10.1	+	+	0	0	0	0	0	0	0	0	+	0	0	0	0	+	+	0	+
11.1	+	+	0	0	0	0	0	0	0	+	0	0	0	0	0	0	+	+	+
12.1	+	+	+	+	+	+	+	+	+	0	0	0	+	0	0	0	0	0	+
13.1	+	+	0	0	+	+	+	+	+	0	+	+	0	0	0	+	+	0	v
14.1	+	+	0	0	+	+	+	+	+	0	+	+	0	0	+	+	0	+	+
15.1	+	+	0	0	+	+	+	+	+	+	+	+	0	0	+	+	+	+	+
16.1	+	+	+	+	+	+	+	+	+	+	+	+	+	0	+	+	+	+	+

125

Partial D Phenotypes[1,3–15]

Partial D phenotype	Associated haplotype	Number of D antigen sites	Number of probands	Ethnic origin	Made anti-D
DII	Ce	3200	One	C	Yes
DIIIa	ce G+	12 300	Many	B	Yes
DIIIb	ce G−		Few	B*	Yes
DIIIc	Ce G+	26 899	Many	C	Yes
DIII type 4		33 255	Few	C	Yes
DIII type 5	ce		One		
DIVa	ce, [(C)−]	9300	Many	B	Yes
DIVb	Ce, cE	4000	Many	C, J	Yes
DIV type 3	Ce	607	One	C	
DIV type 4	Ce		Several		
DVa	ce, Ce, cE	9400	Many	C, J, B	Yes
For DV-like phenotypes see figure					
DVI type 1	cE	300–1000	Many	C	Yes
DVI type 2	Ce	1600–2886	Many	C, J	Yes
DVI type 3	Ce	14 502	Few	C	Yes
DVI type 4	Ce		One	C	
DVII	Ce	3600–8398	Many	C	Yes
DFR	Ce > cE	5300	Many	C	Yes
DBT type 1	Ce > (C)(e) and ce	4300	Several	C, J, B	Yes
DBT type 2	Ce		Several	J	
DHAR (R_0^{Har})	c(e) G−		Many	C	Yes
DHMi	cE	2400	Several	C	Yes
DNB	Ce	6000	Many	C (European) 1 in 292 in Swiss	Yes
DNU	Ce	10 000	Few	C	
DOL	ce	4700	Several	B	Yes
DAR	ce		Many	B	Yes
Weak D Type 4.2.2	ce	1 650	Few	C	Yes
DCS			One	C	
DTI	cE		One	J	
DBS	cE or ce		One	A	
DAL			Several	C	
DFW	Ce		One	C	
DHO	Ce	1300		C	
DHR	cE	3800		C	
DMH	ce			C	Yes
DIM	cE	192	One	C	
Weak D type 15	cE	297	Few	C	Yes
DAU-0	ce	2113	Few	C	
DAU-1	ce	373	One	B	
DAU-2	ce	10 879	One	B	Yes
DAU-3	ce	1909	One	B	Yes

A = Arabic; B = Black; C = Caucasian; J = Japanese.
Some partial D phenotypes in this table are not yet associated with production of alloanti-D; the phenotypes are included here because of their similarity to known partial D phenotypes as determined by molecular analysis or by the D epitope profile.
* The published molecular basis for DIIIb was determined using DNA from Caucasian probands who are G− and probably have a weak D phenotype and thus are not DIIIb as defined by Tippett.[17]

Rhesus Similarity Index[7]

A Rhesus Similarity Index has been devised to characterize the extent of qualitative changes in aberrant D antigens. Based on D epitope density profiles ascertained by using a panel of monoclonal anti-D, this quantitative method aids in the discrimination of normal D from partial D and weak D.

Molecular basis of weak D[7,11,16,18]

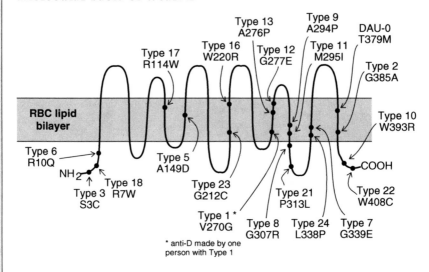

The weak D phenotype is a quantitative, not a qualitative polymorphism and therefore all D epitopes are present. This reduced D antigen expression is usually detected by the indirect antiglobulin test. The molecular basis of weak D phenotypes is heterogeneous. See figure and tables.

The different types of weak D defined at the molecular level, in accordance with ISBT nomenclature, are referred to as 'type' with Arabic numerals, for example, weak D type 1.

Multiple mutations resulting in weak D phenotypes

Weak D type	Amino acid substitution (encoding exon)
14	Ser182Thr, Lys198Asn, Thr201Arg (exon 4)
4.0	Thr201Arg (exon 4); Phe223Val (exon 5)
4.1	Trp16Cys (exon 1); Thr201Arg (exon 4); Phe223Val (exon 5)
4.2.1*	Thr201Arg (exon 4); Phe223Val (exon 5); Ile342Thr (exon 7)

* A silent mutation, 957G>A distinguishes 4.2.1 from partial D phenotypes DAR and type 4.2.2

128

Weak D phenotypes

Weak D phenotype	Associated haplotype	Number of D antigen sites
Type 1	Ce	1285
Type 2	cE	489
Type 3	Ce	1932
Type 4	ce	2288
Type 4.1		3811
Type 5	cE	296
Type 6	Ce	1053
Type 7	Ce	2407
Type 8	Ce	972
Type 9	cE	248
Type 10	cE	1186
Type 11	ce	183
Type 12	Ce	96
Type 13	Ce	956
Type 14	cE	
Type 16	cE	235
Type 17		66
Type 21	Ce	5200

In European populations weak D types 1, 2 and 3 predominate.
Some European guidelines recommend detection of weak D type 2 by routine D typing (without use of indirect antiglobulin test) because weak D type 2 phenotype RBCs have stimulated the production of anti-D[16].

DEL (D$_{el}$) phenotype[19,20]

Weak expression of D, which is detectable only by adsorption and elution of anti-D. Found in Oriental populations.

Molecular basis

Mechanism	Associated haplotype
RHD del exon 9 (1013 bp del including exon 9)	Ce
RHD 1227G>A in exon 9, Lys409Lys, splice site may be affected	Ce
RHD IVS3 + 1 g>a, splice site mutation	Ce
RHD 885G>T in exon 6; Met295Ile*	Ce

* Allele is similar to weak D type 2 but is associated with a different haplotype (ce).

Molecular basis of D-negative phenotypes[3,20,21]

D-negative RBCs lack the RhD protein. Several molecular backgrounds result in the D-negative phenotype: deletion of *RHD* predominates in people of European descent; in Oriental populations an intact but unexpressed *RHD* gene is common; in black African populations two thirds of D-negative people have an inactive *RHD* gene, (the *RHD* pseudogene or *RHD*ψ) with a 37 bp internal duplication resulting in a premature stop codon. Hybrid genes, composed of portions of *RHD* and *RHCE* are also common. (See tables and system pages.)

Mutations in *RHD* encoding the D-negative phenotype

Mechanism	Associated haplotype	Ethnicity
Gene deletion	ce	Caucasians
Duplication of 37 bp at intron 3/exon 4 junction; missense mutations; stop codon in exon 6 (*RHD* pseudogene)	ce	Blacks
48G>A in exon 1; Trp16Stop	Ce	
121C>T in exon 1; Gln41Stop	Ce	
270G>A in exon 2; Trp90Stop	cE	Chinese
990C>G in exon 7; Tyr330Stop	Ce	
711del C in exon 5; fs; Val245Stop	cE	Chinese
906insGGCT in exon 6; fs; IVS6 + 2t>a	Ce	Chinese
5' end of exon 4 delACAG; fs; Met167Stop	Ce	Caucasians
600 delG; Leu228Stop		
635G>T in exon 5; Gly212Val	Ce	
IVS8 + 1 g>a	Ce	Caucasians

Hybrid Rh Genes encoding the D-negative phenotype

Mechanism	Associated haplotype
RHCE(1–9)-*RHD*(10)	cE
RHD(1)-*RHCE*(2–9)-*RHD*(10)*	Ce
RHD(1)-*RHCE*(2–7)-*RHD*(8–10)**	Ce
RHD(1,2)-*RHCE*(3–7)-*RHD*(8–10)	Ce[S]
RHD(1–3)-*RHCE*(4–7)-*RHD*(8–10)[†]	cE
RHD(1–7)-*RHCE*(8)-*RHD*(10)	Ce

Parenthesis indicate exon numbers.
*, **, [†] different breakpoints in two unrelated probands.

Comments

Expression of D may be weakened by a *dCe, dCE* or *d(C)ce^S* complex in trans.
Present on RBCs from chimpanzees and gorillas.

References
1. Tippett, P. et al. (1996) Vox Sang. 70, 123–131.
2. Scott, M. (2002) Transfus. Clin. Biol. 9, 23–29.
3. Avent, N.D. and Reid, M.E. (2000) Blood 95, 375–387.
4. Hemker, M.B. et al. (1999) Blood 94, 4337–4342.
5. Wagner, F.F. et al. (1998) Blood 91, 2157–2168.
6. Wagner, F.F. et al. (1998) Transfusion 38 (Suppl.), 63S (abstract).
7. Wagner, F.F. et al. (2000) Blood 95, 2699–2708.
8. Hyodo, H. et al. (2000) Vox Sang. 78, 122–125.
9. Wagner, F.F. et al. (2001) Transfusion 41, 1052–1058.
10. Faas, B.H.W. et al. (2001) Transfusion 41, 1136–1142.
11. Müller, T.H. et al. (2001) Transfusion 41, 45–52.
12. Noizat-Pirenne, F. et al. (2001) Transfusion 41, 971–972.
13. Omi, T. et al. (2002) Transfusion 42, 481–489.
14. Wagner, F.F. et al. (2002) Blood 100, 2253–2256.
15. Wagner, F.F. et al. (2002) Blood 100, 306–311.
16. Flegel, W.A. and Wagner, F.F. (2002) Clin. Lab. 48, 53–59. (PDF available at: http:/www.uni-ulm.de/%7Ewflegel/Rh/PDF/ClinLab2002.pdf).
17. Reid, M.E. et al. (1998) Immunohematology 14, 89–93.
18. Wagner, F.F. et al. (1999) Blood 93, 385–393.
19. Sun, C.-F. et al. (1998) Vox Sang. 75, 52–57.
20. Shao, C.P. et al. (2002) Vox Sang. 83, 156–161.
21. Wagner, F.F. et al. (2001) BMC Genetics 2 Web/URL: http://www.biomedcentral.com/1471-2156/2/10.

C ANTIGEN

Terminology

ISBT symbol (number)	Rh2 (004.002)
Other names	rh'
History	Reported in 1941 when it was recognized that, in addition to D, the Rh system had four other common antigens. Named because 'C' was the next available letter in the alphabet.

Occurrence

Caucasians	68%
Blacks	27%
Asians	93%

Antithetical antigen

c (**Rh4**)

Expression

Cord RBCs	Expressed
Altered	See system pages for unusual Rh complexes
	Weak on D(C)e carrying **HOFM** (700.050)

Molecular basis associated with C antigen[1]

Amino acid	Ser 103 on RhCe and RhCE protein is critical but the requirements for expression of C antigen are not fully understood; some anti-C require Cys 16 to be present
See system pages for variants	

Effect of enzymes/chemicals on C antigen on intact RBCs

Ficin/papain	Resistant (↑↑)
Trypsin	Resistant (↑)
α-Chymotrypsin	Resistant (↑)
Pronase	Resistant (↑↑)
Sialidase	Resistant
DTT 200 mM	Resistant
Acid	Resistant

In vitro characteristics of alloanti-C

Immunoglobulin class	IgG and IgM
Optimal technique	IAT; enzymes
Complement binding	No

Clinical significance of alloanti-C

Transfusion reaction	Mild to severe/immediate or delayed/hemoglobinuria
HDN	Mild

Autoanti-C

Yes, may be mimicking alloantibody.

Comments

Anti-C is often found in antibody mixtures, especially with anti-G (see RH12) or anti-D (see RH1).

Apparent anti-C in Blacks may be anti-hr^B (see RH31).

Alloanti-C can be made by individuals with the d(C)ce^S (r$'^S$), C^W+, C^X+, and D(C)(e)/ce phenotypes.

C+ RBCs express the G antigen (see RH12).

Not expressed on ape RBCs.

Reference
[1] Cartron, J.-P. (1994) Blood Rev. 8, 199–212.

E ANTIGEN

Terminology

ISBT symbol (number)	Rh3 (004.003)
Other names	rh"
History	Reported in 1943 and named after the next letter in the alphabet when it was realized that the antigen was part the Rh system

Occurrence

Caucasians	29%
Blacks	22%
Asians	39%

Antithetical antigen

e (**Rh5**)

Expression

Cord RBCs	Expressed
Altered	See system pages for unusual Rh complexes

Molecular basis associated with E antigen[1]

Amino acid	Pro 226 on RhcE and RhCE protein is critical but requirements for expression of antigen are not fully understood

See system pages for variants

Molecular basis of E variants[2,3]

Category	Mechanism
EI	Met167Lys
EII	*RHCE(1)-RHD(2,3)-RHCE(4–10)*
EIII	Gln333Glu and Met238Val in cE
EIV	Arg201Tyr in cE

Category EIV RBCs do not lack E epitopes but express them weakly (analogous to weak D).
See table of epitopes expressed by E variant RBCs.

Effect of enzymes/chemicals on E antigen on intact RBCs

Ficin/papain	Resistant (↑↑)
Trypsin	Resistant (↑)
α-Chymotrypsin	Resistant (↑)
Pronase	Resistant (↑↑)
Sialidase	Resistant
DTT 200 mM	Resistant
Acid	Resistant

In vitro characteristics of alloanti-E

Immunoglobulin class	IgG and IgM
Optimal technique	RT; IAT; enzymes
Complement binding	No

Clinical significance of alloanti-E

Transfusion reaction	Mild to moderate/immediate or delayed/hemoglobinuria
HDN	Mild

Autoanti-E

Yes, may be mimicking alloantibody.

Comments

E epitopes expressed by E categories[2].

	Cat EI	Cat EII	Cat EIII	Cat EIV
epE1	+	+	+	+
epE2	+	0	0	+
epE3	0	0	+	+
epE4	0	+	+	+

Many examples of anti-E appear to be naturally-occurring.
Anti-E is often present in sera containing anti-c.
Not found on RBCs from non-human primates.

References
[1] Cartron, J.-P. (1994) Blood Rev. 8, 199–212.
[2] Noizat-Pirenne, F. et al. (1998) Br. J. Haematol. 103, 429–436.
[3] Kashiwase, K. et al. (2001) Transfusion 41, 1408–1412.

c ANTIGEN

Terminology

ISBT symbol (number)	Rh4 (004.004)
Other names	hr′
History	Briefly reported in 1941 when it was recognized that, in addition to D, the Rh system had four other common antigens; named when the antithetical relationship to C was recognized

Occurrence

Caucasians	80%
Blacks	96% [97–98% if d(C)ce^S (r′^{S}) phenotype included]
Asians	47%

Antithetical antigen

C (**RH2**)

Expression

Cord RBCs	Expressed
Altered	See system pages for unusual Rh complexes

Molecular basis associated with c antigen[1,2]

Amino acid

Pro 103 (and Pro 102[3]) on Rhce and RhcE protein are critical but requirements for expression of c antigen are not fully understood.
RhD with a substitution of Ser103Pro expresses a weak c antigen.

Effect of enzymes/chemicals on c antigen on intact RBCs

Ficin/papain   Resistant (↑↑)
Trypsin   Resistant (↑)
α-Chymotrypsin   Resistant (↑)
Pronase   Resistant (↑↑)
Sialidase   Resistant
DTT 200 mM   Resistant
Acid   Resistant

In vitro characteristics of alloanti-c

Immunoglobulin class   Most IgG, some IgM
Optimal technique   IAT; enzymes
Complement binding   No

Clinical significance of alloanti-c

Transfusion reaction   Mild to severe/immediate or delayed/ hemoglobulinuria
HDN   Mild to severe

Autoanti-c

Yes, may be mimicking alloantibody.

Comments

Expressed on RBCs from apes.

References

[1] Faas, B.H.W. et al. (2001) Transfusion 41, 1136–1142.
[2] Cartron, J.-P. (1994) Blood Rev. 8, 199–212.
[3] Westhoff, C.M. et al. (2000) Transfusion 40, 321–324.

e ANTIGEN

Terminology

ISBT symbol (number)	Rh5 (004.005)
Other names	hr″
History	Named in 1945 when its antithetical relationship to E was recognized.

Occurrence

Caucasians	98%
Blacks	98%
Asians	96%

Antithetical antigen

E (**RH3**)

Expression

Cord RBCs	Expressed
Altered	See system pages for unusual Rh complexes
	See table for reactions of monoclonal anti-e with unusual Rh complexes

Molecular basis associated with e antigen[1]

Amino acid	Ala 226 on Rhce and RhCe protein is critical but requirements for expression of e antigen are not fully understood

The presence of Cys 16 in Rhce weakens the expression of e antigen[2].

Effect of enzymes/chemicals on e antigen on intact RBCs

Ficin/papain	Resistant (↑↑)
Trypsin	Resistant (↑)
α-Chymotrypsin	Resistant (↑)
Pronase	Resistant (↑↑)
Sialidase	Resistant
DTT 200 mM	Resistant
Acid	Resistant

In vitro characteristics of alloanti-e

Immunoglobulin class	Most IgG, some IgM
Optimal technique	IAT; enzymes
Complement binding	No

Clinical significance of alloanti-e

Transfusion reaction	Mild to moderate/delayed/hemoglobinuria
HDN	Rare, usually mild

Autoanti-e

Common.

Comments

Alloanti-e-like antibodies may be made by people with e+ RBCs lacking some e epitopes. This occurs more frequently in Blacks than in Caucasians[3].

Several e variants, in people at risk of immunization against lacking Rh epitopes, have been defined with monoclonal anti-e and molecular studies[4,5].

Reaction of monoclonal anti-e with RBCs expressing e-variant haplotypes[2,4]

Haplotype	MS16	MS21	MS62/MS63	MS69	MS70
hr^S- (ceMo)	W	W	0	0	0
hr^S- (ceAR)	+	+	0	W	+
hr^S- (ceEK)	+	+	+	+	+
hr^B- (ce^S)	+	+	+	+	0
$\underline{\underline{C}}$ys16 (ce)	0	+	W	0	NT
R^N	W	0	0	W	0

Not expressed on RBCs from non-human primates.

References
1 Cartron, J.-P. (1994) Blood Rev. 8, 199–212.
2 Westhoff, C.M. et al. (2001) Br. J. Haematol. 113, 666–671.
3 Issitt, P.D. (1991) Immunohematology 7, 29–36.
4 Noizat-Pirenne, F. et al. (2002) Blood 100, 4223–4231.
5 Noizat-Pirenne, F. et al. (2001) Br. J. Haematol. 113, 672–679.

f ANTIGEN

Terminology

ISBT symbol (number)	RH6 (004.006)
Other names	ce, hr
History	Reported in 1953 and named with the next letter of the alphabet when it was thought that c and e *in cis* were required for its expression

Occurrence

Caucasians	65%
Blacks	92%
Asians	12%

Expression

Cord RBCs	Expressed
Altered	See system pages for unusual Rh complexes

Molecular basis associated with f antigen

Requirements for expression of antigen are not understood.

Effect of enzymes/chemicals on f antigen on intact RBCs

Ficin/papain	Resistant
Trypsin	Resistant
α-Chymotrypsin	Resistant
Pronase	Resistant
Sialidase	Resistant
DTT 200 mM	Resistant
Acid	Resistant

In vitro characteristics of alloanti-f

Immunoglobulin class	Most IgG, some IgM
Optimal technique	RT; IAT; enzymes
Complement binding	No

Clinical significance of alloanti-f

Transfusion reaction	Mild/delayed/hemoglobinuria
HDN	Mild

Autoanti-f

Yes.

Comments

The f antigen, an example of a compound antigen, is expressed on RBCs having c (RH4) and e (RH5) antigens in the same haplotype (*in cis*), for example, R_1r (DCe/dce), R_0R_0 (Dce/Dce), etc. The antigen is not expressed when c and e occur on separate haplotypes (*in trans*), e.g., R_1R_2 (DCe/DcE). However, RBCs of some people with the Dc- phenotype express f.

Anti-f is frequently a component of sera containing anti-c or anti-e. Anti-f is useful in distinguishing DCE/dce from DCe/DcE. Apparent anti-f in Blacks may be anti-hr^S (see RH19). Anti-f frequently fade *in vitro* and *in vivo*.

Ce ANTIGEN

Terminology

ISBT symbol (number)	RH7 (004.007)
Other names	rh_i
History	Reported in 1958 when it was observed that C and e *in cis* were required for its expression

Occurrence

Caucasians	68%
Blacks	27%
Asians	92%

Expression

Cord RBCs	Expressed

Molecular basis associated with Ce antigen

Requirements for expression of antigen are not understood.

Effect of enzymes/chemicals on Ce antigen on intact RBCs

Ficin/papain	Resistant ($\uparrow\uparrow$)
Trypsin	Resistant ($\uparrow$)
α-Chymotrypsin	Resistant ($\uparrow$)
Pronase	Resistant ($\uparrow\uparrow$)
Sialidase	Resistant
DTT 200 mM	Resistant
Acid	Resistant

In vitro characteristics of alloanti-Ce

Immunoglobulin class	IgG more common than IgM
Optimal technique	IAT; enzymes
Complement binding	No

Clinical significance of alloanti-Ce

Transfusion reaction	Mild/delayed
HDN	Mild

Comments

Ce is a compound antigen expressed on RBCs with C and e in the same haplotype (*in cis*), for example, on DCe/dce (R_1r) RBCs but not on DCE/dce (R_zr) RBCs.

Anti-Ce is usually found in sera containing anti-C. Apparent anti-Ce in a C+ Black may be anti-hr^B (see RH31).

C^W ANTIGEN

Terminology

ISBT symbol (number)	RH8 (004.008)
Other names	Willis, rh^w
History	Reported in 1946 and named because of the association with C and 'W' from 'Willis', the first proband whose RBCs carried the antigen. For years was thought to be antithetical to C. The weak C antigen on C^W + RBCs is due to an altered expression of C rather than to 'cross-reactivity' of anti-C^W.

Occurrence

Caucasians	2%
Blacks	1%
Finns	4%
Latvians	9%

Antithetical antigens

C^X (**RH9**), MAR (**RH51**)[1]

Expression

Cord RBCs	Expressed
Altered	Weaker on DC^W

See system pages for unusual Rh complexes

Molecular basis associated with C^W antigen[2]

Amino acid	Arg 41
Nucleotide	G at bp 122 in exon 1

C^W- RBCs have Gln 41 and A at bp 122. See system pages.

Effect of enzymes/chemicals on C^W antigen on intact RBCs

Ficin/papain	Resistant ($\uparrow\uparrow$)
Trypsin	Resistant ($\uparrow$)
α-Chymotrypsin	Resistant ($\uparrow$)
Pronase	Resistant ($\uparrow\uparrow$)
Sialidase	Resistant
DTT 200 mM	Resistant
Acid	Resistant

In vitro characteristics of alloanti-C^W

Immunoglobulin class	IgG and IgM
Optimal technique	RT; IAT; enzymes
Complement binding	No

Clinical significance of alloanti-C^W

Transfusion reaction	Mild to severe; immediate/delayed
HDN	Mild to moderate

Comments

Anti-C^W are often naturally-occurring and found in multi-specific sera. Most C^W+ are C+; rare examples are C−. C^W has been associated with D(C)C^We, D(C)C^WE, (C)C^We, (C)C^WE, DC^W- and C^Wce haplotypes.

References

[1] Sistonen, P. et al. (1994) Vox Sang. 66, 287–292.
[2] Mouro, I. et al. (1995) Blood 86, 1196–1201.

C^X ANTIGEN

Terminology

ISBT symbol (number)	RH9 (004.009)
Other names	rh^X
History	Reported in 1954 and named because of the association with C and 'X' because X was the next letter in the alphabet and the antigen had characteristics similar to C^W. Was thought to be antithetical to C. The weak C antigen on C^X+ RBCs is due to an altered expression of C rather than to 'cross-reactivity' of anti-C^X.

Occurrence

Less than 0.01%; more common in Finns.

Antithetical antigens

C^W (**RH8**), MAR (**RH51**)[1]

Expression

Cord RBCs	Expressed

Molecular basis associated with C^X antigen[2]

Amino acid	Thr 36 on RhCe and rarely Rhce
Nucleotide	A at bp 106 in exon 1

C^X- RBCs have Ala 36 and G at bp 106

Effect of enzymes/chemicals on C^X antigen on intact RBCs

Ficin/papain	Resistant (↑↑)
Trypsin	Resistant (↑)
α-Chymotrypsin	Resistant (↑)
Pronase	Resistant (↑↑)
Sialidase	Resistant
DTT 200 mM	Resistant
Acid	Resistant

In vitro characteristics of alloanti-C^X

Immunoglobulin class	IgG and IgM
Optimal technique	37°C; IAT; enzymes
Complement binding	No

Clinical significance of alloanti-C^X

Transfusion reaction	No to moderate; immediate/delayed
HDN	Mild to moderate

Comments

Anti-C^X are often naturally occurring and found in multi-specific sera. C^X+ are C+ except in the rare haplotype dC^Xce^s V – VS+ found in Somalia. C^X has been associated with $D(C)C^Xe$, $(C)C^Xe$ and C^Xce^s haplotypes.

References
[1] Sistonen, P. et al. (1994) Vox Sang. 66, 287–292.
[2] Mouro, I. et al. (1995) Blood 86, 1196–1201.

V ANTIGEN

Terminology

ISBT symbol (number)	RH10 (004.010)
Other names	ce^s, hr^V

History	Reported in 1955 and named after the first letter of the last name of the proband to make anti-V

Occurrence

Caucasians	1%
Blacks	30%

Expression

Cord RBCs	Expressed

Molecular basis associated with V antigen[1,2]

RhceAR (Trp 16; Val 239; Val 245; Gly 263; Lys 267; Val 306) expresses V and not VS.
For haplotypes expressing V together with VS, see system pages.

Effect of enzymes/chemicals on V antigen on intact RBCs

Ficin/papain	Resistant
Trypsin	Presumed resistant
α-Chymotrypsin	Presumed resistant
Pronase	Presumed resistant
Sialidase	Presumed resistant
DTT 200 mM	Resistant
Acid	Resistant

In vitro characteristics of alloanti-V

Immunoglobulin class	IgG
Optimal technique	IAT; enzyme
Complement binding	No

Clinical significance of alloanti-V

Transfusion reaction	Mild/delayed
HDN	No

Comments

Anti-V frequently occurs in multispecific sera, particularly in sera containing anti-D (see RH1).

The V antigen has been associated with Dce^S, ce^S and rarely with DCe^S haplotypes.

Most V+ RBCs are also VS+ (RH20).

References

[1] Hemker, M.B. et al. (1999) Blood 94, 4337–4342.
[2] Daniels, G.L. et al. (1998) Transfusion 38, 951–958.

E^w ANTIGEN

Terminology

ISBT symbol (number)	RH11 (004.011)
Other names	rh^{W2}
History	Reported in 1955 as the cause of HDN and named after the affected family

Occurrence

Less than 0.01%; may be more common in people of German ancestry.

Expression

Cord RBCs	Expressed

Effect of enzymes/chemicals on E^w antigen on intact RBCs

Ficin/papain	Resistant (↑)
Trypsin	Presumed resistant
α-Chymotrypsin	Presumed resistant
Pronase	Presumed resistant
Sialidase	Presumed resistant
DTT 200 mM	Presumed resistant
Acid	Presumed resistant

In vitro characteristics of alloanti-E^w

Immunoglobulin class	IgG
Optimal technique	IAT; enzymes

Clinical significance of alloanti-E^w

HDN   Yes

Comments

Only found associated with the DcE^w haplotype.
Rare specificity. Some, but not all, anti-E detect E^w+ cells.

G ANTIGEN

Terminology

ISBT symbol (number)	RH12 (004.012)
Other names	rh^G
History	Reported in 1958 when a donor's D − C− RBCs were agglutinated by most anti-CD; given the next available letter in the alphabet.

Occurrence

Caucasians	84%
Blacks	92%
Asians	100%

Expression

Cord RBCs	Expressed
Altered	Weak on r^G and r″^{G} RBCs

See system pages for unusual Rh complexes

Molecular basis associated with G antigen[1]

Amino acid   Ser103 on Rh proteins expressing C or D

G− has Pro 103 usually associated with D− phenotype and rarely with D+ phenotype[2]. See system pages.

Effect of enzymes/chemicals on G antigen on intact RBCs

Ficin/papain	Resistant ($\uparrow\uparrow$)
Trypsin	Resistant
α-Chymotrypsin	Resistant
Pronase	Resistant ($\uparrow\uparrow$)
Sialidase	Resistant
DTT 200 mM	Resistant
Acid	Resistant

In vitro characteristics of alloanti-G

Immunoglobulin class	IgG
Optimal technique	IAT; enzymes
Complement binding	No

Clinical significance of alloanti-G

Transfusion reaction	No to severe/delayed
HDN	No to severe

Comments

Anti-G is found as a component in sera from rr people with anti-D (see RH1) (and/or anti-C [see RH2]), D+G− people with anti-C, and some DIIIb people with anti-D.

References

[1] Faas, B.H.W. et al. (1996) Transfusion 36, 506–511.
[2] Faas, B.H.W. et al. (2001) Transfusion 41, 1136–1142.

Hr_0 ANTIGEN

Terminology

ISBT symbol (number)	RH17 (004.017)
History	Anti-Hr_0 reported in 1958 and allocated Rh17 in 1962; defined by absorption/elution studies using sera from D-- probands. Hr_0 was considered to be a high prevalence antigen expressed by all common Rh haplotypes

Occurrence

All populations   100%

Expression

Cord RBCs   Expressed
Altered R_0^{Har}, some e− variants, Rh_{mod}

Effect of enzymes/chemicals on Hr_0 antigen on intact RBCs

Ficin/papain   Resistant ($\uparrow\uparrow$)
Trypsin   Resistant ($\uparrow\uparrow$)
α-Chymotrypsin   Resistant ($\uparrow\uparrow$)
Pronase   Resistant ($\uparrow\uparrow$)
Sialidase   Resistant
DTT 200 mM   Resistant
Acid   Resistant

In vitro characteristics of alloanti-Hr_0

Immunoglobulin class   IgG
Optimal technique   IAT; enzymes

Clinical significance of alloanti-Hr_0

Transfusion reaction   No to severe
HDN   No to severe

Autoanti-Hr_0

Antibody in patients with AIHA called anti-pdl may be anti-Rh17.

Comments

Selected anti-Rh17 may be used to distinguish Rh_{mod} from Rh_{null} phenotypes.
 Anti-Rh17 is made by individuals with D-deletion phenotypes (D--, D.., Dc-,
 DC^W-). Anti-Rh17 are not necessarily mutually compatible.
Hr_0 appears to be composed of several epitopes, some of which may be lacking
 on RBCs with unusual Rh haplotypes, including those with partial e
 expression.

149

Hr ANTIGEN

Terminology

ISBT symbol (number)	RH18 (004.018)
Other names	Hr^S; Shabalala
History	Reported in 1960; two antibodies were distinguished in the serum of Mrs. Shabalala, the Bantu proband. One of the antibodies, anti-Hr, was removed by absorption with R_2R_2 (DcE/DcE) RBCs leaving anti-hr^S.

Occurrence

Most populations: 100%; Hr− only found in Blacks.

Molecular basis associated with Hr antigen[1]

See Rh system pages for unusual Rh complexes.

Clinical significance of alloanti-Hr

Transfusion reaction	No to fatal
HDN	Moderate[2]

Comments

Hr antigen is present on all RBCs except hr^S−, Rh_{null}, and RhCE-depleted phenotypes.

Anti-Hr is made by hr^S− people and may be part of the immune response of people whose RBCs have Rh-depleted phenotypes.

References

1 Noizat-Pirenne, F. et al. (2002) Blood 100, 4223–4231.
2 Moores, P. (1994) Vox Sang. 66, 225–230.

hr^{S}ANTIGEN

Terminology

ISBT symbol (number)	RH19 (004.019)
Other names	Shabalala
History	Reported in 1960. The name 'hr' was from Wiener's terminology for e and superscript 'S' was from Shabalala, the e+ proband who made an apparent alloanti-e. See RH18 (Hr).

Occurrence

All populations, 98% (R$_2$R$_2$ RBCs lack hr^S); RBCs of approximately 1% of e+ Bantu people are hr^S−.

Expression

Cord RBCs	Expressed
Altered	Reduced on DC^{X}e and phenotypes with weak e antigens

Molecular basis associated with hr^S antigen[1]

See Rh system pages for unusual Rh complexes.

Effect of enzymes/chemicals on hr^S antigen on intact RBCs

Ficin/papain	Resistant (↑↑)
Trypsin	Resistant
α-Chymotrypsin	Resistant
Pronase	Resistant (↑↑)
Sialidase	Resistant
DTT 200 mM	Resistant
Acid	Resistant

In vitro characteristics of alloanti-hr^S

Immunoglobulin class	IgG
Optimal technique	IAT; enzymes

Clinical significance of alloanti-hr^S

Generally, not clinically significant but precise information is limited because anti-e-like antibodies are often incorrectly called anti-hr^S. However, the immune response of some hr^S– people may broaden to the clinically significant anti-Hr (anti-Rh18).

Comments

Anti-hr^S reacts preferentially with haplotypes containing ce and on initial testing may be mistaken for anti-f (see RH6) (anti-ce). Antibodies made by hr^S– people are not necessarily anti-hr^S and, unless tested with appropriate rare e variant cells, are more correctly called anti-e-like. cE haplotypes do not express hr^{S2,3}.

References
1 Noizat-Pirenne, F. et al. (2002) Blood 100, 4223–4231.
2 Issitt, P.D. (1991) Immunohematology 7, 29–36.
3 Moores, P. (1994) Vox Sang. 66, 225–230.

VS ANTIGEN

Terminology

ISBT symbol (number)	RH20 (004.020)
Other names	e^s
History	Reported in 1960 and named after the initials of the first lady to make the antibody; the initial of her first name was used because of the association with the V antigen

Occurrence

Blacks, 26–40%; other populations, less than 0.01%.

Expression

Cord RBCs	Expressed
Altered	D(C)(e^S) FPTT+[1]; DC^{W}e/Dce^S (one example, Inkelberger); Dce^S/DCe (one example, Manday) and see system pages

Molecular basis associated with VS antigen[2]

Amino acid	Val 245 in Rhce (several haplotypes)
Nucleotide	G at bp 733 in exon 5

VS− RBCs (wild type) have Leu 245 and C at bp 733.

(C)e^S (r'^{S} haplotype)	*RHD(1, 2 and part of 3)-RHce(part of 3–7)* *RHD(8–10)*[3]
	Val 245 contributed by *RHD*

For haplotypes expressing VS with or without V, see system pages.

Effect of enzymes/chemicals on VS antigen on intact RBCs

Ficin/papain	Resistant
Trypsin	Presumed resistant
α-Chymotrypsin	Presumed resistant
Pronase	Presumed resistant
Sialidase	Presumed resistant
DTT 200 mM	Presumed resistant
Acid	Resistant

In vitro characteristics of alloanti-VS

Immunoglobulin class	IgG
Optimal technique	IAT; enzymes

Clinical significance of alloanti-VS

Transfusion reaction	Mild/delayed
HDN	Positive DAT; no clinical HDN

Comments

Anti-VS is often a component of sera with other specificities. Anti-VS can be naturally-occurring and are heterogeneous.

The majority of V+ RBCs are VS+ (RH:20). The majority of apparent hr^B− (RH:31) RBCs are VS+[4]. The VS antigen has been associated with (C)ce^S, Dce^S and ce^S haplotypes.

References

1 Bizot, M. et al. (1988) Transfusion 28, 342–345.
2 Daniels, G.L. et al. (1998) Transfusion 38, 951–958.
3 Blunt, T. et al. (1994) Vox Sang. 67, 397–401.
4 Reid, M.E. et al. (1997) Vox Sang. 72, 41–44.

C^G ANTIGEN

Terminology

ISBT symbol (number)	RH21 (004.021)
History	Reported in 1961; considered to be the weak C antigen found on r^{G}r^G and r^{G}r RBCs. C^G is also made by all cells expressing C

Occurrence

Caucasians	68%

Comments

There is no monospecific anti-C^G but a minority of anti-C are anti-CC^G. Some consider that the C made by r'^{S} is actually C^{G1}.

Reference

[1] Issitt, P.D. and Anstee, D.J. (1998) Applied Blood Group Serology. 4th Edition, Montgomery Scientific Publications, Durham, NC.

CE ANTIGEN

Terminology

ISBT symbol (number)	RH22 (004.022)
Other names	Jarvis
History	Reported in 1962 and named when it was observed that C and E *in cis* were required for its expression

Occurrence

Less than 1% in most populations; 2% in Asians.

Expression

Cord RBCs   Expressed

Molecular basis associated with CE antigen

Requirements for expression of CE are not understood.

Effect of enzymes/chemicals on CE antigen on intact RBCs

Ficin/papain   Resistant
Trypsin   Presumed resistant
α-Chymotrypsin   Presumed resistant
Pronase   Presumed resistant
Sialidase   Presumed resistant
DTT 200 mM   Presumed resistant
Acid   Presumed resistant

In vitro characteristics of alloanti-CE

Optimal technique   RT [Original anti-CE (Jarvis)]; 37°C

Clinical significance of alloanti-CE

No data are available because only two examples reported.

Comments

The two reported anti-CE appeared to be naturally-occurring and were in sera that also contained anti-C.

This compound antigen is expressed on RBCs with DCE (R_z) and dCE (r_y) haplotypes.

D^W ANTIGEN

Terminology

ISBT symbol (number)   RH23 (004.023)
Other names   Weil

History

Reported in 1962 and named after the first proband whose RBCs had this low prevalence antigen; shown to be an Rh antigen in 1965 and was associated with DVa

Occurrence

Less than 0.01%.

Expression

Cord RBCs   Expressed

Molecular basis associated with D^W antigen

Associated with the partial D antigen of DVa: *RH(D-CE-D)* gene in which all or part of exon 5 of *RHD* is replaced by the same exon from *RHCE*[1].
   See D antigen pages.

Effect of enzymes/chemicals on D^W antigen on intact RBCs

Ficin/papain   Resistant (↑↑)
Trypsin   Presumed resistant
α-Chymotrypsin   Presumed resistant
Pronase   Presumed resistant
Sialidase   Presumed resistant
DTT 200 mM   Presumed resistant
Acid   Presumed resistant

In vitro characteristics of alloanti-D^W

Immunoglobulin class   IgG
Optimal technique   IAT; enzymes

Clinical significance of alloanti-D^W

HDN   Moderate

Comments

D^W is expressed on $D^{Va}Ce$ RBCs in Caucasians, $D^{Va}ce$ in Blacks; the $D^{Va}cE$ haplotype is rare.
Sera containing anti-D^W often contain anti-E.

Anti-D^W (anti-Rh23) is a rare specificity and has been found in multispecific sera. Some examples contain anti-Rh32 and these specificities are not separable. The molecular basis of the Rh haplotype in a person (NR) with D^W-, Rh32− RBCs that were agglutinated by one example of anti-Rh23/Rh32 is given on the system page.

Reference
[1] Rouillac, C. et al. (1995) Blood 85, 2937–2944.

Rh26 (c-LIKE) ANTIGEN

Terminology

ISBT symbol (number)	RH26 (004.026)
Other names	Deal
History	This variant of c was identified in 1964 when the serum of Mrs. Deal, considered to contain a potent anti-c, did not react with some c+ RBCs

Occurrence

Expressed on the majority of c-positive RBCs. The c+ Rh26− phenotype has been found in Italians and Dutch.

Molecular basis associated with Rh26 antigen[1]

Amino acid	Gly 96 on Rhce
Nucleotide	G at bp 286 of *RHce*

Rh26− form has Ser 96 and A at bp 286. See system pages.

Effect of enzymes/chemicals on Rh26 antigen on intact RBCs

Ficin/papain	Resistant (↑)
Trypsin	Presumed resistant
α-Chymotrypsin	Presumed resistant
Pronase	Presumed resistant
Sialidase	Presumed resistant
DTT 200 mM	Presumed resistant
Acid	Presumed resistant

157

In vitro characteristics of alloanti-Rh26

Immunoglobulin class   IgG
Optimal technique   37°C; IAT; enzymes

Clinical significance of alloanti-Rh26

No data are available.

Comments

One c− Rh26+ sample has been described. Rh26− RBCs have weak expression of f antigen.

Reference
1   Faas, B.H.W. et al. (1997) Transfusion 37, 1123–1130.

cE ANTIGEN

Terminology

ISBT symbol (number)   RH27 (004.027)
History   Reported in 1965 and named when it was observed that c and E *in cis* were required for its expression

Occurrence

Caucasians   28%
Blacks   22%
Asians   38%

Molecular basis associated with cE antigen

Requirements for expression of cE are not understood.

Effect of enzymes/chemicals on cE antigen on intact RBCs

Ficin/papain   Resistant
Trypsin   Presumed resistant
α-Chymotrypsin   Presumed resistant

Pronase	Presumed resistant
Sialidase	Presumed resistant
DTT 200 mM	Presumed resistant
Acid	Presumed resistant

In vitro characteristics of alloanti-cE

Immunoglobulin class	IgG
Optimal technique	IAT; enzymes
Complement binding	Yes (one example)

Comments

Few examples of anti-cE have been reported. Expressed on RBCs having c [RH4] and E [RH3] antigens in the same haplotype (*in cis*), for example, R_2r (DcE/dce), r″r (dcE/dce).

The antigen is not expressed when c and E occur on separate haplotypes (*in trans*), for example, R_zr (DCE/dce).

hr^H ANTIGEN

Terminology

ISBT symbol (number)	RH28 (004.028)
History	Reported in 1964. The antigen hr^H, primarily studied among South African Blacks, may be present on some RBCs that type V−VS+. hr^H has a complex relationship with VS (RH20)

Occurrence

Less than 0.01%.

Rh29 ANTIGEN

Terminology

ISBT symbol (number)	RH29 (004.029)
Other names	Total Rh
History	Reported in 1961 and given the next available number. The only Rh29− RBCs are Rh$_{null}$, which were originally called ---/--- when the first proband, an Australian Aboriginal woman, was identified.

Occurrence

All populations	100%

Expression

Cord RBCs	Expressed

Molecular basis of Rh29

Not known. For molecular basis of Rh29− (Rh$_{null}$) see system pages.

Effect of enzymes/chemicals on Rh29 antigen on intact RBCs

Ficin/papain	Resistant (↑↑)
Trypsin	Resistant (↑↑)
α-Chymotrypsin	Resistant (↑)
Pronase	Resistant (↑↑)
Sialidase	Presumed resistant
DTT 200 mM	Presumed resistant
Acid	Resistant

In vitro characteristics of alloanti-Rh29

Immunoglobulin class	IgG and IgM
Optimal technique	37°C; IAT; enzymes

Clinical significance of alloanti-Rh29

Transfusion reaction	No data available but potentially capable
HDN	No to severe

Autoanti-Rh29

Antibody in AIHA called anti-dl may be anti-Rh29.

Comments

Anti-Rh29 is the immune response of some Rh_{null} individuals (both amorph and regulator type). Some anti-Rh29 react with Rh_{mod} cells.

Go^a ANTIGEN

Terminology

ISBT symbol (number)	RH30 (004.030)
Other names	Gonzales, D^{Cor}
History	Briefly reported in 1962 and more extensively in 1967 when Go^a was shown to be an Rh antigen. In 1968 Go^a became associated with D category IV (D^{IVa}ce). It was named after Mrs. Gonzales, the first maker of anti-Go^a. Before partial D phenotypes were categorized, DIVa were called D^{Cor}

Occurrence

Only found in Blacks (2%).

Expression

Cord RBCs	Expressed

Molecular basis associated with Go^a antigen

Go^a is associated with the partial D antigen of category DIVa encoded by a RH(D-CE-D-CE-D) gene in which part of exon 3 and part of exon 7 of RHD are replaced by the equivalent portion RHCE[1].

See D antigen pages.

Effect of enzymes/chemicals on Go^a antigen on intact RBCs

Ficin/papain	Resistant (↑↑)
Trypsin	Resistant
α-Chymotrypsin	Presumed resistant
Pronase	Presumed resistant
Sialidase	Presumed resistant
DTT 200 mM	Presumed resistant
Acid	Resistant

In vitro characteristics of alloanti-Go^a

Immunoglobulin class	IgG
Optimal technique	37°C; IAT; enzymes

Clinical significance of alloanti-Go^a

Transfusion reaction	Moderate/delayed
HDN	Mild to severe

Comments

Go^a is also expressed on D^{IVa}(C)- RBCs.

Anti-Go^a may be immune but are often in multispecific sera, frequently with anti-Rh32 (see RH32) and/or anti-Evans (RH37); these Rh specificities are not separable by absorption/elution.

Reference

[1] Rouillac, C. et al. (1995) Blood 85, 2937–2944.

hr^B ANTIGEN

Terminology

ISBT symbol (number)	RH31 (004.031)
Other name	Bastiaan
History	Reported in 1972. Named 'hr' from Wiener's terminology for e and 'B' from Bastiaan, the first antibody producer. See Hr^B (RH34)

Occurrence

All populations: 98% [R_2R_2 (DcE/DcE) RBCs lack hr^B]

Expression

Cord RBCs	Expressed
Altered	Reduced on phenotypes with weak e antigens. See system pages

Molecular basis associated with hr^B antigen

Not known. See system pages.

Effect of enzymes/chemicals on hr^B antigen on intact RBCs

Ficin/papain	Resistant (↑↑)
Trypsin	Resistant
α-Chymotrypsin	Presumed resistant
Pronase	Presumed resistant
Sialidase	Presumed resistant
DTT 200 mM	Presumed resistant
Acid	Resistant

In vitro characteristics of alloanti-hr^B

Immunoglobulin class	IgG
Optimal technique	37°C; IAT; enzymes

Clinical significance of alloanti-hr^B

Transfusion reaction	Generally, not clinically significant but precise information is limited because anti-e-like antibodies are often incorrectly called anti-hr^B. However, the immune response of some hr^B− people may broaden to the clinically significant anti- Hr^B (Rh34).
HDN	Positive DAT; no clinical HDN

Autoanti-hr^B

Yes, rare (often with transient suppression of antigen).

Comments

The majority of apparent e+ hr^B- RBCs are VS+ (r'^S)[1].
cE haplotypes do not express hr^B.
Anti-hr^B can be mistaken for anti-Ce (see RH7).
Antibodies made by hr^B- people are not necessarily anti-hr^B, and, unless tested with appropriate rare e variant cells, are more correctly labeled anti-e-like.

Reference
1 Beal, C.L. et al. (1995) Immunohematology 11, 74–77.

Rh32 ANTIGEN

Terminology

ISBT symbol (number)	RH32 (004.032)
Other names	$\bar{\bar{R}}^N$
History	Reported in 1971 after several years of investigation and was assigned the next Rh number in 1972. Incorrectly called $\bar{\bar{R}}^N$, which is the name of the original (1960) haplotype with weak C and e antigens later shown to express Rh32

Occurrence

Approximately 1% in Blacks ($\bar{\bar{R}}^N$)
Occurs rarely in Caucasians (mainly DBT phenotype) and in Japanese (DBT phenotype).

Antithetical antigen

Sec (**RH46**)

Expression

Cord RBCs	Expressed
Altered	May be slightly weaker on DBT phenotype RBCs and other rare variants

Molecular basis associated with Rh32 antigen[1,2]

$\bar{\bar{R}}^N$ phenotype: *RH(CE-D-CE)*gene in which exon 4 is replaced by the corresponding exon of *RHD* [with or without nt 445C>A in exon 3 (Thr152Asn)].

Partial D phenotype DBT: *RH(D-CE-D)* gene in which either exons 5–7, or exons 5–9 of *RHD* are replaced by the corresponding exons of *RHCE*.

See system and D antigen pages.

Effect of enzymes/chemicals on Rh32 antigen on intact RBCs

Ficin/papain	Resistant (↑↑)
Trypsin	Resistant
α-Chymotrypsin	Resistant
Pronase	Presumed resistant
Sialidase	Presumed resistant
DTT 200 mM	Presumed resistant
Acid	Resistant

In vitro characteristics of alloanti-Rh32

Immunoglobulin class	IgG
Optimal technique	37°C; IAT; enzymes

Clinical significance of alloanti-Rh32

Transfusion reaction	None reported
HDN	Mild to severe

Comments

The gene complex $\bar{\bar{R}}^N$ produces Rh32 in combination with weakened expression of C (RH2) and e (RH5) antigens and normal or elevated expression of D antigen (RH1). It may be necessary to use sensitive techniques to detect the C antigen on some RBCs.

The RBCs of one proband with the DBT phenotype had weakened expression of C and e; another proband had weakened expression of C only.

Anti-Rh32 may be immune but are often naturally-occurring in multispecific sera. Anti-Rh32 cannot be separated from anti-Goª (see RH30) or anti-Evans (see RH37) by absorption/elution of sera containing these antibodies.

References
1 Beckers, E.A.M. et al. (1996) Br. J. Haematol. 93, 720–727.
2 Rouillac C. et al. (1996) Blood 87, 4853–4861.

Rh33 ANTIGEN

Terminology

ISBT symbol (number)	RH33 (004.033)
Other names	Har; R_0^{Har}; D^{Har}
History	Reported in 1971 and given the next Rh number. Although the complex expressing Rh33 was first detected on RBCs from a German donor, the complex was named R_0^{Har} after the name of an English donor with an informative family

Occurrence

Less than 0.01%. Rh33 may be more common in people of German ancestry.

Expression

Cord RBCs	Presumed expressed
Altered	R_1^{Lisa1}

Molecular basis associated with Rh33 antigen

In the partial D phenotype R_0^{Har}, Rh33 is associated with an *RH(ce-D-ce)* gene in which exon 5 of *RHce* is replaced by exon 5 of *RHD²*. The *RHCeVA* allele, a hybrid *RH(Ce-D-Ce)* gene with exon 5 originating from *RHD* encodes Rh33 and weak C and e antigens. *RHCeVA* may be the allele encoding R_1^{Lisa}.3
    See system pages.

Effect of enzymes/chemicals on Rh33 antigen on intact RBCs

Ficin/papain	Resistant (↑↑)
Trypsin	Presumed resistant
α-Chymotrypsin	Presumed resistant
Pronase	Presumed resistant
Sialidase	Presumed resistant
DTT 200 mM	Presumed resistant
Acid	Presumed resistant

In vitro characteristics of alloanti-Rh33

Immunoglobulin class	IgM
Optimal technique	RT; enzymes
Complement binding	No

Clinical significance of alloanti-Rh33

No data are available.

Comments

Rh33 is encoded by an unusual Rh complex R^{oHar}. This complex also encodes a partial D antigen, normal c (RH4), weak e (RH5), weak f (RH6) and weak Hr_0 (RH17) antigens; it does not encode C (RH2), E (RH3), G (RH12), hr^S (RH19) or Hr (RH18) antigens.

Rh33 is also expressed by the rare complexes $D^{IVa}(C)$-, R_0^{JOH} and R_1^{Lisa}. All Rh33+ RBCs also express the low incidence antigen FPTT (RH50).

Anti-Rh33 is a rare specificity. Two examples were in serum also containing anti-D.

References
[1] Moores, P. et al. (1991) Transfusion 31, 759–761.
[2] Beckers, E.A.M. et al. (1996) Br. J. Haematol. 92, 751–757.
[3] Noizat-Pirenne, F. et al. (2002) Transfusion 42, 627–633.

Hr^B ANTIGEN

Terminology

ISBT symbol (number)	RH34 (004.034)
Other names	Bas; Baas; Bastiaan; Rh34
History	Reported in 1972. Anti-Hr^B initially described the total immune response of Mrs. Bastiaan (hence 'B' in the name), a South African. Later, absorptions showed her serum contained two specificities: anti-hr^B (see **RH31**) and an antibody reacting with RBCs of all common phenotypes that was called anti-Hr^{B1}

Occurrence

All populations	100%

Expression

Cord RBCs	Expressed

Molecular basis associated with Hr^B antigen

See Rh system pages for molecular basis of Hr^B− phenotypes.

Effect of enzymes/chemicals on Hr^B antigen on intact RBCs

Ficin/papain	Resistant (↑↑)
Trypsin	Presumed resistant
α-Chymotrypsin	Presumed resistant
Pronase	Presumed resistant
Sialidase	Presumed resistant
DTT 200 mM	Presumed resistant
Acid	Presumed resistant

In vitro characteristics of alloanti-Hr^B

Immunoglobulin class	IgG
Optimal technique	IAT; enzymes

Clinical significance of alloanti-Hr^B

Transfusion reaction	No data available, presumed to be significant because of similarity to anti-Rh18
HDN	Positive DAT but no clinical HDN[1]

Comments

Weak examples of anti-Hr^B resemble anti-C (see RH3) in that C+ RBCs give the strongest reactions; c+ RBCs give intermediate strength reactions and DcE/DcE (R$_2$R$_2$) cells give the weakest reactions.

Reference
[1] Moores, P. and Smart E. (1991) Vox Sang. 61, 122–129.

Rh35 ANTIGEN

Terminology

ISBT symbol (number)	RH35 (004.035)
Other names	1114

History   Reported in 1971. Rh35 is produced by an
                                 Rh complex that produces weak C and e
                                 antigens and normal D antigen

Occurrence

Less than 0.01%. Rh35 was originally found in people of Danish ancestry.

Expression

Cord RBCs   Presumed expressed

Molecular basis associated with Rh35 antigen

For the molecular basis of a phenotype with weak C and e expression
(CeMA), which may be Rh35+, see system pages.

Effect of enzymes/chemicals on Rh35 antigen on intact RBCs

Ficin/papain   Resistant (↑↑)
Trypsin   Presumed resistant
α-Chymotrypsin   Presumed resistant
Pronase   Presumed resistant
Sialidase   Presumed resistant
DTT 200 mM   Presumed resistant
Acid   Presumed resistant

In vitro characteristics of alloanti-Rh35

Immunoglobulin class   IgG
Optimal technique   Enzymes

Clinical significance of alloanti-Rh35

No data available because only one example of the antibody has been reported.

Be^a ANTIGEN

Terminology

ISBT symbol (number)	RH36 (004.036)
Other names	Berrens
History	Reported in 1953 and named after the family in which HDN occurred. Be^a is produced by a complex that produces weak c, e, and f (ce) antigens and no D antigen. Family studies in 1974 confirmed it as an Rh antigen. Allocated the next Rh number

Occurrence

Less than 0.1%. Propositi were of German/Polish extraction

Expression

Cord RBCs	Expressed

Effect of enzymes/chemicals on Be^a antigen on intact RBCs

Ficin/papain	Resistant (↑↑)
Trypsin	Presumed resistant
α-Chymotrypsin	Presumed resistant
Pronase	Presumed resistant
Sialidase	Presumed resistant
DTT 200 mM	Presumed resistant
Acid	Presumed resistant

In vitro characteristics of alloanti-Be^a

Immunoglobulin class	IgG
Optimal technique	37°C; IAT; enzymes
Complement binding	No

Clinical significance of alloanti-Be^a

Transfusion reaction	None reported
HDN	Moderate to severe

Comments

Anti-Be^a is immune.

Evans ANTIGEN

Terminology

ISBT symbol (number)   RH37 (004.037)

History   Evans, identified in 1968, was named after the family in which HDN occurred. Evans segregated with a D-- like complex (D••) in the family of the second Evans+ proband. Family studies, reported in 1978, confirmed Evans as an Rh antigen

Occurrence

Less than 0.01%. May be more common in Welsh and Scots.

Expression

Cord RBCs   Expressed

Altered   Weak on DIVb RBCs

Molecular basis associated with Evans antigen[1,2]

Dav   *RHD(1–6)-RHCE(7–10)//RHD*

JD   *RHD(1–5 and part 6)-RHCE(part 6 of 6–10)//RHCE(1)-RHD(2–10)*

AT   *RHCE(1)-RHD(2–6)-RHCE(7–10)//RHD*

DIVb   *RHCE//RHD(1–6 and part of 7)-RHCE (part of 7–9)-RHD10*

See system pages.

Effect of enzymes/chemicals on Evans antigen on intact RBCs

Ficin/papain   Resistant (↑↑)

Trypsin   Presumed resistant

α-Chymotrypsin   Presumed resistant

Pronase   Presumed resistant

Sialidase   Presumed resistant

DTT 200 mM   Presumed resistant

Acid   Presumed resistant

In vitro characteristics of alloanti-Evans

Immunoglobulin class	IgM less common than IgG
Optimal technique	37°C; IAT; enzymes
Complement binding	No

Clinical significance of alloanti-Evans

Transfusion reaction	None reported
HDN	Mild and moderate

Comments

The Rh complex D•• produces Evans antigen, elevated expression of D, normal expression of G, the high incidence antigens Rh29 and Dav; C, c, E or e antigens are not produced[3].

Anti-Evans may be naturally-occurring and is often found in multispecific sera. Anti-Evans cannot be separated from anti-Go[a] (see RH30) or anti-Rh32 (see RH32) by absorption/elution of sera containing these antibodies.

References

[1] Huang, C.-H. et al. (2000) Semin. Hematol. 37, 150–165.
[2] Avent, N.D. and Reid M.E. (2000) Blood 95, 375–387.
[3] Tippett, P. et al. (1996) Vox Sang. 70, 123–131.

Rh39 ANTIGEN

Terminology

ISBT symbol (number)	RH39 (004.039)
Other names	C-like
History	Reported in 1979. Anti-Rh39 reacts more strongly with C+ than C− RBCs and can be absorbed to exhaustion by all C+ and C− RBCs with common and uncommon Rh phenotypes except Rh$_{null}$

Occurrence

All populations   100%

Autoanti-Rh39

Yes, always. Made by some C− people.

Comments

One patient with this 'mimicking' anti-C antibody proceeded to make alloanti-C.

Tar ANTIGEN

Terminology

ISBT symbol (number)   RH40 (004.040)
Other names   Targett
History   Reported in 1975 and named after the proband whose RBCs expressed the antigen. When family studies in 1979 showed Tar to be an Rh antigen it was awarded an Rh number. In 1986, Tar was established as a marker for the DVII partial D antigen

Occurrence

Less than 0.01%.

Expression

Cord RBCs   Expressed

Molecular basis associated with Tar antigen[1]

Amino acid   Pro110
The Tar-negative RhD protein has Leu110. See D antigen pages.

Effect of enzymes/chemicals on Tar antigen on intact RBCs

Ficin/papain   Resistant ($\uparrow$)
Trypsin   Presumed resistant
α-Chymotrypsin   Presumed resistant
Pronase   Presumed resistant
Sialidase   Presumed resistant
DTT 200 mM   Resistant
Acid   Presumed resistant

In vitro characteristics of alloanti-Tar

Immunoglobulin class   IgG
Optimal technique   37°C; IAT; enzymes
Complement binding   No

Clinical significance of alloanti-Tar

HDN   Moderate

Comments

In addition to the association with $D^{VII}Ce$, Tar is expressed on a variant RhD protein that also expresses weak c[2]. Tar also was found on a D-- like complex, which produced weaker than usual D antigen.

Anti-Tar is a rare specificity; the antibody has been produced through pregnancy and has been found without known stimulus.

References
[1] Rouillac, C. et al. (1995) Am. J. Hematol. 49, 87–88.
[2] Faas, B.H.W. et al. (2001) Transfusion 41, 1136–1142.

Rh41 ANTIGEN

Terminology

ISBT symbol (number)	RH41 (004.041)
Other names	Ce-like
History	Reported in 1981 and given the next Rh number in 1990. The only example of anti-Rh41 reacted with RBCs that have C and e in the same haplotype. However, unlike anti-Ce, anti-Rh41 reacts with r'^S (dCce^S) RBCs and does not react with C^W and e *in cis*[1]

Occurrence

Caucasians	70%

Expression

Cord RBCs	Presumed expressed

Reference
[1] Svoboda, R.K. et al. (1981) Transfusion 21, 150–156.

Rh42 ANTIGEN

Terminology

ISBT symbol (number)	RH42 (004.042)
Other names	Ce^S; Cce^S; rh^S; Thornton
History	Reported in 1980. It is a marker for the Cce^S V– VS+ haplotype

Occurrence

Less than 0.1% in Caucasians; 2% in Blacks.

Expression

Cord RBCs   Expressed

Effect of enzymes/chemicals on Rh42 antigen on intact RBCs

Ficin/papain   Resistant ($\uparrow$)

In vitro characteristics of alloanti-Rh42

Immunoglobulin class   IgG
Optimal technique   37°C; IAT; enzymes

Clinical significance of alloanti-Rh42

Transfusion reaction   None reported
HDN   Moderate

Comments

At least two examples of anti-Rh42 have been reported.

Crawford ANTIGEN

Terminology

ISBT symbol (number)   RH43 (004.043)
History   Reported in 1980, the only example of anti-Crawford was found in a reagent anti-D. There is little information about the Crawford antigen. RBCs type: VS+, V_+^W

Occurrence

Less than 1% in Blacks.

Expression

Cord RBCs   Expressed

Molecular basis associated with Crawford antigen[1]

Rhce: 16 Cys, 103 Pro, 223 Glu, 226 Ala, 245 Val

Effect of enzymes/chemicals on Crawford antigen on intact RBCs

Ficin/papain   Resistant (↑)

In vitro characteristics of alloanti-Crawford

Immunoglobulin class   IgG
Optimal technique   37°C; IAT; enzymes

Reference
[1] Westhoff, C. Personal Communication.

Nou ANTIGEN

Terminology

ISBT symbol (number)   RH44 (004.044)
History   The antigen was reported in 1969 and
                              named after Mme Nou. from the Ivory
                              Coast who was homozygous for $D^{IVa}(C)$-.
                              Anti-Nou, reported in 1981, is a compo-
                              nent of some anti-Hr$_0$ (see RH17) sera and
                              can be separated by adsorption/elution
                              with $D^{IVa}(C)$-/$D^{IVa}(C)$- cells; the antibody
                              does not react with Rh$_{null}$, D− −, D··
                              DC^W− or Dc− cells

Occurrence

All populations   100%

Expression

Cord RBCs   Expressed

Effect of enzymes/chemicals on Nou antigen on intact RBCs

Ficin/papain   Resistant (↑)

Riv ANTIGEN

Terminology

ISBT symbol (number)   RH45 (004.045)
History   Reported in 1983 and named for the Puerto Rican family in which the antigen and antibody were identified

Occurrence

Six propositi are known.

Expression

Cord RBCs   Expressed

Effect of enzymes/chemicals on Riv antigen on intact RBCs

Ficin/papain   Resistant (↑)
Trypsin   Presumed resistant
α-Chymotrypsin   Presumed resistant
Pronase   Presumed resistant
Sialidase   Presumed resistant
DTT 200 mM   Presumed resistant
Acid   Presumed resistant

In vitro characteristics of alloanti-Riv

Immunoglobulin class   IgG
Optimal technique   37°C; IAT; enzymes
Complement binding   No

Clinical significance of alloanti-Riv

HDN   Mild; caused by the only example of anti-Riv in a serum which also contained anti-Go^a (see RH30)[1]

Comments

The Riv antigen is encoded by the rare gene complex $D^{IVa}(C)-$; this complex also produces Go^a (RH30), Rh33 (RH33), FPTT (RH50), the D antigen (RH1) characteristic of category D^{IVa}, G (RH12), Nou (RH44), and very weak C (RH2), but no c (RH4), E (RH3), e (RH5) or f (RH6) antigen.

Reference

[1] Delehanty, C.L. et al. (1983) Transfusion 23, 410 (abstract).

Sec ANTIGEN

Terminology

ISBT symbol (number)	RH46 (004.046)
History	Described in 1989, given an Rh number in 1990, named after the first antibody producer

Occurrence

All populations: 100%.

Antithetical antigen

Rh32 (**RH32**)

Expression

Cord RBCs	Expressed

Molecular basis associated with the Sec antigen

See system pages.

Effect of enzymes/chemicals on Sec antigen on intact RBCs

Ficin/papain	Resistant ($\uparrow$)
Trypsin	Presumed resistant
α-Chymotrypsin	Presumed resistant
Pronase	Presumed resistant
Sialidase	Presumed resistant
DTT 200 mM	Presumed resistant
Acid	Presumed resistant

In vitro characteristics of alloanti-Sec

Immunoglobulin class	IgG
Optimal technique	37°C; IAT; enzymes
Complement binding	No

Clinical significance of alloanti-Sec

HDN   No to severe

Comments

Immunized D(C)(e)/D(C)(e) people, homozygous for Rh32, make anti-Sec.
Sec is expressed by RBCs of common Rh phenotype but is absent from Rh_{null}
  RBCs and not expressed by the following haplotypes: $\bar{\bar{R}}^N$, D--, Dc-, DC^W-,
  D··.

Dav ANTIGEN

Terminology

ISBT symbol (number)   RH47 (004.047)
History   Reported in 1982 and named after the first
                              donor with D·· RBCs. Anti-Dav is a com-
                              ponent of some anti-Hr_0 (see RH17) sera
                              and can be separated by adsorption/elu-
                              tion with D··/D·· cells

Occurrence

All populations   100%

Expression

Cord RBCs   Expressed

Effect of enzymes/chemicals on Dav antigen on intact RBCs

Ficin/papain   Resistant ($\uparrow$)

Comments

Anti-Dav reacts with cells of all common Rh phenotypes, and with D$\cdot\cdot$ cells but not with Rh$_{null}$, D^{IVa}(C)-, D--, DC^W- and Dc- cells.

JAL ANTIGEN

Terminology

ISBT symbol (number)	RH48 (004.048)
Other names	S.Allen, J.Allen
History	Reported and numbered in 1990 after more than a decade of using the Allen serum; named after J. Allen, whose RBCs possessed the antigen

Occurrence

Less than 0.01%; found in English, French-speaking Swiss, Brazilians and Blacks.

Expression

Cord RBCs	Expressed

Effect of enzymes/chemicals on JAL antigen on intact RBCs

Ficin/papain	Resistant ($\uparrow$)
Trypsin	Presumed resistant
α-Chymotrypsin	Presumed resistant
Pronase	Presumed resistant
Sialidase	Presumed resistant
DTT 200 mM	Presumed resistant
Acid	Presumed resistant

In vitro characteristics of alloanti-JAL

Immunoglobulin class	IgG
Optimal technique	37°C; IAT; enzymes

181

Clinical significance of alloanti-JAL

HDN   No to severe

Comments

The JAL antigen is encoded by two different Rh complexes: in Blacks JAL is associated with weak expression of c antigen (RH4) in a ce complex while in Caucasians JAL is associated with a weak C antigen (RH2) in a Ce complex. In both complexes, JAL is variably associated with weak e antigen (RH5) expression. Three examples of anti-JAL are reported: only one is monospecific[1,2].

References
1 Lomas, C. et al. (1990) Vox Sang. 59, 39–43.
2 Poole, J. et al. (1990) Vox Sang. 59, 44–47.

STEM ANTIGEN

Terminology

ISBT symbol (number)   RH49 (004.049)
Other names   Stemper
History   Reported in 1993 (Rh number was assigned at the 1992 ISBT meeting) and named after the Black family in which the antibody/antigen was first identified

Occurrence

Less than 0.01% in Caucasians; 0.4% in Indians (in South Africa) and 6% in Blacks.

Expression

Cord RBCs   Expressed
Altered   Variable expression among STEM+[1]

Effect of enzymes/chemicals on STEM antigen on intact RBCs

Ficin/papain   Resistant (↑)
Trypsin   Presumed resistant

α-Chymotrypsin	Presumed resistant
Pronase	Presumed resistant
Sialidase	Presumed resistant
DTT 200 mM	Presumed resistant
Acid	Presumed resistant

In vitro characteristics of alloanti-STEM

Immunoglobulin class	IgG
Optimal technique	IAT; enzymes
Complement binding	No

Clinical significance of alloanti-STEM

HDN	Mild

Comments

STEM may be associated with Dce haplotypes that do not produce hr^S (RH19) or hr^B (RH31)[1]. Approximately 65% of hr^S−Hr− RBCs and 30% of hr^B−Hr^B− RBCs are STEM+.

Reference

[1] Marais, I. et al. (1993) Transf. Med. 3, 35–41.

FPTT ANTIGEN

Terminology

ISBT symbol (number)	RH50 (004.050)
Other names	700.048, Mol
History	Reported in 1988 and named after the 'French Post Telegraph and Telecommunication' because several of the original probands worked and donated blood there. Achieved Rh antigen status in 1994

Occurrence

Less than 0.01%.

183

Expression

Cord RBCs	Expressed
Altered	Strength varies with type of FPTT+ Rh complex

Molecular basis associated with FPTT antigen[1-3]

In the partial D phenotype DFR, FPTT is associated with a hybrid *RH(D-CE-D)* gene in which part of exon 4 of *RHD* is replaced by the same part of exon 4 of *RHCE*[4].

In the partial D phenotype, R_0^{Har}, FPTT is associated with a hybrid *RH(ce-D-ce)* and a hybrid *RH(Ce-D-Ce)* in which exon 5 of *RHCE* is replaced by exon 5 of *RHD*.

See system and D antigen pages.

Effect of enzymes/chemicals on FPTT antigen on intact RBCs

Ficin/papain	Resistant (↑)
Trypsin	Presumed resistant
α-Chymotrypsin	Presumed resistant
Pronase	Presumed resistant
Sialidase	Presumed resistant
DTT 200 mM	Presumed resistant
Acid	Presumed resistant

In vitro characteristics of alloanti-FPTT

Immunoglobulin class	IgG
Optimal technique	IAT; enzymes

Clinical significance of alloanti-FPTT

No data are available.

Comments

FPTT antigen is also associated with rare 'depressed' Rh phenotypes that have depressed C (RH2) and/or e (RH5) antigens (one family had weakened expression of VS antigen [RH20])[5].

The rare haplotype $D^{IVa}(C)-$ is FPTT+. All Rh33+ RBCs are FPTT+, but not all FPTT+ are Rh33+.

The only reported example of anti-FPTT was in a multispecific serum (Mol) from a woman who had not been transfused or pregnant.

References
1 Beckers, E.A.M. et al. (1996) Br. J. Haematol. 92, 751–757.
2 Noizat-Pirenne, F. et al. (2002) Transfusion 42, 627–633.
3 Rouillac, C. et al. (1995) Blood 85, 2937–2944.
4 Cartron, J.-P. (1994) Blood Rev. 8, 199–212.
5 Bizot, M. et al. (1988) Transfusion 28, 342–345.

MAR ANTIGEN

Terminology

ISBT symbol (number)	RH51 (004.051)
History	Reported in 1994 and named after the first antibody producer, a Finnish woman with C^W+, C^X+ RBCs

Occurrence

All populations: 100%; Occurrence of MAR− phenotype in Finns is 0.2%.

Antithetical antigen

C^W (**RH8**) and C^X (**RH9**)

Expression

Cord RBCs	Expressed
Altered	Weak on hr^B-; RH:32; DC^We/DC^We; DC^Xe/DC^Xe RBCs

Molecular basis associated with MAR antigen

MAR− RBCs can express C^W (RH8) and C^X (RH9) antigens. MAR is in the vicinity of amino acid residues 36–41 of the RhCe protein[1,2].

See system pages.

Effect of enzymes/chemicals on MAR antigen on intact RBCs

Ficin/papain	Resistant ($\uparrow$)
Trypsin	Resistant
α-Chymotrypsin	Resistant
Pronase	Presumed resistant
Sialidase	Presumed resistant
DTT 200 mM	Resistant
Acid	Presumed resistant

In vitro characteristics of alloanti-MAR

Immunoglobulin class	IgG
Optimal technique	37°C; IAT; enzymes

Clinical significance of alloanti-MAR

The only reported example of anti-MAR was found in the serum of a non-transfused DC^{W}e/DC^{X}e woman upon delivery of her second child[3,4].

Comments

Antibodies made by people with C^W/C^X alleles detect a high prevalence antigen (MAR) and are weakly reactive with C^W+/C^W+ or C^X+/C^X+ RBCs. Antibodies (anti-MAR-like) made by people with C^W/C^W, C^X/C^X are non-reactive with C^W+/C^X+ RBCs[3,4].

References

1 Sistonen, P. et al. (1994) Vox Sang. 66, 287–292.
2 Mouro, I. et al. (1995) Blood 86, 1196–1201.
3 O'Shea, K.P. et al. (2001) Transfusion 41, 53–55.
4 Poole, J. et al. (2001). Transf Med 11 (Suppl. 1), 32 (abstract).

BARC ANTIGEN

Terminology

ISBT symbol (number)   RH52 (004.052)

History   Reported in 1989 as a low prevalence antigen associated with some DVI RBCs. Named after the Badger American Red Cross, where the antibody was first found. Confirmed as an Rh antigen in 1996 and awarded the next number

Occurrence

Less than 0.01%.

Expression

Cord RBCs   Presumed expressed

Altered   Correlation between strength of BARC antigen and partial D antigen. See comments

Molecular basis associated with BARC antigen haplotype[1-3]

BARC is associated with partial D of category VI in a $D^{VI}Ce$ haplotype. There are three types of $D^{VI}Ce$ and each is associated with a hybrid *RH(D-CE-D)* gene.
    See D antigen pages.

Effect of enzymes/chemicals on BARC antigen on intact RBCs

Ficin/papain   Resistant
Trypsin   Presumed resistant
α-Chymotrypsin   Presumed resistant
Pronase   Presumed resistant
Sialidase   Presumed resistant
DTT 200 mM   Presumed resistant
Acid   Presumed resistant

187

In vitro characteristics of alloanti-BARC

Immunoglobulin class	IgG
Optimal technique	IAT; enzymes
Complement binding	No

Clinical significance of alloanti-BARC

No data are available.

Comments

BARC subdivides category DVI[4]. Almost all (76 of 78) $D^{VI}Ce$ complexes express BARC; $D^{VI}cE$ do not express BARC. DVI RBCs with a weak expression of D have a weak expression of BARC. Those with a stronger expression of D have a strong expression of BARC. Anti-BARC is separated from a multi-specific serum (Horowitz) by absorption and elution.

References
[1] Mouro, I. et al. (1994) Blood 83, 1129–1135.
[2] Wagner, F.F. et al. (1998) Blood 91, 2157–2168.
[3] Wagner, F.F. et al. (2001) Transfusion 41, 1052–1058.
[4] Tippett, P. et al. (1996) Vox Sang. 70, 123–131.

JAHK ANTIGEN

Terminology

ISBT symbol (number)	RH53 (004.053)
History	First described in 1995 as a low prevalence antigen associated with the r^G haplotype. Family studies reported in 2002 confirmed Rh antigen status and an Rh number was allocated. Name extracted from the family name of the original antibody producer

Occurrence

Less than 0.01%.

Expression

Cord RBCs   Presumed expressed

Molecular basis associated with JAHK antigen

Not known, but present on RBCs with r^G phenotype but not with the r''^G phenotype or any of the common Rh complexes[1].

See system pages.

Effect of enzymes/chemicals on JAHK antigen on intact RBCs

Ficin/papain   Resistant ($\uparrow$)
Trypsin   Resistant
α-Chymotrypsin   Resistant
Pronase   Presumed resistant
Sialidase   Presumed resistant
DTT 200 mM   Presumed resistant
Acid   Resistant

In vitro characteristics of alloanti-JAHK

Immunoglobulin class   IgG
Optimal technique   37°C; IAT; enzymes

Clinical significance of alloanti-JAHK

Not known.

Comments

Anti-JAHK is found in multispecific sera[1,2].

References
1 Green, C. et al. (2002) Transfus. Med. 12, 55–61.
2 Kosanke, J. et al. (2002) Immunohematology 18, 46–47.

DAK ANTIGEN

Terminology

ISBT symbol (number)   RH54 (004.054)
History   Described in 1999; named 'D' for the D antigen and 'AK' from the name of the original antibody producer. Confirmed as an Rh antigen in 2002

Occurrence

Less than 0.01%.

Expression

Cord RBCs   Presumed expressed
Altered   Weak on $\bar{\bar{R}}^N$

Molecular basis associated with DAK antigen

Not known, but present on RBCs with DIIIa, DOL, and $\bar{\bar{R}}^N$ phenotypes and some other rare Rh variants[1].
    See system and D antigen pages.

Effect of enzymes/chemicals on DAK antigen on intact RBCs

Ficin/papain   Resistant ($\uparrow$)
Trypsin   Resistant
α-Chymotrypsin   Resistant
Pronase   Presumed resistant
Sialidase   Presumed resistant
DTT 200 mM   Presumed resistant
Acid   Resistant

In vitro characteristics of alloanti-DAK

Immunoglobulin class   IgG
Optimal technique   37°C; IAT; enzymes

Clinical significance of alloanti-DAK

Transfusion reaction	Presumed significant
HDN	Presumed significant

Comments

Several examples of anti-DAK exist all in multi-specific sera.

Reference
[1] Sausais, L. et al. (1999). Transfusion 39 (Suppl. 1), 79S (abstract).

LOCR ANTIGEN

Terminology

ISBT symbol (number)	RH55 (004.055)
Other names	700.053
History	Described in 1994 and became part of the Rh blood group system in 2002. Name was derived from two families in which HDN occurred

Occurrence

Less than 0.01%; five probands all European.

Expression

Cord RBCs	Presumed expressed

Effect of enzymes/chemicals on LOCR antigen on intact RBCs

Ficin/papain	Resistant ($\uparrow$)
Trypsin	Resistant
α-Chymotrypsin	Resistant
Pronase	Presumed resistant
Sialidase	Presumed resistant
DTT 200 mM	Presumed resistant
Acid	Resistant

In vitro characteristics of alloanti-LOCR

Immunoglobulin class   IgG
Optimal technique   37°C; IAT; enzymes

Clinical significance of alloanti-LOCR

Transfusion reaction   No data
HDN   Moderate

Comments

Travels with ce and the c or e may be weakened[1].

Reference
[1] Coghlan, G. et al. (1994) Transfusion 34, 492–495.

Number of antigens   19

Terminology

ISBT symbol	LU
ISBT number	005
CD number	CD239
Other name	B-CAM (B-cell adhesion molecule)
History	The first Lutheran antigen was described in 1945 and should have been named Lutteran, after the first Lu(a+) donor, but the writing on the label of the blood sample was misread as Lutheran

Expression

Soluble form	Not described
Other blood cells	Not on lymphocytes, granulocytes, monocytes, platelets
Tissues	Brain, heart, kidney glomeruli, liver, lung, pancreas, placenta, skeletal muscle, arterial wall, tongue, trachea, skin, esophagus, cervix, ileum, colon, stomach, gall bladder[1]

Gene

Chromosome	19q13.2–q13.3
Name	*LU*
Organization	15 exons distributed over approximately 12 kbp
Product	Lutheran glycoprotein (597 amino acids) and B-CAM (557 amino acids) by alternative splicing of exon 13[1,2]
Gene map	

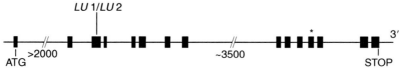

LU 1/LU 2 (229A > G) encode Luᵃ/Luᵇ (His77Arg)

* *LU* 18/*LU* 19 (1614A > G) encode Auᵃ (Lu18)/Auᵇ (Lu19) (Thr539Ala)

├────── 1 kbp

Database accession numbers

GenBank X83425

http://www.bioc.aecom.yu.edu/bgmut/index.htm

Amino acid sequence[1]

```
                                M   EPPDAPAQAR   GAPRLLLLAV   LLAAHPDAQA    -1
EVRLSVPPLV   EVMRGKSVIL   DCTPTGTHDH   YMLEWFLTDR   SGARPRLASA    50
EMQGSELQVT   MHDTRGRSPP   YQLDSQGRLV   LAEAQVGDER   DYVCVVRAGA   100
AGTAEATARL   NVFAKPEATE   VSPNKGTLSV   MEDSAQEIAT   CNSRNGNPAP   150
KITWYRNGQR   LEVPVEMNPE   GYMTSRTVRE   ASGLLSLTST   LYLRLRKDDR   200
DASFHCAAHY   SLPEGRHGRL   DSPTFHLTLH   YPTEHVQFWV   GSPSTPAGWV   250
REGDTVQLLC   RGDGSPSPEY   TLFRLQDEQE   EVLNVNLEGN   LTLEGVTRGQ   300
SGTYGCRVED   YDAADDVQLS   KTLELRVAYL   DPLELSEGKV   LSLPLNSSAV   350
VNCSVHGLPT   PALRWTKDST   PLGDGPMLSL   SSITFDSNGT   YVCEASLPTV   400
PVLSRTQNFT   LLVQGSPELK   TAEIEPKADG   SWREGDEVTL   ICSARGHPDP   450
KLSWSQLGGS   PAEPIPGRQG   WVSSSLTLKV   TSALSRDGIS   CEASNPHGNK   500
RHVFHFGAVS   PQTSQAGVAV   MAVAVSVGLL   LLVVAVFYCV   RRKGGPCCRQ   550
RREKGAPPPG   EPGLSHSGSE   QPEQTGLLMG   GASGGARGGS   GGFGDEC      597
```

LU encodes a leader sequence of 31 amino acids.

Antigen mutation is numbered counting Met as 1.

Carrier molecule

The predicted mature protein has five disulfide-bonded, extracellular, immunoglobulin superfamily (IgSF) domains (two variable-region (V) sets and three constant region (C) sets)[1].

M_r (SDS-PAGE)	Lu: 85 000 (has cytoplasmic tail); B-CAM: 78 000 (no cytoplasmic tail)[3,4]
CHO: N-glycan	five potential sites (residue 290, 346, 352, 388, 408)
CHO: O-glycan	Present
Cysteine residues	10 extracellular
Copies per RBC	1500–4000

Molecular basis of antigens[5–8]

Antigen	Amino acid change	Exon	Nt change	Restriction enzyme
Lu^a/Lu^b	His77Arg	3	230A>G	Aci I (−/+)
Lu4+/Lu4−	Arg175Gln	5	524G>A	
Lu5+/Lu5−	Arg109His	3	326G>A	
Lu6/Lu9	Ser275Phe	7	824C>T	
Lu8/Lu14	Met204Lys	6	611T>A	
Lu12+/Lu12−	Arg34 and Leu35del	2	99del GCGCTT	
Lu13+/Lu13−	Ser447Leu; Gln581Leu	11; 13	1340C>T; 1742A>T	
Lu16+/Lu16−	Arg227Cys	6	679C>T	
Lu17+/Lu17−	Glu114Lys	3	340G>A	
Au^a/Au^b	Ala539Thr	12	1615G>A	
Lu20+/Lu20−	Thr302Met	7	905C>T	
Lu21+/Lu21−	Asp94Glu	3	282C>G	

(−/+) corresponds to presence or absence of restriction enzyme site.

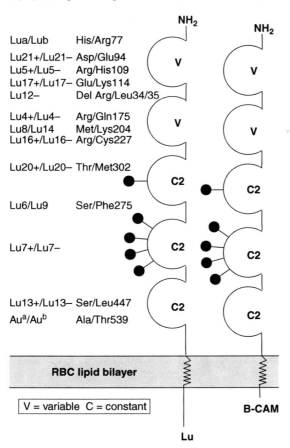

195

Function

Possibly has adhesion properties and may mediate intracellular signaling. The extracellular domains and the cytoplasmic domain contain consensus motifs for the binding of integrin and Src homology 3 domains, respectively, suggesting possible receptor and signal-transduction function[1]. Lutheran glycoprotein binds to laminin (particularly to isoforms that contain α5 chains), strongly suggesting that it is a membrane constituent that is involved in cell–cell and cell–matrix binding events and may function as a laminin receptor during erythropoiesis[9]. IgSF domains 1–3, and possibly domain 5, are involved in laminin binding[10-12].

Disease association

Expression of B-CAM is increased in certain malignant tumors and cells. May mediate adhesion of sickle cells to vascular endothelium.

Phenotypes (% occurrence)

	Most populations
Lu(a+b−)	0.2
Lu(a−b+)	92.4
Lu(a+b+)	7.4
Lu(a−b−)	Rare

Null: Lu(a−b−) recessive type.
Unusual: Lu(a−b−) dominant and X-linked types.

Comparison of three types of Lu(a−b−) phenotypes

Lu(a−b−) phenotype	Lutheran antigens	Make anti-Lu3	CD44	CD75	I/i antigen
Recessive	Absent	Yes	Normal	Normal	Normal/normal
Dominant	Weak	No	Weak (25–39% of normal)	Strong	Normal/weak
X-linked	Weak	No	Normal	Absent	Weak/strong

Molecular basis of recessive Lu(a−b−) phenotype

Missense 733 C>A in exon 6, Cys237Stop, which ablates an *Mwo* I site[13].

Comments

Lu(a−b−) dominant type RBCs [caused by action of *In(Lu)*] have reduced expression of Lutheran, P1, AnWj, Indian, Knops, Cs^a and MER2 blood group antigens. The X-linked type (*XS2*) of Lu(a−b−) has only been found in one family.

Lutheran antigens are not expressed in K562, HEL, HL60, IM9, MOLT4 cells.

Lutheran, along with *Secretor*, provided the first example of autosomal linkage in humans. Some Lu(a−b−) RBCs are acanthocytic[14].

References

1   Parsons, S.F. et al. (1995) Proc. Natl. Acad. Sci. USA 92, 5496–5500.
2   Rahuel, C. et al. (1996) Blood 88, 1865–1872.
3   Parsons, S.F. et al. (1987) Transfusion 27, 61–63.
4   Daniels, G. and Khalid G. (1989) Vox Sang. 57, 137–141.
5   El Nemer, W. et al. (1997) Blood 89, 4608–4616.
6   Crew, V. and Daniels G. (2002) Personal communications.
7   Daniels, G. et al. (2002) Vox Sang. 83 (Suppl. 2), 147 (abstract).
8   Parsons, S.F. et al. (1997) Blood 89, 4219–4225.
9   El Nemer, W. et al. (1998) J. Biol. Chem. 273, 16686–16693.
10   El Nemer, W. et al. (2001) J. Biol. Chem. 276, 23757–23762.
11   Udani, M. et al. (1998) J. Clin. Invest. 101, 2550–2558.
12   Zen, Q. et al. (1999) J. Biol. Chem. 274, 728–734.
13   Mallinson, G. et al. (1997) Transf. Med. 7 (Suppl. 1), 18 (abstract).
14   Udden, M.M. et al. (1987) Blood 69, 52–57.

Lu^a ANTIGEN

Terminology

ISBT symbol (number)	LU1 (005.001)
History	Identified in 1945; named after the donor whose blood stimulated the production of anti-Lu^a in a patient with SLE

Occurrence

Caucasians	8%
Blacks	5%

Antithetical antigen

Lu^b (**LU2**)

Expression

Cord RBCs   Weak

There is considerable variation in the strength of Lu^a expression on RBCs. This variation is inherited.

Molecular basis associated with Lu^a antigen[1,2]

Amino acid   His 77 in IgSF domain 1
Nucleotide   A at bp 230 in exon 3
Restriction enzyme   Ablates an *Aci* I site

Effect of enzymes/chemicals on Lu^a antigen on intact RBCs

Ficin/papain   Resistant (may be weakened)
Trypsin   Sensitive
α-Chymotrypsin   Sensitive
Pronase   Sensitive
Sialidase   Resistant
DTT 200 mM/50 mM   Sensitive/resistant
Acid   Resistant

In vitro characteristics of alloanti-Lu^a

Immunoglobulin class   IgM; IgG
Optimal technique   RT or IAT with characteristic "loose" agglutinates surrounded by unagglutinated RBCs; capillary
Complement binding   Rare

Clinical significance of alloanti-Lu^a

Transfusion reaction   No
HDN   No to mild (rare)

Comments

Anti-Lu^a is not infrequently found in serum from patients following transfusion and also may be naturally-occurring. Sera containing anti-Lu^a often also contain HLA antibodies.

The presence of Lutheran glycoprotein on placental tissue may result in absorption of maternal antibodies to Lutheran antigens.

References
[1] El Nemer, W. et al. (1997) Blood 89, 4608–4616.
[2] Parsons, S.F. et al. (1997) Blood 89, 4219–4225.

Lu^b ANTIGEN

Terminology

ISBT symbol (number)	LU2 (005.002)
History	Named because of its antithetical relationship to Lu^a; anti-Lu^b identified in 1956

Occurrence

All populations	99.8%

Antithetical antigen

Lu^a (**LU1**)

Expression

Cord RBCs	Weak
Altered	Weak on RBCs of the dominant type of Lu(a−b−)

There is considerable variation in the strength of Lu^b expression on RBCs. This variation is inherited.

Molecular basis associated with Lu^b antigen[1,2]

Amino acid	Arg 77 in IgSF domain 1
Nucleotide	G at bp 230 in exon 3
Restriction enzyme	Gains an *Aci* I site

Effect of enzymes/chemicals on Lu^b antigen on intact RBCs

Ficin/papain	Resistant (may be weakened)
Trypsin	Sensitive
α-Chymotrypsin	Sensitive
Pronase	Sensitive
Sialidase	Resistant
DTT 200 mM/50 mM	Sensitive/resistant
Acid	Resistant

In vitro characteristics of alloanti-Lu^b

Immunoglobulin class   IgG; IgM
Optimal technique   RT; IAT; capillary
Complement binding   Rare

Clinical significance of alloanti-Lu^b

Transfusion reaction   Mild to moderate
HDN   Mild

Comments

Weak expression of Lu^b on dominant type Lu(a−b−) RBCs is detectable by absorption/elution.
The presence of Lutheran glycoprotein on placental tissue may result in absorption of maternal antibodies to Lutheran antigens.

References
[1] Parsons, S.F. et al. (1997) Blood 89, 4219–4225.
[2] El Nemer, W. et al. (1997) Blood 89, 4608–4616.

Lu3 ANTIGEN

Terminology

ISBT symbol (number)   LU3 (005.003)
Other names   Lu^{ab}, Lu^{a}Lu^b
History   Reported in 1963; re-named Lu3 to be computer-friendly after Lu(a−b−) RBCs were shown to lack other high-prevalence antigens in the Lutheran system. Dr. Crawford, a blood banker, found her own RBCs to be of the dominant type of Lu(a−b−)

Occurrence

All populations   100%

Expression

Cord RBCs   Weak

Altered   Weak or non-detectable by hemagglutination on RBCs of the dominant type of Lu(a−b−)

Molecular basis associated with Lu3 antigen

Not known. See system pages for molecular basis of Lu:−3 phenotype.

Effect of enzymes/chemicals on Lu3 antigen on intact RBCs

Ficin/papain	Resistant
Trypsin	Sensitive
α-Chymotrypsin	Sensitive
Pronase	Sensitive
Sialidase	Resistant
DTT 200 mM/50 mM	Sensitive/resistant
Acid	Resistant

In vitro characteristics of alloanti-Lu3

Immunoglobulin class	IgG
Optimal technique	IAT
Complement binding	Rare

Clinical significance of alloanti-Lu3

No data are available.

Comments

Immunoblotting showed Lu3 is on the Lutheran glycoprotein[1].

Anti-Lu3 is only made by immunized individuals of the rare recessive type Lu(a−b−). In these cases, Lu(a−b−) blood of the recessive or dominant type should be used for transfusion.

Reference

[1] Daniels, G. and Khalid G. (1989) Vox Sang. 57, 137–141.

Lu4 ANTIGEN

Terminology

ISBT symbol (number)	LU4 (005.004)
Other names	Barnes
History	The first of a series of Lu(a−b+) people who made an antibody compatible only with Lu(a−b−) RBCs. Described in 1971

Occurrence

All populations	100%

Expression

Cord RBCs	Weak
Altered	Weak or non-detectable by hemagglutination on RBCs of the dominant type of Lu(a−b−)

Molecular basis associated with Lu4 antigen[1,2]

Amino acid	Arg175 in IgSF domain 2
Nucleotide	G at bp 524 in exon 5

Lu:−4 has Gln 175 and A at bp 524.

Effect of enzymes/chemicals on Lu4 antigen on intact RBCs

Ficin/papain	Resistant
Trypsin	Sensitive
α-Chymotrypsin	Sensitive
Pronase	Sensitive
Sialidase	Resistant
DTT 200 mM/50 mM	Sensitive/resistant
Acid	Not known

In vitro characteristics of alloanti-Lu4

Immunoglobulin class	IgG
Optimal technique	IAT
Complement binding	No

Clinical significance of alloanti-Lu4

Transfusion reaction	No data are available
HDN	No in two infants born to one Lu:−4 female

Comments

Only one family with the Lu:−4 phenotype has been reported.
The presence of Lutheran glycoprotein on placental tissue may result in absorption of maternal antibodies to Lutheran antigens.

References
1 Daniels, G. et al. (2002) Vox Sang. 83 (Suppl. 2), 147 (abstract).
2 Crew, V. and Daniels, G. Personal communication.

Lu5 ANTIGEN

Terminology

ISBT symbol (number)	LU5 (005.005)
Other names	Beal; Fox
History	Identified in 1972; given the next number in the series of Lu(a−b+) people who made an antibody compatible only with Lu(a−b−) RBCs

Occurrence

All populations	100%

Expression

Cord RBCs	Weak
Altered	Weak or non-detectable by hemagglutination on RBCs of the dominant type of Lu(a−b−)

Molecular basis associated with Lu5 antigen[1,2]

Amino acid	Arg 109 in IgSF domain 1
Nucleotide	G at bp 326 in exon 3
Lu5− has His 109 and A at bp 326.	

Effect of enzymes/chemicals on Lu5 antigen on intact RBCs

Ficin/papain	Resistant
Trypsin	Sensitive
α-Chymotrypsin	Sensitive
Pronase	Sensitive
Sialidase	Resistant
DTT 200 mM/50 mM	Sensitive/resistant
Acid	Not known

In vitro characteristics of alloanti-Lu5

Immunoglobulin class	IgG
Optimal technique	IAT
Complement binding	No

Clinical significance of alloanti-Lu5

Transfusion reaction	No (potentially significant by a chemiluminescent assay)
HDN	No

Comments

Several examples of immune anti-Lu5 have been reported.
The presence of Lutheran glycoprotein on placental tissue may result in absorption of maternal antibodies to Lutheran antigens.

References
[1] Daniels, G. et al. (2002) Vox Sang. 83 (Suppl. 2), 147 (abstract).
[2] Crew, V. and Daniels, G. Personal communication.

Lu6 ANTIGEN

Terminology

ISBT symbol (number)	LU6 (005.006)
Other names	Jan; Jankowski
History	Identified in 1972; given the next number in the series of Lu(a−b+) people who made an antibody compatible only with Lu(a−b−) RBCs

Occurrence

All populations: 100%; three Lu:−6 were Iranian Jews[1].

Antithetical antigen

Lu9 (LU9)

Expression

Cord RBCs	Weak
Altered	Non-detectable by hemagglutination on RBCs of the dominant type of Lu(a−b−)

Molecular basis associated with Lu6 antigen[2,3]

Amino acid	Ser 275 in IgSF domain 3
Nucleotide	C at bp 824 in exon 7

Effect of enzymes/chemicals on Lu6 antigen on intact RBCs

Ficin/papain	Resistant
Trypsin	Sensitive
α-Chymotrypsin	Sensitive
Pronase	Sensitive
Sialidase	Resistant
DTT 200 mM/50 mM	Sensitive/resistant
Acid	Not known

In vitro characteristics of alloanti-Lu6

Immunoglobulin class	IgG
Optimal technique	IAT
Complement binding	No

Clinical significance of alloanti-Lu6

Transfusion reaction	No to moderate
HDN	No

Comments

The presence of Lutheran glycoprotein on placental tissue may result in absorption of maternal antibodies to Lutheran antigens.

References

[1] Yahalom, V. et al. (2002) Transfusion 42, 247–250.
[2] Daniels, G. et al. (2002) Vox Sang. 83 (Suppl. 2), 147 (abstract).
[3] Crew, V. and Daniels, G. (2002) personal communications.

Lu7 ANTIGEN

Terminology

ISBT symbol (number)	LU7 (005.007)
Other names	Gary
History	Identified in 1972; given the next number in the series of Lu(a−b+) people who made an antibody compatible only with Lu(a−b−) RBCs

Occurrence

All populations: 100%; only two probands.

Expression

Cord RBCs	Weak
Altered	Non-detectable by hemagglutination on RBCs of the dominant type of Lu(a−b−)

Molecular basis associated with Lu7 antigen[1]

Not known but Lu7 is located within the fourth IgSF domain.

Effect of enzymes/chemicals on Lu7 antigen on intact RBCs

Ficin/papain	Resistant
Trypsin	Sensitive

α-Chymotrypsin	Sensitive
Pronase	Sensitive
Sialidase	Presumed resistant
DTT 200 mM/50 mM	Presumed sensitive/resistant
Acid	Not known

In vitro characteristics of alloanti-Lu7

Immunoglobulin class	IgG
Optimal technique	IAT
Complement binding	No

Clinical significance of alloanti-Lu7

Transfusion reaction	No to mild
HDN	No

Comments

Only two examples of anti-Lu7 have been described.

The presence of Lutheran glycoprotein on placental tissue may result in absorption of maternal antibodies to Lutheran antigens.

Reference
[1] Parsons, S.F. et al. (1997) Blood 89, 4219–4225.

Lu8 ANTIGEN

Terminology

ISBT symbol (number)	LU8 (005.008)
Other names	Taylor; MT
History	Identified in 1972; given the next number in the series of Lu(a−b+) people who made an antibody compatible only with Lu(a−b−) RBCs

Occurrence

All populations: 100%; several probands have been described.

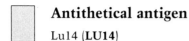

Antithetical antigen

Lu14 (**LU14**)

Expression

Cord RBCs	Weak
Altered	Non-detectable by hemagglutination on RBCs of the dominant type of Lu(a−b−)

Molecular basis associated with Lu8 antigen[1,2]

Amino acid	Met 204 in IgSF domain 2
Nucleotide	T at bp 611 in exon 6

Effect of enzymes/chemicals on Lu8 antigen on intact RBCs

Ficin/papain	Variable
Trypsin	Sensitive
α-Chymotrypsin	Sensitive
Pronase	Sensitive
Sialidase	Resistant
DTT 200 mM/50 mM	Sensitive/resistant
Acid	Not known

In vitro characteristics of alloanti-Lu8

Immunoglobulin class	IgG
Optimal technique	IAT
Complement binding	No

Clinical significance of alloanti-Lu8

Transfusion reaction	No to mild
HDN	No

Comments

The presence of Lutheran glycoprotein on placental tissue may result in absorption of maternal antibodies to Lutheran antigens.

References
[1] Daniels, G. et al. (2002) Vox Sang. 83 (Suppl. 2), 147 (abstract).
[2] Crew, V. and Daniels, G. Personal communication.

Lu9 ANTIGEN

Terminology

ISBT symbol (number)	LU9 (005.009)
Other names	Mull
History	Reported in 1973; given the next available number when its antithetical relationship to Lu6 was recognized

Occurrence

About 1–2%, but probably lower because the orginal anti-Lu9 contained anti-Bg[a].

Antithetical antigen

Lu6 (LU6)

Expression

Cord RBCs	Weak

Molecular basis associated with Lu9 antigen[1,2]

Amino acid	Phe 275 in IgSF domain 3
Nucleotide	T at bp 824 in exon 7

Effect of enzymes/chemicals on Lu9 antigen on intact RBCs

Ficin/papain	Resistant
Trypsin	Sensitive
α-Chymotrypsin	Sensitive
Pronase	Sensitive

Sialidase	Presumed resistant
DTT 200 mM/50 mM	Presumed sensitive/resistant
Acid	Resistant

In vitro characteristics of alloanti-Lu9

Immunoglobulin class	IgG
Optimal technique	IAT and capillary

Clinical significance of alloanti-Lu9

Transfusion reaction	No data; the second example was probably stimulated by transfusion[3].
HDN	Positive DAT but no clinical HDN in case

Comments

The presence of Lutheran glycoprotein on placental tissue may result in absorption of maternal antibodies to Lutheran antigens.

References

1 Daniels, G. et al. (2002) Vox Sang. 83 (Suppl. 2), 147 (abstract).
2 Crew, V. and Daniels, G. Personal communication.
3 Champagne, K. et al. (1999) Immunohematology 15, 113–116.

Lu11 ANTIGEN

Terminology

ISBT symbol (number)	LU11 (005.011)
Other names	Reynolds
History	Identified in 1974; given the next number in the series of Lu(a−b+) people who made an antibody compatible only with Lu(a−b−) RBCs

Occurrence

All populations	100%

Expression

Cord RBCs   Weak
Altered   Non-detectable by hemagglutination on RBCs of
                       the dominant type of Lu(a−b−)

Effect of enzymes/chemicals on Lu11 antigen on intact RBCs

Ficin/papain   Resistant
Trypsin   Presumed sensitive
α-Chymotrypsin   Presumed sensitive
Pronase   Sensitive
Sialidase   Presumed resistant
DTT 200 mM/50 mM   Presumed sensitive/resistant
Acid   Not known

In vitro characteristics of alloanti-Lu11

Immunoglobulin class   IgM and IgG
Optimal technique   RT and IAT
Complement binding   No

Clinical significance of alloanti-Lu11

Transfusion reaction   No to mild (not much data)
HDN   No (not much data)

Comments

Few anti-Lu11 have been reported and are typically very weakly reactive.
No evidence that Lu11 is inherited or carried on the Lutheran glycoprotein.

Lu12 ANTIGEN

Terminology

ISBT symbol (number)	LU12 (005.012)
Other names	Muchowski; Much
History	Identified in 1973; given the next number in the series of Lu(a−b+) people who made an antibody compatible only with Lu(a−b−) RBCs

Occurrence

All populations: 100%; only two probands reported.

Expression

Cord RBCs	Weak
Altered	Non-detectable by hemagglutination on RBCs of the dominant type of Lu(a−b−)

Molecular basis associated with absence of Lu12 antigen[1,2]

Amino acid	Deletion of Arg 34 and Leu 35 in IgSF domain 1
Nucleotide	99delGCGCTT in exon 2

Effect of enzymes/chemicals on Lu12 antigen on intact RBCs

Ficin/papain	Resistant
Trypsin	Sensitive
α-Chymotrypsin	Sensitive
Pronase	Sensitive
Sialidase	Resistant
DTT 200 mM/50 mM	Sensitive/resistant
Acid	Not known

In vitro characteristics of alloanti-Lu12

Immunoglobulin class	IgG
Optimal technique	IAT
Complement binding	No

Clinical significance of alloanti-Lu12

No data are available because only two examples of anti-Lu12 have been reported.

References
1 Crew, V. and Daniels, G. Personal communication.
2 Daniels, G. et al. (2002) Vox Sang. 83 (Suppl. 2), 147 (abstract).

Lu13 ANTIGEN

Terminology

ISBT symbol (number)	LU13 (005.013)
Other names	Hughes
History	Reported in 1983; given the next number in the series of Lu(a−b+) people who made an antibody compatible only with Lu(a−b−) RBCs

Occurrence

All populations: 100%; only a few probands have been reported.

Expression

Cord RBCs	Weak
Altered	Non-detectable by hemagglutination on RBCs of the dominant type of Lu(a−b−)

Molecular basis associated with Lu13 antigen[1,2]

Amino acid	Ser 447 in IgSF domain 5 and Gln 581 in transmembrane domain
Nucleotide	C at bp 1340 in exon 11 and A at bp 1742 in exon 13

Lu:−13 has Leu 447 and Leu 581, and T at bp 1340 and T at bp 1742.

Effect of enzymes/chemicals on Lu13 antigen on intact RBCs

Ficin/papain	Resistant
Trypsin	Sensitive
α-Chymotrypsin	Sensitive
Pronase	Sensitive
Sialidase	Resistant
DTT 200 mM/50 mM	Sensitive/resistant
Acid	Not known

In vitro characteristics of alloanti-Lu13

Immunoglobulin class	IgG
Optimal technique	IAT
Complement binding	No

Clinical significance of alloanti-Lu13

No data are available.

Comments

Only four examples of anti-Lu13 are known.

References
1 Daniels, G. et al. (2002) Vox Sang. 83 (Suppl. 2), 147 (abstract).
2 Crew, V. and Daniels, G. Personal communication.

Lu14 ANTIGEN

Terminology

ISBT symbol (number)	LU14 (005.014)
Other names	Hofanesian
History	Reported in 1977; given the next available number when its antithetical relationship to Lu8 was recognized

Occurrence

English	1.8%
Danes	1.5%

Antithetical antigen

Lu8 (**LU8**)

Expression

Cord RBCs	Presumed weak

Molecular basis associated with Lu14 antigen[1,2]

Amino acid	Lys 204 in IgSF domain 2
Nucleotide	T at bp 611 in exon 6

Effect of enzymes/chemicals on Lu14 antigen on intact RBCs

Ficin/papain	Variable
Trypsin	Presumed sensitive
α-Chymotrypsin	Presumed sensitive
Pronase	Sensitive
Sialidase	Presumed resistant
DTT 200 mM/50 mM	Sensitive/resistant
Acid	Not known

In vitro characteristics of alloanti-Lu14

Immunoglobulin class	IgG
Optimal technique	IAT
Complement binding	No

Clinical significance of alloanti-Lu14

Transfusion reaction	No data
HDN	Positive DAT; HDN in one case

Comments

The presence of Lutheran glycoprotein on placental tissue may result in absorption of maternal antibodies to Lutheran antigens.

References
1 Crew, V. and Daniels, G. Personal communication.
2 Daniels, G. et al. (2002) Vox Sang. 83 (Suppl. 2), 147 (abstract).

Lu16 ANTIGEN

Terminology

ISBT symbol (number)	LU16 (005.016)
History	Reported in 1980 when three Lu(a+b−) Black women were found to have an antibody to a high-prevalence antigen in addition to anti-Lu[b]

Occurrence

All populations	100%

Molecular basis associated with Lu16 antigen[1,2]

Amino acid	Arg 227 in IgSF domain 2
Nucleotide	C at bp 679 in exon 6

LU:−16 has Cys 227 and T at bp 679.

In vitro characteristics of alloanti-Lu16

Immunoglobulin class	IgG
Optimal technique	IAT

References
1 Crew, V. and Daniels, G. Personal communication.
2 Daniels, G. et al. (2002) Vox Sang. 83 (Suppl. 2), 147 (abstract).

Lu17 ANTIGEN

Terminology

ISBT symbol (number)	LU17 (005.017)
Other names	Delcol, nee: Pataracchia
History	Reported in 1979; given the next number in the series of Lu(a−b+) people who made an antibody compatible only with Lu(a−b−) RBCs

Occurrence

All populations: 100%; only found in one Italian proband.

Expression

Cord RBCs	Expressed
Altered	Non-detectable by hemagglutination on RBCs of the dominant type of Lu(a−b−)

Molecular basis associated with Lu17 antigen[1,2]

Amino acid	Glu 114 in IgSF domain 1
Nucleotide	G at bp 340 in exon 3

LU:−17 has Lys 114 and A at bp 340.

Effect of enzymes/chemicals on Lu17 antigen on intact RBCs

Ficin/papain	Resistant
Trypsin	Sensitive
α-Chymotrypsin	Sensitive
Pronase	Sensitive
Sialidase	Resistant
DTT 200 mM/50 mM	Sensitive/resistant
Acid	Not known

In vitro characteristics of alloanti-Lu17

Immunoglobulin class	IgG
Optimal technique	IAT
Complement binding	No

217

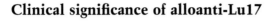

Clinical significance of alloanti-Lu17

In vivo RBC survival studies suggested that anti-Lu17 might cause modest destruction of transfused RBCs.

The only anti-Lu17 was made by a woman with four uneventful pregnancies.

Comments

By immunoblotting, anti-Lu17 reacted with membranes of the dominant type of Lu(a−b−) RBCs.

The presence of Lutheran glycoprotein on placental tissue may result in absorption of maternal antibodies to Lutheran antigens.

References
1 Crew, V. and Daniels, G. Personal communication.
2 Daniels, G. et al. (2002) Vox Sang. 83 (Suppl. 2), 147 (abstract).

Auᵃ ANTIGEN

Terminology

ISBT symbol (number)	LU18 (005.018)
Other names	Auberger; 204.001
History	Named in 1961 after the first producer of anti-Auᵃ, a multi-transfused woman; placed into the Lutheran blood group system in 1990

Occurrence

All populations: 90% (originally the incidence was reported as 82%).

Antithetical antigen

Auᵇ (**LU19**)

Expression

Cord RBCs	Weak
Altered	Non-detectable by hemagglutination on RBCs of the dominant type of Lu(a−b−)

There is considerable variation in the strength of Au^a expression on RBCs. This variation is inherited.

Molecular basis associated with Au^a antigen[1,2]

Amino acid	Ala 539 in IgSF domain 5
Nucleotide	G at bp 1615 in exon 12

Effect of enzymes/chemicals on Au^a antigen on intact RBCs

Ficin/papain	Resistant
Trypsin	Sensitive
α-Chymotrypsin	Sensitive
Pronase	Sensitive
Sialidase	Resistant
DTT 200 mM/50 mM	Sensitive/resistant
Acid	Not known

In vitro characteristics of alloanti-Au^a

Immunoglobulin class	IgG
Optimal technique	IAT
Complement binding	No

Clinical significance of alloanti-Au^a

Transfusion reaction	No to mild
HDN	No

Comments

Only three examples of anti-Au^a have been reported; in sera containing other antibodies[3,4].

The presence of Lutheran glycoprotein on placental tissue may result in absorption of maternal antibodies to Lutheran antigens.

References
1 Crew, V. and Daniels, G. Personal communication.
2 Daniels, G. et al. (2002) Vox Sang. 83 (Suppl. 2), 147 (abstract).
3 Daniels, G.L. et al. (1991) Vox Sang. 60, 191–192.
4 Drachmann, O. et al. (1982) Vox Sang. 43, 259–262.

Au^b ANTIGEN

Terminology

ISBT symbol (number)	LU19 (005.019)
Other names	204.002
History	Reported in 1989 and named because it is antithetical to Au^a

Occurrence

Caucasians	51%
Blacks	68%

Antithetical antigen

Au^a (**LU18**)

Expression

Cord RBCs	Weak
Altered	Non-detectable by hemagglutination on RBCs of the dominant type of Lu(a−b−)

There is considerable variation in the strength of Au^b expression on RBCs. This variation is inherited.

Molecular basis associated with Au^b antigen[1,2]

Amino acid	Thr 539 in IgSF domain 5
Nucleotide	A at bp 1615 in exon 12

Effect of enzymes/chemicals on Au^b antigen on intact RBCs

Ficin/papain	Resistant
Trypsin	Sensitive
α-Chymotrypsin	Sensitive
Pronase	Sensitive

Sialidase	Resistant
DTT 200 mM/50 mM	Presumed sensitive/resistant
Acid	Not known

In vitro characteristics of alloanti-Au^b

Immunoglobulin class	IgG
Optimal technique	IAT
Complement binding	No

Clinical significance of alloanti-Au^b

Transfusion reaction	No to mild
HDN	No

Comments

Four examples of anti-Au^b have been reported, all in sera also containing anti-Lu^a.

The presence of Lutheran glycoprotein on placental tissue may result in absorption of maternal antibodies to Lutheran antigens.

References
[1] Daniels, G. et al. (2002) Vox Sang. 83 (Suppl. 2), 147 (abstract).
[2] Crew, V. and Daniels, G. Personal communication.

Lu20 ANTIGEN

Terminology

ISBT symbol (number)	LU20 (005.020)
History	Reported in 1992, antibody made by an Israeli thalassemic; given the next number in the series of Lu(a−b+) people who made an antibody compatible only with Lu(a−b−) RBCs

Occurrence

All populations: 100%; only one proband, Israeli.

Expression

Cord RBCs	Weak
Altered	Non-detectable by hemagglutination on RBCs of the dominant type of Lu(a−b−)

Molecular basis associated with Lu20 antigen[1,2]

Amino acid	Thr 302 in IgSF domain 3
Nucleotide	C at bp 905 in exon 7

LU:−20 has Met 302 and T at bp 905.

Effect of enzymes/chemicals on Lu20 antigen on intact RBCs

Ficin/papain	Resistant
Trypsin	Sensitive
α-Chymotrypsin	Sensitive
Pronase	Presumed sensitive
Sialidase	Presumed resistant
DTT 200 mM/50 mM	Sensitive/resistant
Acid	Not known

In vitro characteristics of alloanti-Lu20

Immunoglobulin class	IgG
Optimal technique	37°C; IAT
Complement binding	No

Clinical significance of alloanti-Lu20

Not known since only one example has been described.

References
1 Crew, V. and Daniels, G. Personal communication.
2 Daniels, G. et al. (2002) Vox Sang. 83 (Suppl. 2), 147 (abstract).

Lu21 ANTIGEN

Terminology

ISBT symbol (number)   LU21 (005.021)
History   Reported in 2002; given the next number in the series of Lu(a−b+) people who made an antibody compatible only with Lu(a−b−) RBCs

Occurrence

All populations: 100%; only one proband, Israeli.

Expression

Cord RBCs   Weak
Altered   Non-detectable by hemagglutination on RBCs of the dominant type of Lu(a−b−)

Molecular basis associated with Lu21 antigen[1,2]

Amino acid   Asp 94 in IgSF domain 1
Nucleotide   C at bp 282 in exon 3
LU:−21 has Glu 94 and G at bp 282.

Effect of enzymes/chemicals on Lu21 antigen on intact RBCs

Ficin/papain   Resistant
Trypsin   Sensitive
α-Chymotrypsin   Sensitive
Pronase   Presumed sensitive
Sialidase   Presumed resistant
DTT 200 mM/50 mM   Sensitive/resistant
Acid   Not known

In vitro characteristics of alloanti-Lu21

Immunoglobulin class   IgG; IgM
Optimal technique   IAT; RT and 37°C

Clinical significance of alloanti-Lu21

No HDN in the proband's second, third and fourth pregnancies.

Comments

The presence of Lutheran glycoprotein on placental tissue may result in absorption of maternal antibodies to Lutheran antigens.

References

1 Crew, V. and Daniels, G. Personal communication.
2 Daniels, G. et al. (2002) Vox Sang. 83 (Suppl. 2), 147 (abstract).

Number of antigens 24

Terminology

ISBT symbol	KEL
ISBT number	006
CD number	CD238
History	Named in 1946 after the first antibody producer (*Kell*eher) of anti-K that caused HDN

Expression

Other blood cells	Possibly only on erythrocytes, appears early in erythropoiesis
Tissues	Primarily in bone marrow, fetal liver, testes; lesser amounts in other tissues including various parts of the brain, lymphoid organs, heart, and skeletal muscle

Gene

Chromosome	7q33
Name	*KEL*
Organization	19 exons distributed over 21.5 kbp of gDNA
Product	Kell glycoprotein

Gene map

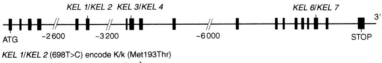

KEL 1/KEL 2 (698T>C) encode K/k (Met193Thr)
KEL 3/KEL 4 (961T>C) encode Kp^a/Kp^b (Trp281Arg)
KEL 6/KEL 7 (1910C>T) encode Js^a/Js^b (Pro597Leu)

⊢——⊣ 1 kbp

Database accession numbers

GenBank   M64934; AH008123
www.bioc.aecom.yu.edu/bgmut/index.htm.

Amino acid sequence[1]

```
MEGGDQSEEE   PRERSQAGGM   GTLWSQESTP   EERLPVEGSR   PWAVARRVLT    50
AILILGLLLC   FSVLLFYNFQ   NCGPRPCETS   VCLDLRDHYL   ASGNTSVAPC   100
TDFFSFACGR   AKETNNSFQE   LATKNKNRLR   RILEVQNSWH   PGSGEEKAFQ   150
FYNSCMDTLA   IEAAGTGPLR   QVIEELGGWR   ISGKWTSLNF   NRTLRLLMSQ   200
YGHFPFFRAY   LGPHPASPHT   PVIQIDQPEF   DVPLKQDQEQ   KIYAQIFREY   250
LTYLNQLGTL   LGGDPSKVQE   HSSLSISITS   RLFQFLRPLE   QRRAQGKLFQ   300
MVTIDQLKEM   APAIDWLSCL   QATFTPMSLS   PSQSLVVHDV   EYLKNMSQLV   350
EEMLLKQRDF   LQSHMILGLV   VTLSPALDSQ   FQEARRKLSQ   KLRELTEQPP   400
MPARPRWMKC   VEETGTFFEP   TLAALFVREA   FGPSTRSAAM   KLFTAIRDAL   450
ITRLRNLPWM   NEETQNMAQD   KVAQLQVEMG   ASEWALKPEL   ARQEYNDIQL   500
GSSFLQSVLS   CVRSLRARIV   QSFLQPHPQH   RWKVSPWDVN   AYYSVSDHVV   550
VFPAGLLQPP   FFHPGYPRAV   NFGAAGSIMA   HELLHIFYQL   LLPGGCLACD   600
NHALQEAHLC   LKRHYAAFPL   PSRTSFNDSL   TFLENAADVG   GLAIALQAYS   650
KRLLRHHGET   VLPSLDLSPQ   QIFFRSYAQV   MCRKPSPQDS   HDTHSPPHLR   700
VHGPLSSTPA   FARYFRCARG   ALLNPSSRCQ   LW                        732
```

Carrier molecule

Single-pass RBC membrane glycoprotein (type II) that is highly folded via disulphide bonds. In the RBC membrane Kell glycoprotein is covalently linked at Cys72 to the Cys347 of the XK protein.

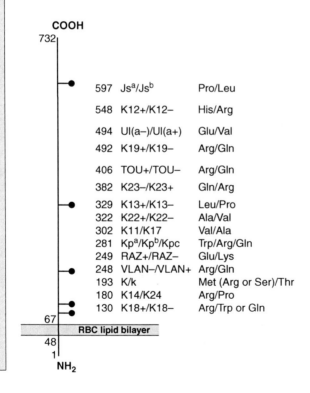

M_r (SDS-PAGE)   93 000; 79 000–80 000 after N-glyconase treatment
CHO: N-glycan   Five sites
Cysteine residues   15 (plus 1 in the membrane)
Copies per RBC   3500–18 000[2]

Molecular basis of antigens[3]

Antigen	Amino acid change	Exon	Nt change	Restriction enzyme
K/k	Met193Thr	6	698T>C	Bsm I (+/−)
Kpᵃ/Kpᵇ	Trp281Arg	8	961T>C	Nla III (+/−)
Kpᵇ/Kpᶜ	Arg281Gln	8	962G>A	Pvu II (−/+)
Jsᵃ/Jsᵇ	Pro597Leu	17	1910C>T	Mnl I (−/+)
	633Leu(silent)	17	2019G>A	Dde I (−/+)
Ul(a−)/Ul(a+)	Glu494Val	13	1601A>T	AccI (−/+)
K11/K17	Val302Ala	8	1025T>C	Hae III (−/+)
K12+/K12−	His548Arg	15	1763A>G	Nla III (+/−)
K13+/K13−	Leu329Pro	9	1106T>C	
K14/K24	Arg180Pro	6	659G>C	Hae III (−/+)
K18+/K18−	Arg130Trp	4	508C>T	Taq II (−/+)
	Arg130Gln	4	509G>A	Eco 57 I (−/+)
K19+/K19−	Arg492Gln	13	1595G>A	
K22+/K22−	Ala322Val	9	1085C>T	Tsp 45 I (−/+)
K23−/K23+	Gln382Arg	10	1265A>G	Bcn I (−/+)
VLAN−/VLAN+	Arg248Gln	8	863G>A	PspG I (−/+)*
TOU+/TOU−	Arg406Gln	11	1337G>A	
RAZ+/RAZ−	Glu249Lys	8	865G>A	EcoR I(−/+)*

* Mutation(s) introduced into primer(s).

Function

Kell glycoprotein is an endothelin-3-converting enzyme, preferentially cleaving big endothelin-3, a 41 amino acid polypeptide, at Trp21-Ile22, creating bioactive endothelin-3, a potent vasoconstrictor. Kell glycoprotein is a member of the Neprilysin (M13) sub-family of zinc endopeptidases and has sequence and structural homology with neutral endopeptidase 24.11 (CALLA, enkephalinase) and endothelin-converting enzyme ECE-1. Kell, in common with all zinc endopeptidases, shares a pentameric sequence, HEXXH, which is central to zinc binding and catalytic activity[4].

Disease association

In one study, 1 in 250 patients with AIHA had autoantibodies directed to Kell system antigens. Transiently depressed Kell system antigens have been associated with the presence of autoantibodies mimicking alloantibodies in AIHA and with microbial infection. The Kell protein was reduced in one case of ITP[5]. Kell antigens are weak on RBCs from McLeod CGD (X-linked type) males (see Kx blood group system 019)

Phenotypes (% occurrence)

	Caucasians	Blacks
K−k+	91	98
K+k−	0.2	Rare
K+k+	8.8	2
Kp(a+b−)	Rare	0
Kp(a−b+)	97.7	100
Kp(a+b+)	2.3	Rare
Js(a+b−)	0	1
Js(a−b+)	100	80
Js(a+b+)	Rare	19

Null: K_0; very rare, but a little less rare in Finland, Japan and Reunion Islands.

Unusual: K_{mod}, McLeod (see Kx blood group system [**XK, 019**] and table below showing comparison of Kell phenotypes), Kp(a+b−), Leach and Gerbich types of Ge-negative.

Comparison of Kell phenotypes[6]

Phenotype	Expression of antigen		Possible antibody in serum	RBC morphology
	Kell system	Kx		
Inherited Kell system phenotypes				
Common	Normal	Weak	Alloantibody	Discocytes
Kp(a+b−)	Slight/ moderate reduction	Slight increase	Anti-Kp^b	Discocytes
K_{mod}	Marked reduction*	Moderate increase	Anti-Ku-like (not mutually compatible)	Discocytes
K_0 heterozygote	Normal	Moderate increase	None	Discocytes
K_0	None†	Marked increase††	Anti-Ku	Discocytes
Inherited Kx system phenotype				
McLeod CGD	Marked reduction	None	Anti-KL (anti-Kx + anti-Km)	Acanthocytes

Phenotype	Expression of antigen Kell system	Kx	Possible antibody in serum	RBC morphology
McLeod non-CGD	Marked reduction	None	Anti-Km (anti-Kx in one case)	Acanthocytes
McLeod carriers‡	Normal to marked reduction	Not known	None	Discocytes and acanthocytes
Other				
Gerbich and Leach phenotypes	Slight decrease	Normal/ weak	Not Kell-related	Discocytes and elliptocytes in Leach phenotype
AIHA (Kell-related)	Normal to marked reduction	Slight increase (when Kell reduced)	"Kell-related" antibodies or non-specific	Discocytes or spherocytes (due to the hemolytic anemia)
Thiol-treated RBCs	Not detectable	Slight increase	Not applicable	Not known

* Will adsorb and elute antibody to inherited antigens in Kell system.
† Do not adsorb and elute.
†† Kx protein is decreased; antigen may be more accessible.
‡ The proportion of normal to McLeod phenotype RBCs varies in different carrier females. Only affected males present with 100% of RBCs having the McLeod phenotype.

Comparison of features of McLeod phenotype with normal and K_0 RBCs

Features	Normal Kell phenotype	K_0	McLeod non-CGD	McLeod CGD
Kell system antigens	++++	0	Weak	Weak
Kx antigen	+	++	0	0
Km antigen	++	0	0	0
Antibodies made	To lacking Kell antigens	Anti-Ku	Anti-Km (anti-Kx in one case)	Anti-Kx + -Km
Creatine kinase level	Normal	Normal	Elevated	Normal or Elevated
Blood for transfusion	Normal antigen-negative phenotype	K_0	McLeod or K_0	McLeod

Features	Normal Kell phenotype	K_0	McLeod non-CGD	McLeod CGD
Gene defect	Not applicable	Mutations in KEL	Mutations in XK	Deletion of XK and CGD
Morphology	Discocytes	Discocytes	Acanthocytes	Acanthocytes
Pathology	None	None	Muscular and neurological defects	Muscular and neurological defects with CGD

For more information on the McLeod syndrome, see web site: www.nefo.med.uni-muenchen.de/~adanek/McLeod.html.

Molecular basis of phenotypes

(Unless otherwise stated found in one proband)

Phenotype	Basis; amino acid change	Proband
K+^{w} k− Kp(a − b+^{w}) Ku+^{w} Js(b+^{w}) K:w14 K:w22	Exon 6 698C>G; Thr193Arg	Japanese[7]
K+^{var}k+	Exon 6 697A>T; Thr193Ser	Swiss[8]
K−k+^{w}	Exon 11 1388C>T; Ala423Val	White[3]
K_0	IVS3 +1 g>c, exon skipping and premature stop codon	Taiwanese[9]
	IVS3 +1 g>a, exon skipping and premature stop codon	Reunion Island (2)[10]
	Exon 4 366 T>A; Cys83Stop	Yugoslavian[10]
	Exon 4 502 C>T; Arg128Stop[10]	African blacks (2)[10]
	Exon 6 694 C>T; Arg192Stop	White[10]
	Exon 9 1162 C>T; Gln348Stop	White[10]
	Exon 10 1208 G>A; Ser363Asn[10]	White (2)[10]
	Exon 18 2147 G>A; Ser676Asn	Isreali[10]
K_{mod}	Exon 10 1208 G>A; Ser363Asn	White[11]
	Exon 10 1208 G>A; Ser363Asn with Exon 18 2150 A>G; Tyr677Cys	White[11]
	Exon 9 1106 T>C; Leu329Pro with Exon 15 G>A; Trp532Stop	(K:−13) White[11]
	Exon 19 G>A; Gly703Arg with Exon 16 C>T; silent	White[11]
McLeod	See Kx blood group system	

Extreme depression of Kell system antigens together with very weak expression of Kx antigen (resembling a McLeod phenotype) on the RBCS of a German proband were caused by homozygosity for Kp^a at the KEL locus with the simultaneous presence of a single base change in the donor splice site of XK[12].

Comments

No Kell system haplotype has been found to have more than one low incidence antigen. The Kp^a antigen *in cis* weakens the expression of Kell antigens (*cis*-modifying effect)[6,13]. K_{mod} is an umbrella term used to describe various phenotypes with very weak expression of Kell antigens and increased expression of Kx antigen. Kell antigens are sensitive to treatment by a mixture of α-chymotrypsin and trypsin or to sequential treatment of antigen-positive RBCs with these enzymes. Antibodies produced by K_{mod} individuals are not necessarily mutually compatible. Antibodies to antigens in the Kell blood group system cause suppression of erythropoiesis in the fetus[14].

References

1 Lee, S. et al. (1991) Proc. Natl. Acad. Sci. USA 88, 6353–6357.
2 Parsons, S.F. et al. (1993) Transf. Med. 3, 137–142.
3 Lee, S. (1997) Vox Sang. 73, 1–11.
4 Lee, S. et al. (2000) Semin. Hematol. 37, 113–121.
5 Williamson, L.M. et al. (1994) Br. J. Haematol. 87, 805–812.
6 Øyen, R. et al. (1997) Immunohematology 13, 75–79.
7 Uchikawa, M. et al. (2000) Vox Sang. 78 (Suppl. 1), 0011 (abstract).
8 Poole, J. et al. (2001) Transfusion 41 (Suppl.), 15S (abstract).
9 Yu, L.C. et al. (2001) J. Biol. Chem. 276, 10247–10252.
10 Lee, S. et al. (2001) J. Biol. Chem. 276, 27281–27289.
11 Lee, S. et al. (2003) Transfusion; (in press).
12 Daniels, G.L. et al. (1996) Blood 88, 4045–4050.
13 Yazdanbakhsh, K. et al. (1999) Blood 94, 310–318.
14 Vaughan, J.I. et al. (1998) N. Engl. J. Med. 338, 798–803.

K ANTIGEN

Terminology

ISBT symbol (number)	KEL1 (006.001)
Other names	Kell; K1
History	Named after first antibody producer (Kelleher) of anti-K, which caused HDN; reported in 1946

Occurrence

Caucasians	9%
Blacks	2%
Orientals	Rare
Iranian Jews	12%
Arabs	May be as high as 25%

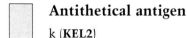

Antithetical antigen

k (**KEL2**)

Expression

Cord RBCs	Expressed
Altered	See table showing comparison of Kell phenotypes on system pages
	Weak with Arg193 or Ser193

Molecular basis associated with K antigen[1-3]

Amino acid	Met 193
Nucleotide	T at bp 698 in exon 6
Restriction enzyme	*Bsm* I site present

Two other mutations encode K antigen, albeit weakly: 698C>G (Thr193Arg) and 697A>T (Thr 193Ser). All three mutations disrupt the N-glycosylation motif, so Asn 191 is not glycosylated

Effect of enzymes/chemicals on K antigen on intact RBCs

Ficin/papain	Resistant
Trypsin	Resistant
α-Chymotrypsin	Resistant*
Pronase	Resistant
Sialidase	Resistant
DTT 200 mM/50 mM	Sensitive/sensitive
Acid	Sensitive

* May be weakened or sensitive if the enzyme preparation is contaminated with trypsin.

In vitro characteristics of alloanti-K

Immunoglobulin class	IgM less common than IgG
Optimal technique	RT or more usually IAT
Complement binding	Rare

Clinical significance of alloanti-K

Transfusion reaction	Mild to severe/delayed/hemolytic
HDN	Mild to severe (rare)

Comments

Some bacteria elicit production of IgM anti-K. Expression of K can be acquired as a result of bacterial activity *in vivo* and *in vitro*.

References
1   Lee, S. et al. (1995) Blood 85, 912–916.
2   Uchikawa, M. et al. (2000) Vox Sang. 78 (Suppl. 1), 0011 (abstract).
3   Poole, J. et al. (2001) Transfusion 41 (Suppl.), 15S (abstract).

k ANTIGEN

Terminology

ISBT symbol (number)	KEL2 (006.002)
Other names	Cellano; K2
History	Identified in 1949 when an antibody was shown to recognize the antithetical antigen to K. Cellano, the original name, was derived by rearranging the proband's last name (Nocella)

Occurrence

Caucasians	99.8%
Blacks	100%

Antithetical antigen

K (**KEL1**)

Expression

Cord RBCs	Expressed
Altered	See table showing comparison of Kell phenotypes on system page. Weakened in rare genetic variants. Weak expression with concomitant 423Val[1].

233

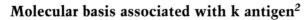

Molecular basis associated with k antigen[2]

Amino acid	Thr 193
Nucleotide	C at bp 698 in exon 6
Restriction enzyme	*Bsm* I site ablated

Effect of enzymes/chemicals on k antigen on intact RBCs

Ficin/papain	Resistant
Trypsin	Resistant
α-Chymotrypsin	Resistant (see K [KEL1])
Pronase	Resistant
Sialidase	Resistant
DTT 200 mM/50 mM	Sensitive/sensitive
Acid	Sensitive

In vitro characteristics of alloanti-k

Immunoglobulin class	IgG more common than IgM
Optimal technique	IAT
Complement binding	No

Clinical significance of alloanti-k

Transfusion reaction	Mild to moderate/delayed
HDN	Mild to severe (rare)

References
1 Lee, S. (1997) Vox Sang. 73, 1–11.
2 Lee, S. et al. (1995) Blood 85, 912–916.

Kp^a ANTIGEN

Terminology

ISBT symbol (number)	KEL3 (006.003)
Other names	Penny; K3
History	Identified in 1957; the antigen, which was shown to be related to the Kell system, took its name from "K" for "Kell" and "p" for the first letter of the antibody producer's name (Penny)

Occurrence

Caucasians	2%
Blacks	Less than 0.01%

Antithetical antigens

Kp^b (**KEL4**), Kp^c (**KEL21**)

Expression

Cord RBCs	Expressed
Altered	See table showing comparison of Kell phenotypes on system pages

Molecular basis associated with Kp^a antigen[1]

Amino acid	Trp 281
Nucleotide	T at bp 961 in exon 8
Restriction enzyme	*Nla* III site present

Effect of enzymes/chemicals on Kp^a antigen on intact RBCs

Ficin/papain	Resistant
Trypsin	Resistant
α-Chymotrypsin	Resistant (see K [KEL1])
Pronase	Resistant
Sialidase	Resistant
DTT 200 mM/50 mM	Sensitive/sensitive
Acid	Sensitive

In vitro characteristics of alloanti-Kp^a

Immunoglobulin class	IgG
Optimal technique	IAT
Complement binding	No

Clinical significance of alloanti-Kp^a

Transfusion reaction	Mild to moderate/delayed
HDN	Mild to severe

Comments

In the presence of Kp^a, other inherited Kell system antigens are suppressed (*cis*-modifier effect) to varying degrees.

To date, in people with K+Kp(a+) RBCs, *K* has always been *in trans* to *Kp^a*.

Reference
[1] Lee, S. et al. (1996) Transfusion 36, 490–494.

Kp^b ANTIGEN

Terminology

ISBT symbol (number)	KEL4 (006.004)
Other names	Rautenberg; K4
History	Identified in 1958 and recognized to be antithetical to Kp^a

Occurrence

All populations	100%

Antithetical antigens

Kp^a (**KEL3**); Kp^c (**KEL21**)

Expression

Cord RBCs	Expressed
Altered	Weak on RBCs from some patients with AIHA
	See table showing comparison of Kell phenotypes on system pages

Molecular basis associated with Kp^b antigen[1]

Amino acid	Arg 281
Nucleotide	C at bp 961, G at bp 962 in exon 8
Restriction enzyme	*Nla* III site ablated

Effect of enzymes/chemicals on Kp^b antigen on intact RBCs

Ficin/papain	Resistant
Trypsin	Resistant
α-Chymotrypsin	Resistant (see K [KEL1])
Pronase	Resistant
Sialidase	Resistant
DTT 200 mM/50 mM	Sensitive/sensitive
Acid	Sensitive

In vitro characteristics of alloanti-Kp^b

Immunoglobulin class	IgG, rarely IgM
Optimal technique	IAT
Complement binding	No

Clinical significance of alloanti-Kp^b

Transfusion reaction	No to moderate/delayed
HDN	Mild to moderate

Autoanti-Kp^b

Yes. May appear as alloantibody on initial presentation due to suppression of Kp^b antigen

Comments

Sera containing anti-Kp^b often contain anti-K (see KEL1).

Reference
1   Lee, S. et al. (1996) Transfusion 36, 490–494.

Ku ANTIGEN

Terminology

ISBT symbol (number)	KEL5 (006.005)
Other names	Peltz; K5
History	Antibody in serum of K_0 [K−k−Kp(a−b−)] person identified in 1957; originally called anti-KkKp^a or anti-Peltz (after the proband); renamed anti-Ku (K for Kell, u for universal) in 1961

Occurrence

All populations	100%

Expression

Cord RBCs	Expressed
Altered	See table showing comparison of Kell phenotypes on system page

Molecular basis associated with Ku antigen

Not known. For the molecular basis associated with an absence of Ku (K_0 phenotype) see Kell system pages.

Effect of enzymes/chemicals on Ku antigen on intact RBCs

Ficin/papain	Resistant
Trypsin	Resistant
α-Chymotrypsin	Resistant (see K [KEL1])
Pronase	Resistant
Sialidase	Resistant
DTT 200 mM/50 mM	Sensitive/sensitive
Acid	Sensitive

In vitro characteristics of alloanti-Ku

Immunoglobulin class	IgG
Optimal technique	IAT
Complement binding	No

Clinical significance of alloanti-Ku

Transfusion reaction   Mild to severe
HDN   No to moderate

Autoanti-Ku

Yes

Comments

Anti-Ku is made by K_0 people who may make additional antibodies directed at other Kell antigens and, rarely, make Kell system specificities without making Ku. K_{mod} people make Ku-like antibodies that are not necessarily mutually compatible.

An antibody detected only in the presence of Cotrimoxazole (CTMX; a drug in the suspension medium of some reagent RBCs) was identified as anti-Ku[1].

Reference
[1] Le Pennec, P. et al. (1999) Transfusion 39 (Suppl.), 81S (abstract).

Jsª ANTIGEN

Terminology

ISBT symbol (number)   KEL6 (006.006)
Other names   Sutter; K6
History   Described in 1958; "J" is from the first name (John) and "s" is from the last name (Sutter) of the first producer of the antibody. Jsª was shown to belong to the Kell system in 1965

Occurrence

Caucasians   0.01%
Blacks   20%

Antithetical antigen

Jsᵇ (**KEL7**)

Expression

Cord RBCs	Expressed
Altered	See table showing comparison of Kell phenotypes on system pages

Molecular basis associated with Js^a antigen[1]

Amino acid	Pro 597
Nucleotide	C at bp 1910 and G at bp 2019 (silent mutation) in exon 17
Restriction enzyme	Mnl I (1910C) and Dde I (2019G) sites ablated

Effect of enzymes/chemicals on Js^a antigen on intact RBCs

Ficin/papain	Resistant
Trypsin	Resistant
α-Chymotrypsin	Resistant (see K [KEL1])
Pronase	Resistant
Sialidase	Resistant
DTT 200 mM/50 mM	Sensitive/sensitive (see Comments)
Acid	Sensitive

In vitro characteristics of alloanti-Js^a

Immunoglobulin class	IgG more common than IgM
Optimal technique	IAT
Complement binding	No

Clinical significance of alloanti-Js^a

Transfusion reaction	No to moderate/delayed
HDN	Mild to severe[2]

Comments

At least one example of "naturally-occurring" anti-Js^a has been reported in a Japanese woman.

Js^a is extremely sensitive to thiol reagents (it is sensitive to 2 mM DTT) most likely because it is located between two cysteine residues.

Chimpanzee RBCs type KEL:−1, 2, −3, 4, 5, 6, −7.

References
1   Lee, S. et al. (1995) Transfusion 35, 822–825.
2   Gordon, M.C. et al. (1995) Vox Sang. 69, 140–141.

Js^b ANTIGEN

Terminology

ISBT symbol (number)	KEL7 (006.007)
Other names	Matthews; K7
History	Named in 1963 when it was found to be antithetical to Js^a; joined the Kell blood group system in 1965

Occurrence

Caucasians	100%
Blacks	99%

Antithetical antigen

Js^a (**KEL6**)

Expression

Cord RBCs	Expressed
Altered	See table showing comparison of Kell phenotypes on system pages

Molecular basis associated with Js^b antigen[1]

Amino acid	Leu 597
Nucleotide	T at bp 1910 and A at bp 2019 (silent mutation) in exon 17
Restriction enzyme	*Mnl* I (1910T) and *Dde* I (2019A) sites present

Effect of enzymes/chemicals on Js^b antigen on intact RBCs

Ficin/papain	Resistant (some enhanced)
Trypsin	Resistant
α-Chymotrypsin	Resistant (see K [KEL1])
Pronase	Resistant
Sialidase	Resistant
DTT 200 mM/50 mM	Sensitive/sensitive (see Comments)
Acid	Sensitive

In vitro characteristics of alloanti-Js^b

Immunoglobulin class	IgG
Optimal technique	IAT
Complement binding	No

Clinical significance of alloanti-Js^b

Transfusion reaction	Mild to moderate/delayed
HDN	Mild to severe (one fatality)[2]

Comments

Js^b is extremely sensitive to thiol reagents (it is sensitive to 2 mM DTT) most likely because it is located between two cysteine residues.
Chimpanzee RBCs type KEL:−1, 2, −3, 4, 5, 6, −7.

References
[1] Lee, S. et al. (1995) Transfusion 35, 822–825.
[2] Stanworth, S. et al. (2001) Vox Sang. 81, 134–135.

Ul^a ANTIGEN

Terminology

ISBT symbol (number)	KEL10 (006.010)
Other names	Karhula; K10
History	Described in 1968 and shown to be part of the Kell system in 1969. Named after the last letters of the antibody producer (Karhula)

Occurrence

Less than 0.01% in most populations; 2.6% in Finns (higher in some regions); 0.46% in Japanese.

Expression

Cord RBCs   Expressed

Molecular basis associated with Ul[a] antigen[1]

Amino acid   Val 494
Nucleotide   T at bp 1601 in exon 13
Restriction enzyme   *Acc* I site present
Wild type *KEL* has A at bp 1601 and Ul(a−) Kell glycoprotein has Glu 494.

Effect of enzymes/chemicals on Ul[a] antigen on intact RBCs

Ficin/papain   Resistant
Trypsin   Resistant
α-Chymotrypsin   Resistant (see K [KEL1])
Pronase   Presumed resistant
Sialidase   Presumed resistant
DTT 200 mM/50 mM   Presumed sensitive
Acid   Presumed sensitive

In vitro characteristics of alloanti-Ul[a]

Immunoglobulin class   IgG
Optimal technique   IAT

Clinical significance of alloanti-Ul[a]

Transfusion reaction   No data but anti-Ul[a] has been stimulated by transfusion
HDN   1 case[2]. 19 Ul(a−) mothers with Ul(a+) children did not make anti-Ul[a]

Comments

Only a few examples of anti-Ul[a] reported: two in Finland and two in Japan.

References
1 Lee, S. et al. (1996) Transfusion 36, 490–494.
2 Sakuma, K. et al. (1994) Vox Sang. 66, 293–294.

K11 ANTIGEN

Terminology

ISBT symbol (number)	KEL11 (006.011)
Other names	Côté
History	Found in 1971 in the serum of Mrs. Côté; the first of a series of K−k+ people who made an antibody compatible only with K_0 RBCs; a para-Kell antigen until proven to belong to Kell in 1974

Occurrence

All populations	100%

Antithetical antigen

K17 (**KEL17**)

Expression

Cord RBCs	Expressed
Altered	See table showing comparison of Kell phenotypes on system pages

Molecular basis associated with K11 antigen[1]

Amino acid	Val 302
Nucleotide	T at bp 1025 in exon 8
Restriction enzyme	*Hae* III site ablated

Effect of enzymes and chemicals on K11 antigen on intact RBCs

Ficin/papain	Resistant
Trypsin	Resistant

α-Chymotrypsin	Resistant (see K [KEL1])
Pronase	Resistant
Sialidase	Resistant
DTT 200 mM/50 mM	Sensitive/sensitive
Acid	Sensitive

In vitro characteristics of alloanti-K11

Immunoglobulin class	IgG
Optimal technique	IAT

Clinical significance of alloanti-K11

Transfusion reaction	Mild to moderate (not much data)
HDN	No to mild (not much data)

Reference
1   Lee, S. et al. (1996) Transfusion 36, 490–494.

K12 ANTIGEN

Terminology

ISBT symbol (number)	KEL12 (006.012)
Other names	Bøc (Bockman); Spears
History	Described in 1973; given the next number in the series of K−k+ people who made an antibody compatible only with K_0 RBCs

Occurrence

All populations: 100%; K:−12 phenotype has been identified in four Caucasian families.

Expression

Cord RBCs	Presumed expressed
Altered	See table showing comparison of Kell phenotypes on system pages

Molecular basis associated with K12 antigen[1]

Amino acid	His 548
Nucleotide	A at bp1763 in exon 15, K:−12 has G at bp 1763 (548 Arg)
Restriction enzyme	*Nla* III site ablated in K:−12

Effect of enzymes/chemicals on K12 antigen on intact RBCs

Ficin/papain	Resistant
Trypsin	Resistant
α-Chymotrypsin	Resistant (see K [KEL1])
Pronase	Resistant
Sialidase	Resistant
DTT 200 mM/50 mM	Sensitive/sensitive
Acid	Presumed sensitive

In vitro characteristics of alloanti-K12

Immunoglobulin class	IgG
Optimal technique	IAT

Clinical significance of alloanti-K12

Transfusion reaction	K:12 blood transfused to two patients (DL, MS) did not cause a transfusion reaction.
HDN	No data although Mrs. Bøc had at least two children

Reference
[1] Lee, S. (1997) Vox Sang. 73, 1–11.

K13 ANTIGEN

Terminology

ISBT symbol (number)	KEL13 (006.013)
Other names	SGRO
History	Described in 1974, given the next Kell system number. The K:−13 proband is a K_{mod}, thereby

explaining the weak expression of Kell antigens in this phenotype[1]

Occurrence

All populations: 100%; K:−13 found in one family.

Expression

Cord RBCs   Presumed expressed
Altered   See table showing comparison of Kell phenotypes on system pages

Molecular basis associated with K13 antigen[1]

Amino acid   Leu 329
Nucleotide   1106 T in exon 9
K:−13 proband has C at bp 1106 (Pro 329) in one gene, and 1716G>A in exon 15 (Trp532Stop) in the other gene.

Effect of enzymes/chemicals on K13 antigen on intact RBCs

Ficin/papain   Resistant
Trypsin   Resistant
α-Chymotrypsin   Resistant (see K [KEL1])
Pronase   Resistant
Sialidase   Resistant
DTT 200 mM/50 mM   Sensitive/sensitive
Acid   Presumed sensitive

In vitro characteristics of alloanti-K13

Immunoglobulin class   IgG
Optimal technique   IAT

Clinical significance of alloanti-K13

No data are available. The K:−13 sister of the proband had seven children without making anti-K13.

Reference
[1] Lee, S. et al. (2003) Transfusion (in press).

K14 ANTIGEN

Terminology

ISBT symbol (number)	KEL14 (006.014)
Other names	San; Santini; Dp
History	Described in 1973, given the next number in the series of K−k+ people who made an antibody compatible only with K_0 RBCs

Occurrence

All populations: 100%; K:−14 found in three French-Cajun families.

Antithetical antigen

K24 (**KEL24**)

Expression

Cord RBCs	Expressed
Altered	See table showing comparison of Kell phenotypes on system pages

Molecular basis associated with K14 antigen[1]

Amino acid	Arg 180
Nucleotide	G at bp 659 in exon 6
Restriction enzyme	*Hae* III site ablated

Effect of enzymes/chemicals on K14 antigen on intact RBCs

Ficin/papain	Resistant
Trypsin	Resistant
α-Chymotrypsin	Resistant (see K [KEL1])
Pronase	Resistant
Sialidase	Resistant
DTT 200 mM/50 mM	Sensitive/sensitive
Acid	Presumed sensitive

In vitro characteristics of alloanti-K14

Immunoglobulin class IgG
Optimal technique IAT

Clinical significance of alloanti-K14

Clinical significance is largely unknown since only three families have been reported.
HDN one case[2]

References
[1] Lee, S. (1997) Vox Sang. 73, 1–11.
[2] Wallace, M.E. et al. (1976) Vox Sang. 30, 300–304.

K16 ANTIGEN

Terminology

ISBT symbol (number) KEL16 (006.016)
Other names Weak k; k-like
History When anti-k was absorbed with McLeod RBCs, an antibody remained that was non-reactive with McLeod RBCs and reactive with all other k+ RBCs. In 1976, this antibody was named anti-K16 and the antigen K16. No further studies have been performed

K17 ANTIGEN

Terminology

ISBT symbol (number) KEL17 (006.017)
Other names Wk^a; Weeks
History Reported in 1974; given a Kell system number because the antigen had linkage disequilibrium to K, and in 1975 was shown to be antithetical to K11

Occurrence

All populations   0.3%

Antithetical antigen

K11 (**KEL11**)

Expression

Cord RBCs   Presumed expressed
Altered   See table showing comparison of Kell phe-
                             notypes on system pages

Molecular basis associated with K17 antigen[1]

Amino acid   Ala 302
Nucleotide   C at bp 1025 in exon 8
Restriction enzyme   *Hae* III site gained

Effect of enzymes/chemicals on K17 antigen on intact RBCs

Ficin/papain   Resistant
Trypsin   Resistant
α-Chymotrypsin   Presumed resistant (see K [KEL1])
Pronase   Presumed resistant
Sialidase   Presumed resistant
DTT 200 mM/50 mM   Sensitive/sensitive
Acid   Presumed sensitive

In vitro characteristics of alloanti-K17

Immunoglobulin class   IgG
Optimal technique   IAT

Clinical significance of alloanti-K17

Transfusion reaction   No data
HDN   No data

Reference
1   Lee, S. et al. (1996) Transfusion 36, 490–494.

K18 ANTIGEN

Terminology

ISBT symbol (number)	KEL18 (006.018)
Other names	V.M.; Marshall
History	Described in 1975, given the next number in the series of K−k+ people who made an antibody compatible only with K_0 RBCs

Occurrence

All populations: 100%; K:−18 found in three families.

Expression

Cord RBCs	Presumed expressed
Altered	See table showing comparison of Kell phenotypes on system pages

Molecular basis associated with K18 antigen[1]

Amino acid   Arg 130
Type 1

Nucleotide	C at bp 508 in exon 4. K:−18 has T (130 Trp)
Restriction enzyme	K:−18 gains a *Taq* II site

Type 2

Nucleotide	G at bp 509 in exon 4. K:−18 has A (130 Gln)
Restriction enzyme	K:−18 gains *Eco* 57 I site

Effect of enzymes/chemicals on K18 antigen on intact RBCs

Ficin/papain	Resistant
Trypsin	Resistant
α-Chymotrypsin	Resistant (see K [KEL1])
Pronase	Resistant
Sialidase	Resistant
DTT 200 mM/50 mM	Sensitive/sensitive
Acid	Presumed sensitive

251

In vitro characteristics of alloanti-K18

Immunoglobulin class   IgG
Optimal technique   IAT

Clinical significance of alloanti-K18

Clinical significance is largely unknown. Chromium survival studies showed accelerated RBC destruction in one case[2].

References

1. Lee, S. (1997) Vox Sang. 73, 1–11.
2. Barrasso, C. et al. (1983) Transfusion 23, 258–259.

K19 ANTIGEN

Terminology

ISBT symbol (number)   KEL19 (006.019)
Other names   Sub; Sublett
History   Described in 1979; given the next number in the series of K−k+ people who made an antibody compatible only with K_0 RBCs

Occurrence

All populations   100%

Expression

Cord RBCs   Presumed expressed
Altered   See table showing comparison of Kell phenotypes on system pages

Molecular basis associated with K19 antigen[1]

Amino acid   Arg 492
Nucleotide   G at bp 1595 in exon 13
K:−19 has A at bp 1595 and Gln 492

Effect of enzymes/chemicals on K19 antigen on intact RBCs

Ficin/papain	Resistant
Trypsin	Resistant
α-Chymotrypsin	Resistant (see K [KEL1])
Pronase	Resistant
Sialidase	Resistant
DTT 200 mM/50 mM	Sensitive/sensitive
Acid	Presumed sensitive

In vitro characteristics of alloanti-K19

Immunoglobulin class	IgG
Optimal technique	IAT

Clinical significance of alloanti-K19

Transfusion reaction	In one case moderate/delayed/hemolytic
HDN	No data

Comments

Only two examples of anti-K19 have been described, one made by a woman (race unknown) probably as a result of pregnancy, the other made by a Black, multiply transfused man.

Reference
[1]   Lee, S. (1997) Vox Sang. 73, 1–11.

Km ANTIGEN

Terminology

ISBT symbol (number)	KEL20 (006.020)
Other names	K20
History	The suffix "m" denotes the association with the McLeod phenotype[1]. Reported in 1979

Occurrence

All populations   100%

Expression

Cord RBCs   Presumed expressed
Altered   See table showing comparison of Kell phe-
                              notypes on system pages

Molecular basis associated with Km antigen

Unknown

Effect of enzymes/chemicals on Km antigen on intact RBCs

Ficin/papain   Resistant
Trypsin   Resistant
α-Chymotrypsin   Not known
Pronase   Not known
Sialidase   Not known
DTT 200 mM/50 mM   Not known (sensitive to AET)
Acid   Not known

In vitro characteristics of alloanti-Km

Immunoglobulin class   IgG
Optimal technique   IAT

Clinical significance of alloanti-Km

Transfusion reaction   Delayed/hemolytic in one case[2]
HDN   Not applicable; anti-Km has been only
                              made by McLeod males

Comments

Anti-Km is made by non-CGD McLeod males. Both McLeod and K_0
phenotype blood will be compatible. Anti-Km + anti-Kx (sometimes called
anti-KL) is made by CGD McLeod males and only McLeod blood will be
compatible[3].

References
1 Marsh, W.L. (1979) Vox Sang. 36, 375.
2 Marsh, W.L. et al. (1979) Transfusion 19, 604–608.
3 Marsh, W.L. and Redman, C.M. (1990) Transfusion 30, 158–167.

Kp^c ANTIGEN

Terminology

ISBT symbol (number)	KEL21 (006.021)
Other names	Levay; K21
History	First reported in 1945; joined the Kell system in 1979 when it was shown to be antithetical to Kp^a and Kp^b. Anti-Levay was the first antibody to a low prevalence ('private') antigen found

Occurrence

Less than 0.01%; Japanese, up to 0.32% (several *Kp^c* homozygotes).

Antithetical antigens

Kp^a (**KEL3**), Kp^b (**KEL4**)

Expression

Cord RBCs	Expressed
Altered	See table showing comparison of Kell phenotypes on system pages

Molecular basis associated with Kp^c antigen[1]

Amino acid	Gln 281
Nucleotide	A at bp 962 in exon 8
Restriction enzyme	*Pvu* II site is gained

Effect of enzymes/chemicals on Kp^c antigen on intact RBCs

Ficin/papain	Resistant
Trypsin	Resistant
α-Chymotrypsin	Not known
Pronase	Presumed resistant
Sialidase	Presumed resistant
DTT 200 mM/50 mM	Sensitive/sensitive
Acid	Presumed sensitive

In vitro characteristics of alloanti-Kp^c

Immunoglobulin class	IgG; IgM
Optimal technique	IAT; saline RT

Clinical significance of alloanti-Kp^c

No data are available.

Comments

A Japanese Kp(a−b−) blood donor with anti-Kp^b was found to be Levay positive.

Reference
[1] Lee, S. et al. (1996) Transfusion 36, 490–494.

K22 ANTIGEN

Terminology

ISBT symbol (number)	KEL22 (006.022)
Other names	N.I.; Ikar
History	Described in 1982; given the next number in the series of K−k+ people who made an antibody compatible only with K$_0$ RBCs

Occurrence

All populations: 100%; K:−22 found in two Iranian Jewish families.

Expression

Cord RBCs   Expressed
Altered   See table showing comparison of Kell phe-
                             notypes on system pages

Molecular basis associated with K22 antigen[1]

Amino acid   Ala 322
Nucleotide   C at bp 1085 in exon 9
Restriction enzyme   *Tsp* 45 I site gained in K:−22
K:−22 has T at bp 1085 and 322 Val.

Effect of enzymes/chemicals on K22 antigen on intact RBCs

Ficin/papain   Resistant
Trypsin   Resistant
α-Chymotrypsin   Resistant (see K [KEL1])
Pronase   Resistant
Sialidase   Resistant
DTT 200 mM/50 mM   Sensitive/sensitive
Acid   Presumed sensitive

In vitro characteristics of alloanti-K22

Immunoglobulin class   IgG
Optimal technique   IAT

Clinical significance of alloanti-K22

Transfusion reaction   No data
HDN   Mild to severe in one case

Reference
1   Lee, S. (1997) Vox Sang. 73, 1–11.

K23 ANTIGEN

Terminology

ISBT symbol (number)	KEL23 (006.023)
Other names	Centauro
History	Reported in 1987; antibody identified in the serum of a pregnant woman; assigned to Kell because the serum precipitated Kell glycoprotein from her husband's RBCs

Occurrence

Less than 0.01%. Found in one Italian family.

Expression

Cord RBCs	Expressed

Molecular basis associated with K23 antigen[1]

Amino acid	Arg 382
Nucleotide	G at bp 1265 in exon 10
Restriction enzyme	*Bcn* I site gained

K:−23 has A at bp 1265 and Gln 383

Effect of enzymes/chemicals on K23 antigen on intact RBCs

Ficin/papain	Resistant
Trypsin	Resistant
α-Chymotrypsin	Not known
Pronase	Presumed resistant
Sialidase	Presumed resistant
DTT 200 mM/50 mM	Sensitive/sensitive
Acid	Presumed sensitive

In vitro characteristics of alloanti-K23

Immunoglobulin class	IgG
Optimal technique	IAT

Clinical significance of alloanti-K23

Transfusion reaction   No data
HDN   Positive DAT; no clinical HDN

Reference
[1] Lee, S. (1997) Vox Sang. 73, 1–11.

K24 ANTIGEN

Terminology

ISBT symbol (number)   KEL24 (006.024)
Other names   CL; Callais; Cls
History   Described in 1985 when it was shown to be antithetical to K14

Occurrence

Found in three French-Cajun families.

Antithetical antigen

K14 (**KEL14**)

Expression

Cord RBCs   Expressed
Altered   See table showing comparison of Kell phenotypes on system pages

Molecular basis associated with K24 antigen[1]

Amino acid   Pro 180
Nucleotide   C at bp 659 in exon 6
Restriction enzyme   *Hea* III site gained

Effect of enzymes/chemicals on K24 antigen on intact RBCs

Ficin/papain	Resistant
Trypsin	Resistant
α-Chymotrypsin	Presumed resistant (see K [KEL1])
Pronase	Presumed resistant
Sialidase	Presumed resistant
DTT 200 mM/50 mM	Sensitive/sensitive
Acid	Presumed sensitive

In vitro characteristics of alloanti-K24

Immunoglobulin class	IgG
Optimal technique	IAT

Clinical significance of alloanti-K24

Transfusion reaction	No data, only one example of anti-K24 has been described
HDN	Positive DAT; no clinical HDN

Reference
[1] Lee, S. (1997) Vox Sang. 73, 1–11.

VLAN ANTIGEN

Terminology

ISBT symbol (number)	KEL25 (006.025)
History	Named in 1996 after the last name of the proband whose RBCs possessed the antigen

Occurrence

Found in one Dutch family[1].

Expression

Cord RBCs	Presumed expressed

Molecular basis associated with VLAN antigen[2]

Amino acid	Gln 248
Nucleotide	A at bp 863 in exon 8
Restriction enzyme	None [*Psp* GI (*Eco* RII) site created by introducing a 861 T>C mutation in one of the primers]

VLAN– has G at bp 863 and 248 Arg.

Effect of enzymes/chemicals on VLAN antigen on intact RBCs

Ficin/papain	Resistant
Trypsin	Presumed resistant
α-Chymotrypsin	Presumed resistant (see K [**KEL1**])
Pronase	Presumed resistant
Sialidase	Presumed resistant
DTT 200 mM/50 mM	Sensitive/sensitive
Acid	Presumed sensitive

In vitro characteristics of alloanti-VLAN

Immunoglobulin class	IgG
Optimal technique	IAT

Clinical significance of alloanti-VLAN

No data. The only example of the antibody was found in serum BUS by an incompatible crossmatch[1].

References
1 Jongerius, J.M. et al. (1996) Vox Sang. 71, 43–47.
2 Lee, S. et al. (2001) Vox Sang. 81, 259–263.

TOU ANTIGEN

Terminology

ISBT symbol (number)	KEL26 (006.026)
History	Named in 1995 after the last name of the proband whose serum contained an antibody to a high incidence antigen; provisional assignment to Kell system ratified in 1998

Occurrence

All populations: 100%; TOU– found in two families, one Native American and one Hispanic[1].

Expression

Cord RBCs	Presumed expressed
Altered	See table showing comparison of Kell phenotypes on system pages

Molecular basis associated with TOU antigen[2]

Amino acid	Arg 406
Nucleotide	G at bp 1337 in exon 11

TOU– has A at bp 1337 and 406 Gln.

Effect of enzymes/chemicals on TOU antigen on intact RBCs

Ficin/papain	Resistant
Trypsin	Resistant
α-Chymotrypsin	Presumed resistant
Pronase	Presumed resistant
Sialidase	Presumed resistant
DTT 200 mM/50 mM	Sensitive/sensitive
Acid	Presumed sensitive

In vitro characteristics of alloanti-TOU

Immunoglobulin class	IgG
Optimal technique	IAT

Clinical significance of alloanti-TOU

No data.

References
1 Jones, J. et al. (1995) Vox Sang. 69, 53–60.
2 Lee, S. (1997) Vox Sang. 73, 1–11.

RAZ ANTIGEN

Terminology

ISBT symbol (number)   KEL27 (006.027)
Other name   K27
History   Named in 1994 after the proband whose serum contained an antibody to a high incidence antigen; provisional Kell system assignment ratified in 2002

Occurrence

All populations: 100%; KEL:−27 found in one Indian (from Gujarat) family[1].

Expression

Cord RBCs   Presumed expressed
Altered   See table showing comparison of Kell phenotypes on system pages

Molecular basis associated with RAZ antigen[2]

Amino acid   Glu 249
Nucleotide   G at bp 865 in exon 8
Restriction enzyme None; (*Eco* RI site created in KEL:−27 phenotype by introducing a 869 A>T mutation in one of the primers)
KEL:−27 has A at bp 865 and 249 Lys.

Effect of enzymes/chemicals on RAZ antigen on intact RBCs

Ficin/papain   Resistant
Trypsin   Resistant
α-Chymotrypsin   Weakened
Pronase   Presumed resistant
Sialidase   Presumed resistant
DTT 200 mM/50 mM   Sensitive/sensitive
Acid   Presumed sensitive

In vitro characteristics of alloanti-RAZ

Immunoglobulin class   IgG
Optimal technique   IAT

Clinical significance of alloanti-RAZ

Transfusion reaction   No data
HDN   No data

References

[1] Daniels, G.L. et al. (1994) Transfusion 34, 818–820.
[2] Lee, S. et al. (2001) Vox Sang. 81, 259–263.

Number of antigens   6

Terminology

ISBT symbol	LE
ISBT number	007
History	Discovered by Mourant in 1946; named after one of the two original donors in whom anti-Le^a was identified

Expression

Soluble form	Saliva and all body fluids except CSF (in secretors)
Other blood cells	Lymphocytes, platelets
Tissues	Pancreas, mucosa of stomach, small and large intestine, skeletal muscle, renal cortex, adrenal glands

Gene

Chromosome	19p13.3
Name	*LE (FUT3)*
Organization	3 exons distributed over approximately 8 kbp of gDNA
Product	α(1,3/1,4) fucosyltransferase

Gene map

Database accession numbers

GenBank   X53578
http://www.bioc.aecom.yu.edu/bgmut/index.htm.

Amino acid sequence of α(1,3/1,4) fucosyltransferase[1]

```
MDPLGAAKPQ   WPWRRCLAAL   LFQLLVAVCF   FSYLRVSRDD   ATGSPRAPSG    50
SSRQDTTPTR   PTLLILLWTW   PFHIPVALSR   CSEMVPGTAD   CHITADRKVY   100
PQADTVIVHH   WDIMSNPKSR   LPPSPRPQGQ   RWIWFNLEPP   PNCQHLEALD   150
RYFNLTMSYR   SDSDIFTPYG   WLEPWSGQPA   HPPLNLSAKT   ELVAWAVSNW   200
KPDSARVRYY   QSLQAHLKVD   VYGRSHKPLP   KGTMMETLSR   YKFYLAFENS   250
LHPDYITEKL   WRNALEAWAV   PVVLGPSRSN   YERFLPPDAF   IHVDDFQSPK   300
DLARYLQELD   KDHARYLSYF   RWRETLRPRS   FSWALDFCKA   CWKLQQESRY   350
QTVRSIAAWF   T                                                  361
```

Carrier molecule[2]

Lewis antigens are not the primary gene product.

Lewis antigens are not intrinsic to RBCs but are located on type 1 glycosphingolipids that are adsorbed onto blood cells from the plasma. The biosynthesis of Lewis antigens results from the interaction of two independent loci *FUT3* (*LE*) and *FUT2* (*SE*). See Section III.

```
Gal
  | β1–3
GlcNAc
  |
 Gal
  |
 Glc
  |
 Cer
```

Function

Sialylated forms of Le^a and Le^b may serve as ligands for E-selectins

Disease association

Lewis antigens may be lost from RBCs as a result of infectious mononucleosis complicated with hemolysis, severe alcoholic cirrhosis, and alcoholic pancreatitis.

Patients with fucosidosis may have increased expression of Lewis antigens in their saliva and on their RBCs.

Glycoconjugates with Le^b activity mediate attachment of *Helicobacter pylori*, a major causative agent of gastric ulcers, to gastric mucosal epithelium.

RBCs from patients with leukocyte adhesion deficiency (LADII) syndrome are Le(a−b−), and are Bombay phenotype, due to a mutation in the GDP-fucose transporter[3,4].

Severity of SCD and risk of ischemic heart disease may be increased in patients with the Le(a−b−) phenotype[5].

Renal graft survival is inferior in patients lacking Lewis antigens, suggesting that Lewis antibodies may play a role in graft rejection.

Phenotypes (% occurrence)

	Caucasians	Blacks	Japanese
Le(a+b−)	22	23	0.2
Le(a−b+)	72	55	73
Le(a+b+^{w})	Rare	Rare	16.8
Le(a−b−)	6	22–30	10

Null: Le(a−b−)
Unusual: Le(a+b+^{w}) (sej/sej) is in 16.8% of Japanese
Le(a+b+), rare in European populations, is found in some Asian populations (e.g. Australian Aborigines, Chinese in Taiwan, Japanese, and Polynesians) with an incidence of 10–40%. The phenotype is the result of a mutation in the FUT2 (SE) gene: 385A>T; Ile129Phe. (See **H [018]** chapter.)

Le(a−b−) RBC phenotypes due to FUT3 mutations

Information taken from http://www.bioc.aecom.yu.edu/bgmut/index.htm.

cDNA change	Amino acid change	Origin
59T>G	Leu20Arg	Indonesia, Japan, Sweden, Denmark
59T>G; 445C>A	Leu20Arg; Leu149Met	Denmark, Sweden
59T>G; 508G>A	Leu20Arg; Gly170Ser	Japan, Denmark
59T>G; 1067T>A	Leu20Arg; Ile356Lys	Indonesia, Japan, Denmark
202T>C; 314C>T	Trp68Arg; Thr105Met	Denmark, Sweden
202T>C; 314C>T; 484G>A	Trp68Arg; Thr105Met; Asp162Asn	S. Africa, Caucasian
202T>C; 1067T>A	Trp68Arg; Ile356Lys	Denmark
484G>A; 667G>A	Asp162Asn; Gly227Arg	Africa (Xhosa)
484G>A; 667G>A; 808G>A	Asp162Asn; Gly227Arg; Val270Met	Africa (Xhosa)
760G>A	Asp254Asn	Indonesia
1067T>A	Ile356Lys	Indonesia, Denmark

Le(a−b−) RBC phenotypes due to other mutations

For mutations in FUT2 (SE) and the gene encoding the GDP-fucose transporter see **H [018]** system. In addition, 95% of FUT6-deficient people have Le(a−b−) RBCs. 9% of people in Java do not express FUT6. For mutations in FUT6 that cause the Le(a−b−) phenotype see http://www.bioc.aecom.yu.edu/bgmut/index.htm.

267

Comments

Saliva is a good source of soluble Lewis antigens and should be made isotonic before use in hemagglutination tests.

During pregnancy, expression of Lewis antigens on RBCs is often greatly reduced[6].

Le^x (SSEA-1; CD15) and Le^y, products of *FUT3* on type 2 precursor chains, are not associated with the RBC surface and are not part of the Lewis blood group system[7]. Le^x and Le^y are isomers of, respectively, Le^a and Le^b and often occur in sialylated forms. Sialyl Le^x is a major neutrophil ligand for E-selectin[8]. Le^a, Le^x, their sialyl derivates, and also Le^b and Le^y, accumulate in tumor tissues. Evidence indicates that adhesion of tumor cells to endothelial cells is mediated between the sialylated Le^a and Le^x antigens and E-selectin and represents an important factor in hematogenous metastasis of tumor cells.

Anti-Le^x and anti-Le^y reagents are murine monoclonal antibodies made in response to immunization with various tumor cells.

Le^x

Gal
| β1–4
GlcNAc $\xrightarrow{\alpha1-3}$ Fuc
|
Gal
|
Glc
|
Cer

Sialyl Le^x

Gal $\xrightarrow{\alpha2-3}$ NeuAc
| β1–4
GlcNAc $\xrightarrow{\alpha1-3}$ Fuc
|
Gal
|
Glc
|
Cer

References

1 Kukowska-Latallo, J.F. et al. (1990) Genes Dev. 4, 1288–1303.
2 Hauser, R. (1995) Transfusion 35, 577–581.
3 Hirschberg, C.B. (2001) J. Clin. Invest. 108, 3–6.
4 Luhn, K. et al. (2001) Nat. Genet. 28, 69–72.
5 Fisher, T.C. et al. (2002) Blood 100, 449a–450a.
6 Henry, S. et al. (1996) Vox Sang. 70, 21–25.
7 Lowe, J.B. (1995) In: Molecular Basis of Human Blood Group Antigens (Cartron, J.-P. and Rouger, P. eds) Plenum Press, New York, pp. 75–115.
8 Walz, G. et al. (1990) Science 250, 1132–1135.

Le^a ANTIGEN

Terminology

ISBT symbol (number)	LE1 (007.001)
History	Identified in 1946; named Lewis after one of the two original producers of anti-Le^a

Occurrence

Caucasians	22%
Blacks	23%

Expression

Cord RBCs	Not expressed; although some cord RBCs will react with anti-Le^a by IAT
Altered	Weak in Le(a+b+); often weakened during pregnancy

Molecular basis associated with Le^a antigen[1]

```
Gal
 | β1–3
GlcNAc  α1–4  Fuc
 |
Gal
 |
Glc
 |
Cer
```

Effect of enzymes/chemicals on Le^a antigen on intact RBCs

Ficin/papain	Resistant (↑↑)
Trypsin	Resistant (↑↑)
α-Chymotrypsin	Resistant (↑↑)
Pronase	Resistant (↑↑)
Sialidase	Resistant
DTT 200 mM	Resistant
Acid	Resistant

In vitro characteristics of alloanti-Le^a

Immunoglobulin class	IgM more frequent than IgG
Optimal technique	RT; 37 °C; IAT; enzymes
Neutralization	Saliva from secretors
Complement binding	Yes; some hemolytic

Clinical significance of alloanti-Le^a

Transfusion reaction	Hemolytic (rare)
HDN	No (1 mild case)

Comments

Anti-Le^a (in conjunction with anti-Le^b [see LE2]) is a frequent naturally occurring antibody made by Le(a−b−) people, especially in pregnancy.

Reference
1 Henry, S. et al. (1995) Vox Sang. 69, 166–182.

Le^b ANTIGEN

Terminology

ISBT symbol (number)	LE2 (007.002)
History	Anti-Le^b was identified in 1948; initially appeared to detect an antigen antithetical to Le^a

Occurrence

Caucasians	72%
Blacks	55%

Expression

Cord RBCs	Not expressed
Altered	Weak in Le(a+b+); often weakened during pregnancy

Molecular basis associated with Le^b antigen[1]

Gal $\underline{\alpha1-2}$ Fuc
| β1–3
GlcNAc $\underline{\alpha1-4}$ Fuc
|
Gal
|
Glc
|
Cer

Effect of enzymes/chemicals on Le^b antigen on intact RBCs

Ficin/papain	Resistant (↑↑)
Trypsin	Resistant (↑↑)
α-Chymotrypsin	Resistant (↑↑)
Pronase	Resistant (↑↑)
Sialidase	Resistant
DTT 200 mM	Resistant
Acid	Resistant

In vitro characteristics of alloanti-Le^b

Immunoglobulin class	IgM more frequent than IgG
Optimal technique	RT; 37°C; IAT; enzymes
Neutralization	Saliva from secretors
Complement binding	Yes; some hemolytic

Clinical significance of alloanti-Le^b

Transfusion reaction	No
HDN	No (1 mild case)

Comments

Le^b is the receptor for *Helicobacter pylori* in gastric mucosal epithelium.

Anti-Le^b (in conjunction with anti-Le^a [see LE1]) is a frequent naturally occurring antibody made by Le(a−b−) people. There are two kinds of anti-Le^b: anti-Le^{bH} (LE4), reacting with group O and A$_2$ Le(b+) RBCs, and anti-Le^{bL}, reacting with all Le(b+) RBCs. Other antibodies react specifically with the compound antigens, for example, ALe^b (LE5) and BLe^b (LE6).

Reference
[1]   Henry, S. et al. (1995) Vox Sang. 69, 166–182.

Le^{ab} ANTIGEN

Terminology

ISBT symbol (number)	LE3 (007.003)
Other names	X; Le^x; Le^{abx}
History	Described in 1949 as the antigen reacting with anti-X; referred to as Le^x from the mid-1950s; formally assigned to Lewis and renamed Le^{ab} by ISBT in 1998

Occurrence

All populations: on Le(a+b−) and Le(a−b+) RBCs from adults and on 90% of cord samples.

Expression

Cord RBCs	Expressed

Molecular basis associated with Le^{ab} antigen[1]

The binding site for anti-Le^{ab} comprises the disaccharide Fucα1>4GlcNAc>R which is within the Le^a and Le^b active structures.

```
        Leᵃ                              Leᵇ

       Gal                             Gal  α1–2  Fuc
        |β1–3                           |β1–3
     ┌GlcNAc α1–4 Fuc┐              ┌GlcNAc α1–4 Fuc┐
        |      ↑                        |      ↑
       Gal    Leᵃᵇ                     Gal    Leᵃᵇ
        |                               |
       Glc                             Glc
        |                               |
       Cer                             Cer
```

In vitro characteristics of alloanti-Le^{ab}

Immunoglobulin class	Presumed IgM
Optimal technique	Presumed RT

Comments

The reactive cord samples [serologically Le(a−b−)] are from babies with a *FUT3* (*LE*) gene. It is possible that a distinct Le^{ab} determinant (with only weak affinity for anti-Le^a) is formed early in embryonic development. The reactivity with Le(a+b−), Le(a−b+) and cord RBCs cannot be separated.

Anti-Le^{ab} is a fairly common specificity and is frequently found with anti-Le^a and/or anti-Le^b; it occurs mainly in serum from Le(a−b−) secretors of blood group A$_1$, B, or A$_1$B. Anti-Le^{ab} is inhibited by saliva that contains Le^a, and is weakly inhibited by saliva that contains Le^b. Saliva from Le(a−) non-secretors may also have a very weak inhibitory effect.

Reference
1   Schenkel-Brunner, H. (2000) Human Blood Groups: Chemical and Biochemical Basis of Antigen Specificity, 2nd Edition, Springer-Verlag Wien, New York.

Le^{bH} ANTIGEN

Terminology

ISBT symbol (number)	LE4 (007.004)
History	Antigen detected by the original anti-Le^b in 1948; named Le^{bH} in 1959 upon recognition of heterogeneity of anti-Le^b; allocated ISBT number in 1998

Occurrence

Present on group O and A$_2$ Le(b+) RBCs, that is, those with strong expression of H antigen. Group A$_1$ or B RBCs react weakly or not at all.

Expression

Cord RBCs	Not expressed
Altered	Weak in Le(a+b+), often weakened during pregnancy

Molecular basis associated with Le^{bH} antigen[1]

Not fully understood. Anti-Le^{bH} appears to react with type 1 H (see H [018] system) on group O RBCs, where the structure is not blocked by the immunodominant blood group A or B sugars. The determinant must also involve the L-fucose added by the FUT3 (LE) specified transferase because Le(a−b−) RBCs from secretors (which carry type 1 H) do not react with anti-Le^{bH}.

Effect of enzymes/chemicals on Le^{bH} antigen on intact RBCs

Refer to Le^b antigen (**LE2**).

In vitro characteristics of alloanti-Le^{bH}

Immunoglobulin class   IgM predominates
Optimal technique   RT; 37°C; enzymes
Neutralization   Saliva (inhibited by saliva that contains H, or H and Le^b)
Complement binding   Yes; some hemolytic

Clinical significance of alloanti-Le^{bH}

Transfusion reaction   No
HDN   No

Comments

Anti-Le^{bH} is a more common specificity that anti-Le^{bL}. Anti-Le^{bH} is unlikely to cause incompatible crossmatches if ABO identical blood is selected.

Reference
[1] Henry, S. et al. (1995) Vox Sang. 69, 166–182.

ALe^b ANTIGEN

Terminology

ISBT symbol (number)   LE5 (007.005)
Other name A_1Le^b
History   Anti-A_1Le^b identified in 1967 during crossmatching; name derived from the unusual antibody reactivity; received an ISBT number in 1998; name amended to ALe^b

Occurrence

On all A_1 Le(b+) and A_1B Le(b+) RBCs, that is, on RBCs of secretors of A who have a *FUT3* (*LE*) gene.

Expression

Cord RBCs   Not expressed

Molecular basis associated with ALe^b antigen[1]

```
GalNAc
  | β1–3
 Gal  α1–2  Fuc
  | β1–3
GlcNAc  α1–4  Fuc
  |
 Gal
  |
 Glc
  |
 Cer
```

ALe^b is expressed when Le^b is modified by the addition of A-gene specified N-Acetyl-D-galactosamine

Effect of enzymes and chemicals on ALe^b antigen on intact RBCs

Refer to Le^b (**LE2**).

In vitro characteristics of alloanti-ALe^b

Refer to Le^b (**LE2**).

Clinical significance of alloanti-ALe^b

Few examples of anti-ALe^b reported; may be lymphocytotoxic.

Comments

ALe^b is often referred to as A$_1$Le^b because the transferase produced by the A_1 gene is more efficient than the A$_2$ transferase at adding sufficient quantities of N-Acetyl-D-galactosamine for detection by anti-ALe^b. ALe^b can also be adsorbed onto lymphocytes.

Anti-ALe^b is a single specificity that cannot be separated into anti-A$_1$ and anti-Le^b. Monoclonal anti-ALe^b have been produced; some cross-react with the isomer ALe^y.

Reference
[1] Henry, S. et al. (1995) Vox Sang. 69, 166–182.

BLe^b ANTIGEN

Terminology

ISBT symbol (number)	LE6 (007.006)
History	Allocated a Lewis system number by ISBT in 1998

Occurrence

On all B Le(b+ and A$_1$B Le(b+) RBCs, that is, on RBCs of secretors of B who have a *FUT3* (*LE*).

Expression

Cord RBCs	Not expressed

Molecular basis associated with BLe^b antigen[1]

```
Gal
 │ α1–3
Gal  α1–2  Fuc
 │ β1–3
GlcNAc  α1–4  Fuc
 │
Gal
 │
Glc
 │
Cer
```

BLe^b is expressed when Le^b is modified by the addition of *B*-gene specified D-galactose

Effect of enzymes/chemicals on BLe^b antigen on intact RBCs

Refer to Le^b antigen (**LE2**).

In vitro characteristics of alloanti-BLe^b

Refer to Le^b (**LE2**).

Clinical significance of alloanti-BLe^b

Few examples of anti-BLe^b reported; unlikely to complicate transfusion or HDN; may be lymphocytotoxic.

Comments

BLe^b can be adsorbed onto lymphocytes. Anti-BLe^b is a single specificity; it cannot be separated into anti-B and anti-Le^b.

Reference
[1] Henry, S. et al. (1995) Vox Sang. 69, 166–182.

Number of antigens

6

Terminology

ISBT symbol	FY
ISBT number	008
CD number	CD234
History	Named after the family of the first proband who made anti-Fy^a

Expression

Soluble form	No
Other blood cells	Not on granulocytes, lymphocytes, monocytes, platelets
Tissues	Brain, colon, endothelium, lung, spleen, thyroid, thymus, kidney. Not on liver, placenta

Gene

Chromosome	1q22–q23
Name	*FY*
Organization	2 exons distributed over 1.521 kbp of gDNA
Product	Major (β) Duffy glycoprotein
	Minor (α) Duffy glycoprotein

Gene map

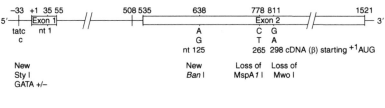

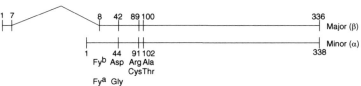

Database accession numbers

GenBank   U01839

http://www.bioc.aecom.yu.edu/bgmut/index.htm.

Amino acid sequence of Fy^b[1,2]

Major product:

```
MGNCLHRAEL SPSTENSSQL DFEDVWNSSY GVNDSFPDGD YDANLEAAAP  50
CHSCNLLDDS ALPFFILTSV LGILASSTVL FMLFRPLFRW QLCPGWPVLA 100
QLAVGSALFS IVVPVLAPGL GSTRSSALCS LGYCVWYGSA FAQALLLGCH 150
ASLGHRLGAG QVPGLTLGLT VGIWGVAALL TLPVTLASGA SGGLCTLIYS 200
TELKALQATH TVACLAIFVL LPLGLFGAKG LKKALGMGPG PWMNILWAWF 250
IFWWPHGVVL GLDFLVRSKL LLLSTCLAQQ ALDLLLNLAE ALAILHCVAT 300
PLLLALFCHQ ATRTLLPSLP LPEGWSSHLD TLGSKS                336
```

Antigen mutation is numbered by counting Met as 1.

The minor product of 338 amino acids has MASSGYVLQ in place of the first seven amino acids (MGNCLHR) at the N-terminus.

Carrier molecule

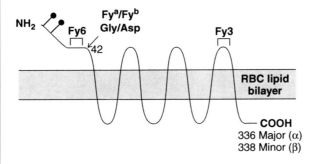

M_r (SDS-PAGE)	35 000–45 000
CHO: N-glycan	Two potential sites in N-terminal domain
CHO: O-glycan	No sites
Copies per RBC	6000–13 000

Molecular basis of Fy antigens

Antigen	Amino acid change	Exon	Nt change	Restriction enzyme
Fy^a/Fy^b	Gly42Asp	2	125G>A	Ban I (+/−)

Function

Receptor for both C-X-C (acute inflammation) and C-C (chronic inflammation) proinflammatory chemokines: IL-8 (interleukin-8), MGSA (melanoma growth stimulatory activity); MCP-1 (monocyte chemotactic protein 1); RANTES (regulated on activation, normal T-expressed and secreted). Clears proinflammatory peptides[3,4]. Fy(a−b−) RBCs do not bind chemokines and RBCs with suppressed Fy^b antigen expression encoded by Fy^x bind reduced levels of chemokines.

Disease association

Receptor for *Plasmodium vivax, P. knowlesi* malarial parasites: Fy(a−b−) RBCs resist invasion.

Phenotypes (% occurrence)[5]

	Caucasian	Blacks	Chinese	Japanese	Thai
Fy(a+b−)	17	9	90.8	81.5	69
Fy(a−b+)	34	22	0.3	0.9	3
Fy(a+b+)	49	1	8.9	17.6	28
Fy(a−b−)	Very rare	68	0	0	0

25% of Israeli Arabs and 4% of Israeli Jews have Fy(a−b−) RBCs.
Null: Fy(a−b−).
Unusual: Fy^x haplotype expresses weak Fy^b antigen not detected by all anti-Fy^b.

Molecular basis of Fy(a−b−) phenotypes

−33T>C in GATA-1 binding motif of *FYB*[6]
−33T>C in GATA-1 binding motif of *FYA*[7]
287G>A in exon 2 of *FYA*; Trp96Stop[8]
407G>A in exon 2 of *FYB* Trp136Stop[8]
408G>A in exon 2 of *FYB* Trp136Stop[8]

Molecular basis of Fy^x [Fy(b+^{w})] phenotype

265C>T; Arg89Cys (invariably with 298G>A; Ala100Thr)[1].

Comments

Fy3 [FY3] may be an earlier evolutionary development than Fy^b [FY2], while Fy^a may have arisen during human ontogeny.

Fy(a−b−) is present in Arabs, Jews, Brazilians and Gypsies.
The mutation 298G>A; Ala100Thr occurs in Whites without 265C>T and does not reduce expression of Fy^b.

References
1 Pogo, A.O. and Chaudhuri, A. (2000) Semin. Hematol. 37, 122–129.
2 Iwamoto, S. et al. (1996) Blood 87, 378–385.
3 Darbonne, W.C. et al. (1991) J. Clin. Invest. 88, 1362–1369.
4 Neote, K. et al. (1994) Blood 84, 44–52.
5 Pierce, S.R. and Macpherson, C.R. eds. (1988) Blood Group Systems: Duffy, Kidd and Lutheran, American Association of Blood Banks, Arlington, VA.
6 Tournamille, C. et al. (1995) Nat. Genet. 10, 224–228.
7 Zimmerman, P.A. et al. (1999) Proc. Natl. Acad. Sci. USA 96, 13973–13977.
8 Rios, M. et al. (2000) Br. J. Haematol. 108, 448–454.

Fy^a ANTIGEN

Terminology

ISBT symbol (number)	FY1 (008.001)
History	Antibody identified in 1950 in the serum of Mr. Duffy, who was a transfused hemophiliac. The last two letters of his name were used for the antigen name

Occurrence

Caucasians	66%
Blacks	10%
Asians	99%
Thai	97%

Antithetical antigen

Fy^b (**FY2**)

Expression

Cord RBCs	Expressed

Molecular basis associated with Fyª antigen[1-4]

Amino acid   Gly at residue 42 (44 in minor isoform)
Nucleotide   G at bp 125 in exon 2
Restriction enzyme   *Ban* I site present

Effect of enzymes/chemicals on Fyª antigen on intact RBCs

Ficin/papain   Sensitive
Trypsin   Resistant
α-Chymotrypsin   Sensitive
Pronase   Sensitive
Sialidase   Resistant
DTT 200 mM   Resistant
Acid   Resistant

In vitro characteristics of alloanti-Fyª

Immunoglobulin class   IgG; IgM rarely
Optimal technique   IAT
Complement binding   Rare

Clinical significance of alloanti-Fyª

Transfusion reaction   Mild to severe; immediate/delayed
HDN   Mild to severe (rare)

Autoanti-Fyª

Autoantibodies mimicking alloantibodies have been reported[5].

Comments

Fyª has been demonstrated on fetal RBCs as early as 6 weeks gestation. Adult
    level of Fyª expression attained approximately 12 weeks after birth.
Fyª antigen found on RBCs of baboon, but not chimpanzee, gorilla, gibbon,
    rhesus, cynomolgus, squirrel monkey, capuchin, douroncoli.

References
1   Chaudhuri, A. et al. (1995) Blood 85, 615–621.
2   Mallinson, G. et al. (1995) Br. J. Haematol. 90, 823–829.
3   Iwamoto, S. et al. (1996) Blood 87, 378–385.
4   Tournamille, C. et al. (1995) Hum. Genet. 95, 407–410.
5   Harris, T. (1990) Immunohematology 6, 87–91.

Fy^b ANTIGEN

Terminology

ISBT symbol (number)	FY2 (008.002)
History	Named in 1951 when the antigen was shown to be antithetical to Fy^a

Occurrence

Caucasians	83%
Blacks	23%
Chinese	9.2%
Asians	18.5%
Thai	31%

Antithetical antigen

Fy^a (**FY1**)

Expression

Cord RBCs	Expressed
Altered	Weak on Fy^x

Molecular basis associated with Fy^b antigen[1-4]

Amino acid	Asp at residue 42 (44 in minor isoform)
Nucleotide	A at bp 125 in exon 2
Restriction enzyme	*Ban* I site absent

Effect of enzymes/chemicals on Fy^b antigen on intact RBCs

Ficin/papain	Sensitive
Trypsin	Resistant (weakened)
α-Chymotrypsin	Sensitive
Pronase	Sensitive
Sialidase	Resistant
DTT 200 mM	Resistant
Acid	Resistant

In vitro characteristics of alloanti-Fy^b

Immunoglobulin class   IgM rare; most IgG
Optimal technique   IAT
Complement binding   Rare

Clinical significance of alloanti-Fy^b

Transfusion reaction   Mild to severe; immediate rare; delayed
HDN   Mild (rare)

Autoanti-Fy^b

Several examples of autoanti-Fy^b mimicking alloantibody reported; one caused AIHA.

Comments

Poor immunogen.
Black individuals who have Fy(a−b−) RBCs invariably possess the *Fy^b* gene.
Fy^b found on RBCs from chimpanzee, gorilla, gibbon, rhesus, cynomolgus, baboon; not found on RBCs from squirrel monkey, capuchin, dourocoli.

References

1   Chaudhuri, A. et al. (1995) Blood 85, 615–621.
2   Mallinson, G. et al. (1995) Br. J. Haematol. 90, 823–829.
3   Iwamoto, S. et al. (1996) Blood 87, 378–385.
4   Tournamille, C. et al. (1995) Hum. Genet. 95, 407–410.

Fy3 ANTIGEN

Terminology

ISBT symbol (number)   FY3 (008.003)
Other names   Fy^{ab}; Fy^{a}Fy^b
History   Anti-Fy3, found in 1971, was made by a pregnant Australian Fy(a−b−) woman who had been transfused. The specificity was named anti-Fy3 (and not anti-Fy^{ab}) because the antigenic determinant was resistant to enzyme treatment

Occurrence

Caucasians	100% [Fy(a−b−) found in 4 Caucasians and one Cree Indian]
Blacks	32%
Asians	99.9%
Yeminite Jews	99%
Israeli Jews	96%
Israeli Arabs	75%

Expression

Cord RBCs	Expressed; increases after birth
Altered	Weak on Fy^x

Molecular basis associated with Fy3 antigen

Not known. See system pages for molecular basis associated with an absence of Fy3. The third extracellular loop contains sequences necessary for binding of monoclonal anti-Fy3.

Effect of enzymes/chemicals on Fy3 antigen on intact RBCs

Ficin/papain	Resistant
Trypsin	Resistant
α-Chymotrypsin	Resistant
Pronase	Resistant
Sialidase	Resistant
DTT 200 mM	Resistant
Acid	Resistant

In vitro characteristics of alloanti-Fy3

Immunoglobulin class	IgG
Optimal technique for detection	IAT, enzymes
Complement binding	Rare

Clinical significance of alloanti-Fy3

Transfusion reaction	Mild to moderate; delayed
HDN	Mild (rare)

Comments

The anti-Fy3 made by three non-Black women reacted strongly with cord RBCs whereas the anti-Fy3 made by Black people does not react or reacts very weakly with cord RBCs.

Formation of anti-Fy3 is usually preceded by formation of anti-Fy^a (see FY1). In spite of the high percentage of the Fy:-3 phenotype among Blacks, anti-Fy3 is a rare specificity. To date, no Black Fy(a$-$b$-$) individual has made anti-Fy^b (see FY2), which is probably due to the presence of an Fy^b gene that is silent only in the erythroid lineage.

Anti-Fy3 agglutinates Rh$_{null}$ RBCs while anti-Fy5 does not (see FY5).

Fy:-3 [Fy(a$-$b$-$)] RBCs resist invasion by *P. vivax* and *P. knowlesi* malarial parasites.

Fy3 is expressed on RBCs from chimpanzees, gorilla, gibbon, rhesus, cynomolgus, baboon, dourocoli.

Fy4 ANTIGEN

Terminology

ISBT symbol (number)	FY4 (008.004)
Other name	Fy^c
History	Reported in 1973 and given the next Fy number. The only example of anti-Fy4 reacted with RBCs of Blacks with the Fy(a+b$-$), Fy(a$-$b+) and Fy(a$-$b$-$) phenotypes but mostly not with those of the Fy(a+b+) phenotype nor with RBCs from many whites, regardless of Fy phenotype. The reactivity was enhanced with papain treated RBCs. Inconsistent results in tests performed in different laboratories and the instability of the antibody on storage throws doubts upon the existence of Fy4

Fy5 ANTIGEN

Terminology

ISBT symbol (number)	FY5 (008.005)
History	Reported in 1973 and given the next Fy number when the antibody was shown to detect a novel antigen. Antibody was made by a Black Fy(a$-$b$-$) Fy:-3 boy with leukemia

Occurrence

Blacks	32%
Most populations	99.9%

Expression

Cord RBCs	Expressed
Altered	Weak on Fy^x and D-- RBCs; absent from Rh$_{null}$ RBCs

Molecular basis associated with Fy5 antigen

Not known; possible interaction between Duffy and Rh proteins[1].

Effect of enzymes/chemicals on Fy5 antigen on intact RBCs

Ficin/papain	Resistant
Trypsin	Not tested
α-Chymotrypsin	Not tested
Pronase	Not tested
Sialidase	Resistant
DTT 200 mM	Resistant
Acid	Presumed resistant

In vitro characteristics of alloanti-Fy5

Immunoglobulin class	IgG
Optimal technique	IAT
Complement binding	Not known

Clinical significance of alloanti-Fy5

Transfusion reaction	Mild, delayed in one case
HDN	No data

Comments

Several examples of anti-Fy5 have been found. All are in Black, multiply transfused (mostly because of sickle cell disease) Fy(a−b−) patients.

Fy(a−b−) RBCs from Black individuals are FY:−3,−5.

Fy(a−b−) RBCs from 1 Caucasian (AZ; AKA Findlay) are FY:−3,5.

287

Rh_{null} RBCs are FY:3,−5.

Expressed on chimpanzee and gorilla RBCs, not on rhesus and cynomolgus RBCs.

Reference
[1] Colledge, K.I. et al. (1973) Vox Sang. 24, 193–199.

Fy6 ANTIGEN

Terminology

ISBT symbol (number)	FY6 (008.006)
History	Reported in 1987. This antigen, although numbered by the ISBT, has only been defined by murine monoclonal antibodies. No human anti-Fy6 has been described

Occurrence

Blacks	32%
Most populations	100%

Expression

Cord RBCs	Expressed
Altered	Weak on Fy^x

Molecular basis associated with Fy6 antigen

Anti-Fy6 bind epitopes within amino acid residues [19]Gln-Leu-Asp-Phe-Glu-Asp-Val-Trp[26] [1].

Effect of enzymes/chemicals on Fy6 antigen on intact RBCs

Ficin/papain	Sensitive
Trypsin	Resistant
α-Chymotrypsin	Sensitive
Pronase	Sensitive

Sialidase   Resistant
DTT 200 mM   Resistant
Acid   Resistant

Comments

Amino acid residues 8(Ala) to 43(Asp) are critical for *Plasmodim vivax* binding[2].

Not on K562, Raja, MOLT4, HPBall (B-cell line), HL60, U937 (myelomono-cytic line).

Fy6 is expressed on RBCs from chimpanzee, gorilla, gibbon, squirrel monkey, dourocoli. Fy6 is not expressed on RBCs from rhesus, cynomolgus, baboon, capuchin.

References
[1] Wasniowska, K. et al. (2002) Transf. Med. 12, 205–211.
[2] Pogo, A.O. and Chaudhuri, A. (2000) Semin. Hematol. 37, 122–129.

Number of antigens 3

Terminology

ISBT symbol   JK
ISBT number   009
History   Named in 1951 after the family of the first proband to make anti-Jk^a

Expression

Soluble form   No
Other blood cells   Not on lymphocytes, granulocytes, monocytes or platelets
Tissues   Kidney

Gene

Chromosome   18q11-q12
Name   *JK (SLC14A1, HUTII)*
Organization   11 exons distributed over 30 kbp of gDNA. Exons 4–11 encode the mature protein
Product   Urea transporter

Gene map

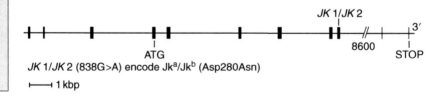

JK 1/JK 2 (838G>A) encode Jk^a/Jk^b (Asp280Asn)

Database accession numbers

GenBank   Y19039
http://www.bioc.aecom.yu.edu/bgmut/index.htm.

Amino acid sequence[1]

```
MEDSPTMVRV DSPTMVRGEN QVSPCQGRRC FPKALGYVTG DMKKLANQLK   50
DKPVVLQFID WILRGISQVV FVNNPVSGIL ILVGLLVQNP WWALTGWLGT  100
VVSTLMALLL SQDRSLIASG LYGYNATLVG VLMAVFSDKG DYFWWLLLPV  150
CAMSMTCPIF SSALNSMLSK WDLPVFTLPF NMALSMYLSA TGHYNPFFPA  200
KLVIPITTAP NISWSDLSAL ELLKSIPVGV GQIYGCDNPW TGGIFLGAIL  250
LSSPLMCLHA AIGSLLGIAA GLSLSAPFED IYFGLWGFNS SLACIAMGGM  300
FMALTWQTHL LALGCALFTA YLGVGMANFM AEVGLPACTW PFCLATLLFL  350
IMTTKNSNIY KMPLSKVTYP EENRIFYLQA KKRMVESPL              389
```

Antigen mutation is numbered counting Met as 1

Carrier molecule

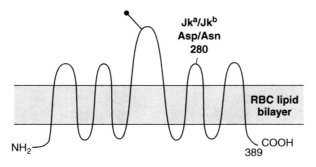

M_r (SDS-PAGE)	Predicted 43 000
CHO: N-glycan	Two potential sites
Cysteine residues	10
Copies per RBC	14 000

Molecular basis of antigens

Antigen	Amino acid change	Exon	Nt change	Restriction enzyme
Jk^a/Jk^b	Asp280Asn	9	838G > A	*MnlI* (+/−)

Function

Red cell urea transporter[2].

Disease association

Jk(a−b−) individuals are unable to maximally concentrate urine.

Phenotypes (% occurrence)

	Caucasians	Blacks	Asians
Jk(a+b−)	26.3	51.1	23.2
Jk(a−b+)	23.4	8.1	26.8
Jk(a+b+)	50.3	40.8	49.1
Jk(a−b−)	Rare	Rare	0.9 (Polynesians)

Null: Jk(a−b−).
Unusual: Jk(a−b−) [In(Jk)].

Molecular basis of Jk(a−b−) phenotypes[3-7]

Basis	Ethnicity
I5−1g>a; skips exon 6 (7,8,9)	(BS) French
I7+1g>t; skips exon 7 (6,8,9; 6,7,8,9)	(LP) Polynesians and Chinese
871T>C in exon 9; Ser293Pro	Finns
582C>G in exon 7; Tyr194Stop	Swiss
Deletion exons 4 and 5	
Deletion exons 4 and 5 in gDNA, and exons 3–5 in cDNA	English

Comments

Jk(a−b−) RBCs resist lysis by 2 M urea[8].
Dominant type Jk(a−b−) [In(Jk)] RBCs have been found in Japanese.
Two transient Jk(a−b−) donors have been described[9,10]. One was a Russian woman with myleofibrosis who made anti-Jk3 while her RBCs typed Jk(a−b−).

References
1 Lucien, N. et al. (2002) J. Biol. Chem. 227, 34101–34108.
2 Sands, J.M. (2002) J. Am. Soc. Nephrol. 13, 2795–2806.
3 Irshaid, N.M. et al. (2000) Transfusion 40, 69–74.
4 Lucien, N. et al. (1998) J. Biol. Chem. 273, 12973–12980.
5 Sidoux-Walter, F. et al. (2000) Blood 96, 1566–1573.
6 Irshaid, N.M. et al. (2002) Br. J. Haematol. 116, 445–453.
7 Lucien, N. et al. (2002) Blood 99, 1079–1081.
8 Mougey, R. (1990) Immunohematology 6, 1–8.
9 Issitt, P.D. et al. (1990) Transfusion 30, 46–50.
10 Obarski, G. et al. (1987) Transfusion 27, 548 (abstract).

Jk^a ANTIGEN

Terminology

ISBT symbol (number)	JK1 (009.001)
History	Reported in 1951. Name derived from the initials of the sixth child of the antibody maker, Mrs. Kidd

Occurrence

Caucasians	77%
Blacks	92%
Asians	73%

Antithetical antigen

Jk^b (**JK2**)

Expression

Cord RBCs	Expressed

Molecular basis associated with Jk^a antigen

Amino acid	Asp at residue 280
Nucleotide	G at bp 838 in exon 9
Restriction enzyme	*Mnl*I site present

Effect of enzymes/chemicals on Jk^a antigen on intact RBCs

Ficin/papain	Resistant (↑↑)
Trypsin	Resistant (↑↑)
α-Chymotrypsin	Resistant (↑↑)
Pronase	Resistant (↑↑)
Sialidase	Resistant
DTT 200 mM	Resistant
Acid	Resistant

In vitro characteristics of alloanti-Jk^a

Immunoglobulin class   IgG; many IgG plus IgM; IgM
Optimal technique   IAT; enzymes; PEG
Complement binding   Yes, provided IgM is present; some hemolytic[1]

Clinical significance of alloanti-Jk^a

Transfusion reaction   No to severe; immediate or delayed/ hemolytic
HDN   Mild to moderate (rare)

Autoanti-Jk^a

Yes.

Comments

Jk^a has been demonstrated on fetal RBCs as early as 11 weeks gestation.
Anti-Jk^a fades *in vitro* and *in vivo*. Often found in multi-specific sera.
Anti-Jk^a may react more strongly with Jk(a+b−) than Jk(a+b+) RBCs (i.e. show dosage).

Reference
[1] Yates, J. et al. (1998) Transf. Med. 8, 133–140.

Jk^b ANTIGEN

Terminology

ISBT symbol (number)   JK2 (009.002)
History   Found in 1953 and named for its antithetical relationship to Jk^a

Occurrence

Caucasians   74%
Blacks   49%
Asians   76%

Antithetical antigen

Jk^a (**JK1**)

Expression

Cord RBCs   Expressed

Molecular basis associated with Jk^b antigen

Amino acid   Asn at residue 280
Nucleotide   A at bp 838 in exon 9
Restriction enzyme   *Mnl*I site lost

Effect of enzymes/chemicals on Jk^b antigen on intact RBCs

Ficin/papain   Resistant (↑↑)
Trypsin   Resistant (↑↑)
α-Chymotrypsin   Resistant (↑↑)
Pronase   Resistant (↑↑)
Sialidase   Resistant
DTT 200 mM   Resistant
Acid   Resistant

In vitro characteristics of alloanti-Jk^b

Immunoglobulin class IgG; many IgG plus IgM; IgM
Optimal technique   IAT; enzymes; PEG
Complement binding   Yes, provided IgM is present; some hemolytic[1]

Clinical significance of alloanti-Jk^b

Transfusion reaction No to severe; immediate or delayed/hemolytic
HDN   No to mild (rare)

Autoanti-Jk^b

Yes.

Comments

Jk^b has been demonstrated on fetal RBCs as early as 7 weeks gestation. Anti-Jk^b fade *in vitro* and *in vivo*. Often found in multi-specific sera.

Reference
1   Yates, J. et al. (1998) Transf. Med 8, 133–140.

Jk3 ANTIGEN

Terminology

ISBT symbol (number)	JK3 (009.003)
Other names	Jk^{ab}; Jk^{a}Jk^b
History	Anti-Jk^{a}Jk^b was identified in 1959 and renamed anti-Jk3 by ISBT when numbers became popular

Occurrence

Most populations	100%
Polynesians, Finns	>99%

Expression

Cord RBCs	Expressed
Altered	Weak on Jk(a−b−) of the *In(Jk)* type (detected by absorption/elution)

Molecular basis associated with Jk3 antigen

Unknown. See Jk system pages for molecular basis of Jk(a−b−) phenotype.

Effect of enzymes/chemicals on Jk3 antigen on intact RBCs

Ficin/papain	Resistant (↑↑)
Trypsin	Resistant (↑↑)
α-Chymotrypsin	Resistant (↑↑)
Pronase	Resistant (↑↑)

Sialidase	Resistant
DTT 200 mM	Resistant
Acid	Resistant

In vitro characteristics of alloanti-Jk3

Immunoglobulin class	IgG more common than IgM
Optimal technique	IAT; PEG
Complement binding	Yes; some hemolytic

Clinical significance of alloanti-Jk3

Transfusion reaction	No to severe/immediate or delayed
HDN	No to mild

Autoanti-Jk3

Rare.

Comments

Anti-Jk3 has been found in a non-transfused male.

People with *In(Jk)* do not make anti-Jk3 and the molecular basis of this phenotype is unknown.

Number of antigens   20 (plus 1 provisionally assigned)

Terminology

ISBT symbol	DI
ISBT number	010
CD number	CD233
History	Named after the producer of the first anti-Di^a, discovered during the investigation of a case of HDN in a Venezuelan family. Diego was described in 1955 by Layrisse et al.; it had been mentioned briefly by Levine et al. in 1954

Expression

Soluble form	No
Tissues	Intercalated cells of the distal and collecting tubules of the kidney. An isoform of AE1 is expressed in the distal nephron of the kidney

Gene

Chromosome	17q21-q22
Name	DI (SLC4A1, AE1, EPB3)
Organization	20 exons distributed over 20 kbp of gDNA
Product	Band 3 (Anion exchanger 1; anion transport protein)

Gene map

DI 1/DI 2 (2561T→C) encode Di^a/Di^b (Leu854Pro)

DI 3/DI 4 (1972A→G) encode Wr^a/Wr^b (Lys658Glu)

For other mutations, see table.

├─────┤ 1 kbp

Database accession numbers

GenBank   X77738; M27819

http://www.bioc.aecom.yu.edu/bgmut/index.htm

Amino acid sequence

```
MEELQDDYED  MMEENLEQEE  YEDPDIPESQ  MEEPAAHDTE  ATATDYHTTS   50
HPGTHKVYVE  LQELVMDEKN  QELRWMEAAR  WVQLEENLGE  NGAWGRPHLS  100
HLTFWSLLEL  RRVFTKGTVL  LDLQETSLAG  VANQLLDRFI  FEDQIRPQDR  150
EELLRALLLK  HSHAGELEAL  GGVKPAVLTR  SGDPSQPLLP  QHSSLETQLF  200
CEQGDGGTEG  HSPSGILEKI  PPDSEATLVL  VGRADFLEQP  VLGFVRLQEA  250
AELEAVELPV  PIRFLFVLLG  PEAPHIDYTQ  LGRAAATLMS  ERVFRIDAYM  300
AQSRGELLHS  LEGFLDCSLV  LPPTDAPSEQ  ALLSLVPVQR  ELLRRRYQSS  350
PAKPDSSFYK  GLDLNGGPDD  PLQQTGQLFG  GLVRDIRRRY  PYYLSDITDA  400
FSPQVLAAVI  FIYFAALSPA  ITFGGLLGEK  TRNQMGVSEL  LISTAVQGIL  450
FALLGAQPLL  VVGFSGPLLV  FEEAFFSFCE  TNGLEYIVGR  VWIGFWLILL  500
VVLVVAFEGS  FLVRFISRYT  QEIFSFLISL  IFIYETFSKL  IKIFQDHPLQ  550
KTYNYNVLMV  PKPQGPLPNT  ALLSLVLMAG  TFFFAMMLRK  FKNSSYFPGK  600
LRRVIGDFGV  PISILIMVLV  DFFIQDTYTQ  KLSVPDGFKV  SNSSARGWVI  650
HPLGLRSEFP  IWMMFASALP  ALLVFILIFL  ESQITTLIVS  KPERKMVKGS  700
GFHLDLLLVV  GMGGVAALFG  MPWLSATTVR  SVTHANALTV  MGKASTPGAA  750
AQIQEVKEQR  ISGLLVAVLV  GLSILMEPIL  SRIPLAVLFG  IFLYMGVTSL  800
SGIQLFDRIL  LLFKPPKYHP  DVPYVKRVKT  WRMHLFTGIQ  IICLAVLWVV  850
KSTPASLALP  FVLILTVPLR  RVLLPLIFRN  VELQCLDADD  AKATFDEEEG  900
RDEYDEVAMP  V                                               911
```

Antigen mutations are numbered by counting Met as 1.

Carrier molecule[1,2]

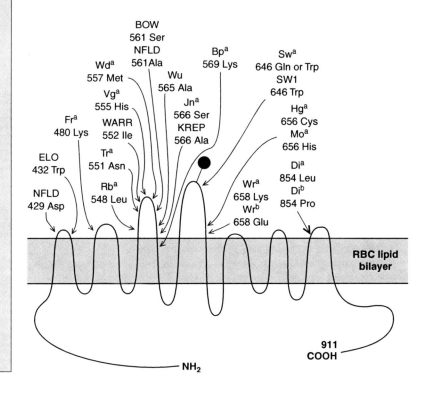

BOW
561 Ser
NFLD
561 Ala
Wd^a
557 Met
Vg^a
555 His
Fr^a
480 Lys
WARR
552 Ile
ELO
432 Trp
Tr^a
551 Asn
NFLD
429 Asp
Rb^a
548 Leu
Wu
565 Ala
Jn^a
566 Ser
KREP
566 Ala
Bp^a
569 Lys
Sw^a
646 Gln or Trp
SW1
646 Trp
Hg^a
656 Cys
Mo^a
656 His
Di^a
854 Leu
Di^b
854 Pro
Wr^a
658 Lys
Wr^b
658 Glu

RBC lipid bilayer

NH₂

911
COOH

M_r (SDS-PAGE)   95 000–105 000
CHO: N-glycan   One site at Asn 642 in the 4th extracellular loop (carries more than half the ABH antigens on the RBC)
Copies per RBC   1 000 000
α-Chymotrypsin cleaves band 3 on intact RBCs at residues 553 and 555 in the third extracellular loop.

Molecular basis of antigens[3]

Antigen	Amino acid change	Exon	Nt change	Restriction enzyme
Di^b/Di^a	Pro854Leu	19	2561C>T	Nae I (−/+)
Wr^b/Wr^a	Glu658Lys	16	1972G>A	
Wd(a−)/Wd(a+)	Val557Met	14	1669G>A	Msl I (−/+)
Rb(a−)/Rb(a+)	Pro548Leu	14	1643C>T	EcoN I (−/+)
WARR−/WARR+	Thr552Ile	14	1654C>T	Bbs I (+/−)
ELO−/ELO+	Arg432Trp	12	1294C>T	Msp I (+/−)
Wu−/Wu+	Gly565Ala	14	1694G>C	Apa I (+/−)
Bp(a−)/Bp(a+)	Asn569Lys	14	1707C>A	Tth 2 (+/−)
Mo(a−)/Mo(a+)	Arg656His	16	1967G>A	Bsm I (−/+)
Hg(a−)/Hg(a+)	Arg656Cys	16	1966C>T	Cac8 I (−/+)
Vg(a−)/Vg(a+)	Tyr555His	14	1663T>C	Dra III (−/+)
Sw(a−)/Sw(a+)	Arg646Gln or Trp	16	1937G>A or 1936 C>T	
BOW−/BOW+	Pro561Ser	14	1681C>T	Ban I(+/−), BstE II(−/+)
NFLD−/NFLD+	Pro561Ala	14	1681C>G	
	Glu429Asp	12	1287A>T	
Jn(a−)/Jn(a+)	Pro566Ser	14	1696C>T	
KREP−/KREP+	Pro566Ala	14	1696C>G	
Tr(a−)/Tr(a+)	Lys551Asn	14	1653C>G	Bbs I (+/−)
Fr(a−)/Fr(a+)	Glu480Lys	13	1438G>A	Bsa I, BsmA I (+/−)
SW1−/SW1+	Arg646Trp	16	1936C>T	

Function

Band 3 makes up 20% of the RBC membrane proteins, it has two functionally independent domains and numerous roles.

N-terminus cytoplasmic domain (residues 1–359): Anchored to the membrane skeleton via ankyrin and protein 4.2 and contributes to maintaining the structural integrity of the RBC; interacts with several glycolytic enzymes, hemoglobin, catalase and hemichromes (the oxidation products of denatured hemoglobin).

C-terminus membrane domain (residues 360–911): Anion exchange (HCO_3/Cl) across the RBC membrane; contributes to the stability of the lipid bilayer through interaction with adjacent phospholipid molecules.

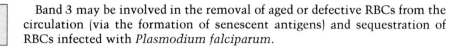

Band 3 may be involved in the removal of aged or defective RBCs from the circulation (via the formation of senescent antigens) and sequestration of RBCs infected with *Plasmodium falciparum*.

Disease association[1]

A severely hydropic baby, who lacked band 3 and protein 4.2, had to be resuscitated and kept alive by transfusion. Products of variant alleles of band 3 have been implicated in the pathogenesis of South East Asian ovalocytosis, congenital acanthocytosis, hereditary spherocytosis and distal renal tubular acidosis. None of the alleles encoding variants of the Diego blood group system are associated with disease. Band 3 has a role in the attachment of malarial parasites to the surface of RBCs and in adhesion of parasitized cells to the vascular epithelium.

Phenotypes (% occurrence)

	Caucasians	Blacks	Asians	South American Indians
Di(a+b−)	<0.01	<0.01	<0.01	<0.1
Di(a−b+)	>99.9	>99.9	90	64
Di(a+b+)	<0.1	<0.1	10	36

Null: Di(a−b−) in one baby.
Unusual: Weak Di(b+).

Comments

Band 3 and glycophorin A (GPA) interact during biosynthesis and within the RBC membrane: GPA appears to be a chaperone to aid the correct folding and efficient transport of band 3 to the RBC membrane. Lack of GPA in the RBC membrane results in failure to express Wr^b (DI4) antigen. Altered form of band 3 present in South East Asian ovalocytes due to a deletion of amino acid residues 400–408[4].

Band 3 Memphis has a Lys56 to Glu change.

References

1 Bruce, L.J. and Tanner, M.J. (1999) Baillieres Best. Pract. Res. Clin. Haematol. 12, 637–654.
2 Schofield, A.E. et al. (1994) Blood 84, 2000–2012.
3 Zelinski, T. (1998) Transf. Med. Rev. 12, 36–45.
4 Tanner, M.J. (1993) Semin. Hematol. 30, 34–57.

Di^a ANTIGEN

Terminology

ISBT symbol (number)	DI1 (010.001)
Other names	Diego
History	Named after Mrs. Diego, producer of the first example of anti-Di^a; reported in detail in 1955; identified as a result of HDN

Occurrence

Most populations	0.01%	
South American Indians	36%	(2% in Caracas Indians to 54% in Carajas Indians)
Japanese	12%	
Chippewa Indians (Canada)	11%	
Chinese	5%	
Poles	0.47%	

Antithetical antigen

Di^b (**DI2**)

Expression

Cord RBCs	Expressed

Molecular basis associated with Di^a antigen[1]

Amino acid	Leu 854
Nucleotide	T at bp 2561 in exon 19
Restriction enzyme	Gains a *Nae* I site

Di^a is predominantly associated with band 3 Memphis variant II, which has a mutation of Lys 56 to Glu in the cytoplasmic domain. RBCs with the band 3 Memphis variant II bind stilbene disulfonate (H$_2$DIDS) more readily than do RBCs with Memphis variant I or common type band 3. Di^a with Lys 56 has been found in Amazonian Indians[2].

Effect of enzymes/chemicals on Di^a antigen on intact RBCs

Ficin/papain	Resistant
Trypsin	Resistant
α-Chymotrypsin	Resistant

Pronase	Resistant
Sialidase	Resistant
DTT 200 mM	Resistant
Acid	Resistant

In vitro characteristics of alloanti-Di^a

Immunoglobulin class	IgG (often IgG1 and IgG3)
Optimal technique	IAT
Complement binding	Some

Clinical significance of alloanti-Di^a

Transfusion reaction	None to severe/delayed
HDN	Mild to severe

Comments

In contrast to many of the Diego system antibodies, anti-Di^a is usually found as a single specificity; only occasionally does it occur in sera containing multiple antibodies to low incidence antigens. Examples of agglutinating anti-Di^a and naturally occurring anti-Di^a exist, but are rare.

References
[1] Bruce, L.J. et al. (1994) J. Biol. Chem. 269, 16155–16158.
[2] Baleotti, W, Jr. et al. (2003) Vox Sang. 84, 326–330.

Di^b ANTIGEN

Terminology

ISBT symbol (number)	DI2 (010.002)
Other names	Luebano
History	Anti-Di^b identified in 1967; detected an antigen of high incidence antithetical to Di^a

Occurrence

Most populations	100%
Native Americans	96%

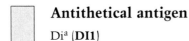

Antithetical antigen

Di^a (**DI1**)

Expression

Cord RBCs	Expressed
Altered	Weak on South East Asian ovalocytes and on the Di(a−b+) and Di(a+b+) RBCs of some Hispanic-Americans

Molecular basis associated with Di^b antigen[1]

Amino acid	Pro 854
Nucleotide	C at bp 2561 in exon 19

Effect of enzymes/chemicals on Di^b antigen on intact RBCs

Ficin/papain	Resistant
Trypsin	Resistant
α-Chymotrypsin	Resistant
Pronase	Resistant
Sialidase	Resistant
DTT 200 mM	Resistant
Acid	Resistant

In vitro characteristics of alloanti-Di^b

Immunoglobulin class	IgG
Optimal technique	IAT
Complement binding	No

Clinical significance of alloanti-Di^b

Transfusion reaction	None to moderate/delayed
HDN	Mild

Autoanti-Di^b

Yes.

Comments

Anti-Di^b demonstrates dosage, reacting more strongly with Di(a−b+) than with Di(a+b+) RBCs.

Reference
[1] Bruce, L.J. et al. (1994) J. Biol. Chem. 269, 16155–16158.

Wr^a ANTIGEN

Terminology

ISBT symbol (number)	DI3 (010.003)
Other names	Wright; 700.001; 211.001
History	Identified in 1953 as the cause of HDN in the Wright family; assigned to the Diego blood group system in 1995

Occurrence

Less than 0.01%.

Antithetical antigen

Wr^b (**DI4**)

Expression

Cord RBCs	Expressed

Molecular basis associated with Wr^a antigen[1]

Amino acid	Lys 658
Nucleotide	A at bp 1972 in exon 16

Effect of enzymes/chemicals on Wr^a antigen on intact RBCs

Ficin/papain	Resistant
Trypsin	Resistant
α-Chymotrypsin	Resistant

Pronase	Resistant
Sialidase	Resistant
DTT 200 mM	Resistant
Acid	Resistant

In vitro characteristics of alloanti-Wr^a

Immunoglobulin class	IgM; IgG
Optimal technique	RT; IAT
Complement binding	No

Clinical significance of alloanti-Wr^a

Transfusion reaction	None to severe/immediate or delayed/hemolytic
HDN	Mild to severe

Comments

Alloanti-Wr^a is often an apparent naturally-occurring antibody and is found in the serum of 1% to 2% of blood donors. It is frequently found in multi-specific sera and is a common specificity in patients with AIHA.

Reference
[1] Bruce, L.J. et al. (1995) Blood 85, 541–547.

Wr^b ANTIGEN

Terminology

ISBT symbol (number)	DI4 (010.004)
Other names	Fritz; MF; 901.010; 900.024; 211.002
History	Anti-Wr^b identified in 1971; detected an antigen of high incidence antithetical to Wr^a; assigned to Diego blood group system in 1995

Occurrence

All populations	100%

Antithetical antigen

Wr^a (**DI3**)

Expression

Cord RBCs	Expressed
Altered	On ENEP− (HAG+) and ENAV− (MARS+) RBCs[1]

Molecular basis associated with Wr^b antigen[2]

Amino acid	Glu 658
Nucleotide	G at bp 1972 in exon 16

For expression, Wr^b antigen requires the presence of amino acid residues 59–76 of GPA[3].

Effect of enzymes/chemicals on Wr^b antigen on intact RBCs

Ficin/papain	Resistant (one example); sensitive (one example)[4]
Trypsin	Resistant
α-Chymotrypsin	Resistant
Pronase	Resistant
Sialidase	Resistant
DTT 200 mM	Resistant
Acid	Resistant

In vitro characteristics of alloanti-Wr^b

Immunoglobulin class	IgM plus IgG
Optimal technique	IAT
Complement binding	No

Clinical significance of alloanti-Wr^b

Transfusion reaction	Not known because only three individuals with Wr(a+b−) RBCs and alloanti-Wr^b have been described. However, a strong anti-Wr^b made by an untransfused, previously pregnant woman, with the Mi.V/Mi.V, Wr(a-b-) phenotype, gave results in the chemiluminescence assay that suggested the anti-Wr^b was likely to cause accelerated destruction of transfused incompatible RBCs[5].
HDN	Positive DAT but not clinical HDN

Autoanti-Wr^b

Yes. Fairly common specificity in patients with AIHA. Some autoanti-Wr^b appear to be benign while others are not: in two cases autoanti-Wr^b, reactive at 37°C, resulted in fatal intravascular hemolysis.

Comments

Alloanti-Wr^b can occur as separable components in the sera of immunized En(a−) people.

RBCs with En(a−) or M^{k}M^k phenotypes (which lack GPA) type as Wr(a−b−). Wr^b is not expressed on En^a.UK, GP.Hil, GP.TSEN, GP.SAT, GP.TK or GP.Dantu hybrid glycophorin molecules. All have glutamic acid at residue 658 of band 3 but lack the required amino acids from GPA[6]. GP.HAG (Gln63Lys of GPA) and GP.MARS (Ala65Pro of GPA) have an altered expression of Wr^b.

References

[1] Poole, J. (2000) Blood Rev. 14, 31–43.
[2] Bruce, L.J. et al. (1995) Blood 85, 541–547.
[3] Reid, M.E. (1999) Immunohematology 15, 5–9.
[4] Storry, J.R. et al. (2001) Transfusion 41 (Suppl.), 23S (abstract).
[5] Poole, J. et al. (1997) Transf. Med. 7 (Suppl.), 27.
[6] Huang, C.-H. et al. (1996) Blood 87, 3942–3947.

Wd^a ANTIGEN

Terminology

ISBT symbol (number)	DI5 (010.005)
Other names	Waldner; 700.030
History	Reported in 1983; first described in the Waldner family; identified when RBCs were being typed with a serum known to contain anti-Fr^a; assigned to Diego blood group system in 1996

Occurrence

Found in two Schmiedeleut Hutterite families, one family in Holland with probable Hutterite connections and two Namibian sisters.

Expression

Cord RBCs   Expressed

Molecular basis associated with Wd^a antigen[1,2]

Amino acid   Met 557
Nucleotide   A at bp 1669 in exon 14
Restriction enzyme   Gains a *Msl* I site
Wd(a−)   Val 557 and G at bp 1669

Effect of enzymes/chemicals on Wd^a antigen on intact RBCs

Ficin/papain   Resistant
Trypsin   Resistant
α-Chymotrypsin   Sensitive
Pronase   Presumed sensitive
Sialidase   Presumed resistant
DTT 200 mM   Resistant
Acid   Resistant

In vitro characteristics of alloanti-Wd^a

Immunoglobulin class   IgM; IgG (at least one IgG example reported)
Optimal technique   RT
Complement binding   No

Clinical significance of alloanti-Wd^a

Anti-Wd^a was not made by any of six Wd(a−) women, who between them gave birth to 30 Wd(a+) children[3]. In the same study, anti-Wd^a was found in the serum of 1 of 358 pregnant women. No other data are available.

Comments

Anti-Wd^a is a common specificity in multi-specific sera.

References
1   Bruce, L.J. et al. (1996) Vox Sang. 71, 118–120.
2   Jarolim, P. et al. (1997) Transfusion 37, 607–615.
3   Lewis, M. and Kaita H. (1981) Am. J. Hum. Genet. 33, 418–420.

Rb^a ANTIGEN

Terminology

ISBT symbol (number)	DI6 (010.006)
Other names	Redelberger; 700.027
History	Reported in 1978; found on the RBCs of Mr. Redelberger, a donor and donor recruiter; assigned to Diego blood group system in 1996.

Occurrence

Found in three families.

Expression

Cord RBCs	Expressed

Molecular basis associated with Rb^a antigen[1]

Amino acid	Leu 548
Nucleotide	T at bp 1643 in exon 14
Restriction enzyme	Gains an *EcoN* I site
Rb(a−)	Pro 548 and C at bp 1643

Effect of enzymes/chemicals on Rb^a antigen on intact RBCs

Ficin/papain	Variable
Trypsin	Resistant
α-Chymotrypsin	Variable
Pronase	Variable
Sialidase	Presumed resistant
DTT 200 mM	Presumed resistant
Acid	Presumed resistant

In vitro characteristics of alloanti-Rb^a

Immunoglobulin class	IgM (predominantly); IgG from the original report
Optimal technique	RT
Complement binding	No

Clinical significance of alloanti-Rbª

Five Rb(a−) women, who gave birth to Rb(a+) children, did not make anti-Rbª. No other data are available.

Comments

Common specificity in multi-specific sera.

Reference
[1] Jarolim, P. et al. (1997) Transfusion 37, 607–615.

WARR ANTIGEN

Terminology

ISBT symbol (number)	DI7 (010.007)
Other names	Warrior; 700.055
History	Identified in 1991 as a result of HDN in the Warrior family; assigned to Diego blood group system in 1996

Occurrence

Found in two kindred, both with Native American heritage. The eldest WARR+ member in the Warrior kindred was of pure Absentee Shawnee ancestry[1].

Expression

Cord RBCs	Expressed

Molecular basis associated with WARR antigen[2]

Amino acid	Ile 552
Nucleotide	T at bp 1654 in exon 14
Restriction enzyme	Ablates *Bbs* I site
WARR−	Thr 552 and C at bp 1654

Effect of enzymes/chemicals on WARR antigen on intact RBCs

Ficin/papain	Resistant
Trypsin	Resistant
α-Chymotrypsin	Sensitive
Pronase	Sensitive
Sialidase	Presumed resistant
DTT 200 mM	Resistant
Acid	Presumed resistant

In vitro characteristics of alloanti-WARR

Immunoglobulin class	IgG
Optimal technique	IAT
Complement binding	No data

Clinical significance of alloanti-WARR

Transfusion reaction	No data are available
HDN	Mild

Comments

It is a common specificity in multi-specific sera and immune anti-WARR exists.

References
[1] Coghlan, G. et al. (1995) Vox Sang. 68, 187–190.
[2] Jarolim, P. et al. (1997) Transfusion 37, 398–405.

ELO ANTIGEN

Terminology

ISBT symbol (number)	DI8 (010.008)
Other names	700.051
History	Antigen recognized in 1979, reported in detail in 1993; named after the first name of the original proband; assigned to Diego blood group system in 1998

Occurrence

Less than 0.01%.

Expression

Cord RBCs   Expressed

Molecular basis associated with ELO antigen[1]

Amino acid	Trp 432
Nucleotide	T at bp 1294 in exon 12
Restriction enzyme	*Msp* I and *Bst* NI sites ablated
ELO−	Arg 432 and C at bp 1294

Effect of enzymes/chemicals on ELO antigen on intact RBCs

Ficin/papain	Resistant
Trypsin	Resistant
α-Chymotrypsin	Variable
Pronase	Variable
Sialidase	Presumed resistant
DTT 200 mM	Resistant
Acid	Presumed resistant

In vitro characteristics of alloanti-ELO

Immunoglobulin class	IgG
Optimal technique	IAT
Complement binding	No

Clinical significance of alloanti-ELO

Transfusion reaction	No data are available
HDN	Severe

Comments

Several examples of immune monospecific anti-ELO exist and it is often found in multi-specific sera.

Reference
[1] Zelinski, T. (1998) Transf. Med. Rev. 12, 36–45.

Wu ANTIGEN

Terminology

ISBT symbol (number)	DI9 (010.009)
Other names	Wulfsberg (700.013); Hov (700.038); Haakestad
History	Identified in 1967; named after the original Wu+ donor; assigned to Diego blood group system in 1998

Occurrence

Less than 0.01% (Dutch, Danish, and Norwegian ancestry; also in one Black proband).

Expression

Cord RBCs	Expressed

Molecular basis associated with Wu antigen[1]

Amino acid	Ala 565
Nucleotide	C at bp 1694 in exon 14
Restriction enzyme	*Apa* I site ablated
Wu−	Gly 565 and G at bp 1694

Effect of enzymes/chemicals on Wu antigen on intact RBCs

Ficin/papain	Resistant
Trypsin	Resistant
α-Chymotrypsin	Sensitive
Pronase	Sensitive
Sialidase	Presumed resistant
DTT 200 mM	Resistant
Acid	Presumed resistant

In vitro characteristics of alloanti-Wu

Immunoglobulin class	IgM; IgG (few)
Optimal technique	RT; IAT
Complement binding	No

Clinical significance of alloanti-Wu

No data are available.

Comments

Anti-Wu is often found in multi-specific sera and may be naturally-occurring. Several members in one family of Dutch descent are likely to be homozygous for *Wu*. The serological relationship with NFLD and BOW[2] cannot be explained by the molecular knowledge.

References
[1] Zelinski, T. (1998) Transf. Med. Rev. 12, 36–45.
[2] Kaita, H. et al. (1992) Transfusion 32, 845–847.

Bp^a ANTIGEN

Terminology

ISBT symbol (number)	DI10 (010.010)
Other names	Bishop; 700.010
History	Antigen discovered in 1964 on the RBCs of Mr. Bishop; assigned to Diego blood group system in 1998

Occurrence

Two probands (English and Italian).

Expression

Cord RBCs	Presumed expressed

Molecular basis associated with Bp^a antigen[1]

Amino acid	Lys 569
Nucleotide	A at bp 1707 in exon 14
Restriction enzyme	T*th* 2 site ablated
Bp(a−)	Asn 569 and C at bp 1707

Effect of enzymes/chemicals on Bp^a antigen on intact RBCs[2]

Ficin/papain	Sensitive
Trypsin	Sensitive
α-Chymotrypsin	Sensitive
Pronase	Sensitive
Sialidase	Presumed resistant
DTT 200 mM	Presumed resistant
Acid	Presumed resistant

In vitro characteristics of alloanti-Bp^a

Immunoglobulin class	IgM
Optimal technique	RT
Complement binding	No data

Clinical significance of alloanti-Bp^a

Transfusion reaction	No data are available
HDN	No

Comments

Alloanti-Bp^a is often found in sera containing multiple naturally-occurring antibodies (including anti-Wr^a) and in sera from patients with AIHA. The band 3 amino acid substitution associated with expression of Bp^a antigen is likely to be located within the RBC lipid bilayer; thus, the enzyme sensitivity of Bp^a is somewhat surprising and may indicate the interaction with another enzyme sensitive component in the formation of the Bp^a epitope.

References
[1] Zelinski, T. (1998) Transf. Med. Rev. 12, 36–45.
[2] Jarolim, P. et al. (1998) Blood 92, 4836–4843.

Mo^a ANTIGEN

Terminology

ISBT symbol (number)	DI11 (010.011)
Other names	Moen; 700.022
History	Antigen found in 1972 when random donors were screened for Jn^a; assigned to Diego blood group system in 1998

Occurrence

Three probands have been reported: one from Norway and two from Belgium.

Antithetical antigen

Hg^a (DI12)

Expression

Cord RBCs   Presumed expressed

Molecular basis associated with Mo^a antigen[1]

Amino acid	His 656
Nucleotide	A at bp 1967 in exon 16
Restriction enzyme	*Bsm* I site gained
Mo(a−)	Arg 656 and G at bp 1967

Effect of enzymes/chemicals on Mo^a antigen on intact RBCs

Ficin/papain	Resistant
Trypsin	Resistant
α-Chymotrypsin	Resistant
Pronase	Resistant
Sialidase	Presumed resistant
DTT 200 mM	Resistant
Acid	Presumed resistant

In vitro characteristics of alloanti-Mo^a

Immunoglobulin class	IgM; IgG
Optimal technique	RT; IAT
Complement binding	Presumed no

Clinical significance of alloanti-Mo^a

No data are available.

Comment

Anti-Mo^a may be naturally-occurring and is found in multi-specific sera.

Reference
1   Zelinski, T. (1998) Transf. Med. Rev. 12, 36–45.

Hgᵃ ANTIGEN

Terminology

ISBT symbol (number)	DI12 (010.012)
Other names	Hughes; 700.034; Tarplee; Tarp
History	The Hgᵃ antigen was described in 1983. Its name was derived from the maiden name (Hughes) of the original Hg(a+) panel donor identified during routine antibody screening tests. Hgᵃ joined the Diego blood group system in 1998

Occurrence

Rare: reported in three Welsh families and an Australian donor from New South Wales.

Antithetical antigen

Moᵃ (**DI11**)

Expression

Cord RBCs	Expressed

Molecular basis associated with Hgᵃ antigen[1]

Amino acid	Cys 656
Nucleotide	T at bp 1966 in exon 16
Restriction enzyme	Gains a *Cac8* I site
Hg(a−)	Arg 656 and G at bp 1966

Effect of enzymes/chemicals on Hgᵃ antigen on intact RBCs[2]

Ficin/papain	Resistant
Trypsin	Resistant
α-Chymotrypsin	Resistant
Pronase	Resistant

Sialidase   Presumed resistant
DTT 200 mM   Presumed resistant
Acid   Presumed resistant

In vitro characteristics of alloanti-Hg[a]

Immunoglobulin class   IgM; IgG
Optimal technique   RT; IAT
Complement binding   No data

Clinical significance of alloanti-Hg[a]

No data are available.

Comments

Anti-Hg[a] is found in multispecific sera; anti-Hg[a] as a single specificity has not been reported.

References

[1] Zelinski, T. (1998) Transf. Med. Rev. 12, 36–45.
[2] Jarolim, P. et al. (1998) Blood 92, 4836–4843.

Vg[a] ANTIGEN

Terminology

ISBT symbol (number)   DI13 (010.013)
Other names   VanVugt; 700.029
History   Antigen reported in 1981; found while screening Australian donors with anti-Wu; named after the first antigen positive donor Miss Van Vugt; assigned to Diego blood group system in 1998

Occurrence

Only one family reported.

Expression

Cord RBCs   Presumed expressed

Molecular basis associated with Vg^a antigen[1]

Amino acid	His 555
Nucleotide	C at bp 1663 in exon 14
Restriction enzyme	Gains a *Dra* III site
Vg(a−)	Tyr 555 and T at bp 1663

Effect of enzymes/chemicals on Vg^a antigen on intact RBCs[2]

Ficin/papain	Resistant
Trypsin	Resistant
α-Chymotrypsin	Sensitive
Pronase	Sensitive
Sialidase	Presumed resistant
DTT 200 mM	Presumed resistant
Acid	Presumed resistant

In vitro characteristics of alloanti-Vg^a

Immunoglobulin class	IgM and rarely IgG
Optimal technique	RT (IAT)
Complement binding	No data

Clinical significance of alloanti-Vg^a

No data are available.

Comments

Anti-Vg^a is a relatively common antibody (11 examples of anti-Vg^a were found among 1669 donor sera) and is found in multi-specific sera that frequently also contain anti-Wr^a.

References
[1] Zelinski, T. (1998) Transf. Med. Rev. 12, 36–45.
[2] Jarolim, P. et al. (1998) Blood 92, 4836–4843.

Sw^a ANTIGEN

Terminology

ISBT symbol (number)	DI14 (010.014)
Other names	Swann; 700.004
History	Antigen identified in 1959, when serum from an AIHA patient was crossmatched against RBCs from donor Donald Swann; assigned to Diego blood group system in 1998

Occurrence

Less than 0.01%.

Expression

Cord RBCs	Presumed expressed

Molecular basis associated with Sw^a antigen[1]

Amino acid	Gln or Trp 646
Nucleotide	A or T at bp 1937 in exon 16
Sw(a−)	Arg 646 and G at bp 1937

Effect of enzymes/chemicals on Sw^a antigen on intact RBCs

Ficin/papain	Resistant
Trypsin	Resistant
α-Chymotrypsin	Resistant
Pronase	Presumed resistant
Sialidase	Presumed resistant
DTT 200 mM	Presumed resistant
Acid	Presumed resistant

In vitro characteristics of alloanti-Sw^a

Immunoglobulin class	IgM; IgG
Optimal technique	RT (IAT)
Complement binding	No data

Clinical significance of alloanti-Sw^a

No data are available.

Comments[2,3]

Anti-Sw^a is often found in AIHA and in multi-specific sera. Anti-Sw^a and anti-Fr^a when present in the same serum show cross-reactivity and cannot be separated by absorption.

Anti-Sw^a also react with SW1+ (see DI21) RBCs: RBCs may be Sw(a+), SW1− (Gln 646, the more common type), or Sw(a+) SW1+ (Trp 646). Sw(a−), SW1+ RBCs have not been found.

References
1 Zelinski, T. et al. (2000) Vox Sang. 79, 215–218.
2 Lewis, M. et al. (1988) Vox Sang. 54, 184–187.
3 Contreras, M. et al. (1987) Vox Sang. 52, 115–119.

BOW ANTIGEN

Terminology

ISBT symbol (number)	DI15 (010.015)
Other names	Bowyer; 700.046
History	Antigen reported in 1988; identified on the RBCs of a donor (Bowyer) during an incompatible crossmatch; assigned to Diego blood group system in 1998

Occurrence

Only a few probands have been reported.

Antithetical antigen

NFLD (**DI16**) [at amino acid 561, see Comments]

Expression

Cord RBCs	Presumed expressed

Molecular basis associated with BOW antigen[1]

Amino acid	Ser 561
Nucleotide	T at bp 1681 in exon 14
Restriction enzyme	*Ban* I site ablated; gains a *BstE* II site
BOW−, NFLD−	Pro 561 and C at bp 1681

Effect of enzymes/chemicals on BOW antigen on intact RBCs

Ficin/papain	Resistant
Trypsin	Resistant
α-Chymotrypsin	Sensitive
Pronase	Sensitive
Sialidase	Presumed resistant
DTT 200 mM	Resistant
Acid	Presumed resistant

In vitro characteristics of alloanti-BOW

Immunoglobulin class	IgG; some IgM
Optimal technique	IAT; RT
Complement binding	No data

Clinical significance of alloanti-BOW

No data are available.

Comments

Several examples of immune monospecific anti-BOW exist and it is often found in multi-specific sera. Molecular analysis showed that both BOW and NFLD are associated with a substitution at amino acid residue 561: serine is present when BOW is expressed and alanine is present when NFLD is expressed. Thus, BOW and NFLD can be considered antithetical even though NFLD has a second mutation of Glu429Asp.

The serological relationship to Wu[2] cannot be explained by the molecular knowledge, although the critical residue (565 Ala) for Wu is relatively close to residue 561 of band 3.

References
[1] McManus, K. et al. (2000) Transfusion 40, 325–329.
[2] Kaita, H. et al. (1992) Transfusion 32, 845–847.

NFLD ANTIGEN

Terminology

ISBT symbol (number)	DI16 (010.016)
Other names	Newfoundland; 700.037
History	Antigen reported in 1984; was found on the RBCs of a French Canadian in Newfoundland; assigned to Diego system in 1998

Occurrence

Only a few probands (two French Canadian and two Japanese families) have been reported.

Antithetical antigen

BOW (**DI15**) [at amino acid 561, see Comments]

Expression

Cord RBCs   Presumed expressed

Molecular basis associated with NFLD antigen[1]

Amino acid   Asp 429 and Ala 561
Nucleotide   T at bp 1287 in exon 12 and G at bp 1681
                              in exon 14
NFLD−, BOW−   Glu 429 with A at bp 1287, and Pro 561
                              with C at bp 1681

Effect of enzymes/chemicals on NFLD antigen on intact RBCs

Ficin/papain   Resistant
Trypsin   Resistant
α-Chymotrypsin   Sensitive
Pronase   Sensitive
Sialidase   Presumed resistant
DTT 200 mM   Resistant
Acid   Presumed resistant

In vitro characteristics of alloanti-NFLD

Immunoglobulin class   IgM and IgG
Optimal technique   RT (with albumin); IAT
Complement binding   No data

Clinical significance of alloanti-NFLD

A Japanese NFLD− woman gave birth to three NFLD+ children without making anti-NFLD. No other data are available.

Comments

Anti-NFLD is found in multi-specific sera. Molecular analysis showed that both NFLD and BOW are associated with a substitution at amino acid residue 561: alanine is present when NFLD is expressed and serine is present when BOW is expressed. Thus, NFLD and BOW can be considered antithetical at this residue. However, NFLD has a second mutation of Glu429Asp. The epitope defining NFLD may be created through an association and/or interaction between the first (residue 429) and third (residue 561) of extracellular loops of band 3.

The serological relationship to Wu[2] cannot be explained by the molecular knowledge although the critical residue (565Ala) for expression of Wu is relatively close to residue 561 of band 3.

References
1 McManus, K. et al. (2000) Transfusion 40, 325–329.
2 Kaita, H. et al. (1992) Transfusion 32, 845–847.

Jn^a ANTIGEN

Terminology

ISBT symbol (number)	DI17 (010.017)
Other names	Nunhart; JN; 700.014
History	Antigen described in 1967; identified on the RBCs of Mr. J.N. during a study of the incidence of Wr^a in the Prague population; assigned to Diego blood group system in 1998

Occurrence

Two probands (one of Polish, the other of Slovakian descent)[1].

Antithetical antigen

KREP (**DI18**)

Expression

Cord RBCs	Presumed expressed

Molecular basis associated with Jn^a antigen[1]

Amino acid	Ser 566
Nucleotide	T at bp 1696 in exon 14
Jn(a−)	Pro 566 and C at bp 1696

Effect of enzymes/chemicals on Jn^a antigen on intact RBCs

Ficin/papain	Resistant
Trypsin	Resistant
α-Chymotrypsin	Sensitive
Pronase	Presumed sensitive
Sialidase	Presumed resistant
DTT 200 mM	Resistant
Acid	Presumed sensitive

In vitro characteristics of alloanti-Jn^a

Immunoglobulin class	IgM (no data regarding presence of an IgG component)
Optimal technique	RT
Complement binding	No data

Clinical significance of alloanti-Jn^a

No data are available.

Comments

Anti-Jn^a is found in multi-specific sera that also contain anti-KREP. The majority are naturally-occurring.

Reference
[1] Poole J. (1999) Immunohematology 15, 135–143.

KREP ANTIGEN

Terminology

ISBT symbol (number)	DI18 (010.018)
Other names	IK
History	Found in 1997 during investigation of the second Jn(a+) proband; named after the antigen-positive donor; assigned to Diego blood group system in 1998

Occurrence

One Polish proband (IK)[1].

Antithetical antigen

Jn^a (DI17)

Expression

Cord RBCs   Presumed expressed

Molecular basis associated with KREP antigen[1]

Amino acid   Ala 566
Nucleotide   G at bp 1696 in exon 14
KREP−   Pro 566 and C at bp 1696

Effect of enzymes/chemicals on KREP antigen on intact RBCs

Ficin/papain   Resistant
Trypsin   Resistant
α-Chymotrypsin   Sensitive
Pronase   Presumed sensitive
Sialidase   Presumed resistant
DTT 200 mM   Resistant
Acid   Presumed resistant

In vitro characteristics of alloanti-KREP

Immunoglobulin class   IgM (no data regarding presence of an IgG
                             component)
Optimal technique   RT
Complement binding   No data

Clinical significance of alloanti-KREP

No data are available.

Comments

Anti-KREP is naturally-occurring and is present in multi-specific sera. Among 13 sera tested, 12 contained anti-Jn^a and anti-KREP and only one serum contained anti-KREP without anti-Jn^a.

Reference
[1] Poole, J. (1999) Immunohematology 15, 135–143.

Tr^a ANTIGEN

Terminology

ISBT symbol (number)	DI19 (010.019) (provisional)
Other names	Traversu; Lanthois; 700.008
History	Antigen found in the 1960s during random testing of English blood donors with a multi-specific serum, that also contained anti-Wr^a; named after the first positive donor, Traversu. Remains provisionally numbered by the ISBT because only one Tr(a+) person has been studied in detail and because serologic discrepancies were observed in earlier testing[1].

Occurrence

Only found in two English probands.

Expression

Cord RBCs	Presumed expressed

Molecular basis associated with Tr^a antigen[2]

Amino acid	Asn 551
Nucleotide	G at bp 1653 in exon 14
Restriction enzyme	*Bbs* I site ablated
Tr(a−)	Lys 551 and C at bp 1653

Effect of enzymes/chemicals on Tr^a antigen on intact RBCs

Ficin/papain	Resistant
Trypsin	Resistant
α-Chymotrypsin	Sensitive
Pronase	Presumed sensitive
Sialidase	Presumed resistant
DTT 200 mM	Presumed resistant
Acid	Presumed resistant

In vitro characteristics of alloanti-Tr^a

Immunoglobulin class	IgM and IgG
Optimal technique	RT; IAT
Complement binding	No data

Clinical significance of alloanti-Tr^a

No data are available.

Comments

Anti-Tr^a was found as a separable specificity in 12 of 18 sera that contained anti-Wr^a.

Anti-Tr^a is found in multi-specific sera and in serum from patients with AIHA.

References
1 Jarolim, P. et al. (1997) Transfusion 37, 607–615.
2 Zelinski, T. (1998) Transf. Med. Rev. 12, 36–45.

Fr^a ANTIGEN

Terminology

ISBT symbol (number)	DI20 (010.020)
Other names	Froese; 700.026
History	Reported in 1978 and named after the family (Froese) in which it was first recognized; assigned to Diego blood group system in 2000

Occurrence

The reported Fr(a+) probands originate from three Mennonite kindred in Manitoba, Canada.

Expression

Cord RBCs	Expressed

Molecular basis associated with Fr^a antigen[1]

Amino acid	Lys 480
Nucleotide	A at bp 1438 in exon 13
Restriction enzyme	Ablates a *Bsa* I site and a *BsmA* I site
Fr(a−)	Glu 480 and G at bp 1438

Effect of enzymes/chemicals on Fr^a antigen on intact RBCs

Ficin/papain	Resistant
Trypsin	Presumed resistant
α-Chymotrypsin	Presumed resistant
Pronase	Presumed resistant
Sialidase	Presumed resistant
DTT 200 mM	Presumed resistant
Acid	Presumed resistant

In vitro characteristics of alloanti-Fr^a

Immunoglobulin class	IgG; IgM (few)
Optimal technique	IAT; RT
Complement binding	No data

Clinical significance of alloanti-Fr^a

Transfusion reaction	No data are available
HDN	Positive DAT, but no clinical HDN

Comments

Several examples of immune monospecific anti-Fr^a exist and it is often found in multi-specific sera. Anti-Fr^a and anti-Sw^a, when present in the same serum show cross-reactivity and cannot be separated by absorption[2].

References
[1] McManus, K. et al. (2000) Transfusion 40, 1246–1249.
[2] Contreras, M. et al. (1987) Vox Sang. 52, 115–119.

SW1 ANTIGEN

Terminology

ISBT symbol (number)	DI21 (010.021)
Other names	700.041
History	SW1 was documented in 1987; revealed by heterogeneity among sera containing anti-Sw^a; assigned to Diego blood group system in 2000

Occurrence

Less than 0.01%.

Expression

Cord RBCs   Presumed expressed

Molecular basis associated with SW1 antigen[1]

Amino acid	Trp 646
Nucleotide	T at bp 1936 in exon 16
SW1−, Sw(a+)	Gln 646 and A at bp 1936
SW1−, Sw(a−)	Arg 646 and C at bp 1936

Effect of enzymes/chemicals on SW1 antigen on intact RBCs

Ficin/papain	Resistant
Trypsin	Resistant
α-Chymotrypsin	Resistant
Pronase	Presumed resistant
Sialidase	Presumed resistant
DTT 200 mM	Presumed resistant
Acid	Presumed resistant

In vitro characteristics of alloanti-SW1

Immunoglobulin class	IgM; IgG
Optimal technique	RT; IAT
Complement binding	No data

Clinical significance of alloanti-SW1

No data are available.

Comments

Examples of anti-SW1 exist that do not react with Sw(a+) RBCs but all anti-Sw[a] react with SW1+ RBCs[1]. See Sw[a] (DI14) for more details.

Reference

[1] Zelinski, T. et al. (2000) Vox Sang. 79, 215–218.

Number of antigens   2

Terminology

ISBT symbol	YT
ISBT number	011
Other name	Cartwright
History	Named after the high prevalence antigen, Yt^a; became a system in 1964 after discovery of the antithetical antigen

Expression

Other blood cells	Not on lymphocytes, granulocytes or monocytes
Tissues	Brain, muscle, nerves

Gene

Chromosome	7q22.1
Name	*YT (ACHE)*
Organization	6 exons distributed over 2.2 kpb of gDNA (Exons 5 and 6 are alternatively spliced)
Product	Acetylcholinesterase (AChE)

Gene map

```
         *
  ■──■■■■──────■────────■────────────────■■──■■──■──3'
  ATG                                          STOP
```

* *YT 1/YT 2* (1057C>A) encode Yt^a/Yt^b (H353N)

⊢—⊣ 1 kbp

Database accession numbers

GenBank   M55040

www.bioc.aecom.yu.edu/bgmut/index.htm

Amino acid sequence[1]

```
                 M RPPQCLLHTP SLASPLLLLL LWLLGGGVGA   -1
EGREDAELLV TVRGGRLRGI RLKTPGGPVS AFLGIPFAEP PMGPRRFLPP   50
EPKQPWSGVV DATTFQSVCY QYVDTLYPGF EGTEMWNPNR ELSEDCLYLN  100
VWTPYPRPTS PTPVLVWIYG GGFYSGASSL DVYDGRFLVQ AERTVLVSMN  150
YRVGAFGFLA LPGSREAPGN VGLLDQRLAL QWVQENVAAF GGDPTSVTLF  200
GESAGAASVG MHLLSPPSRG LFHRAVLQSG APNGPWATVG MGEARRRATQ  250
```

```
LAHLVGCPPG  GTGGNDTELV  ACLRTRPAQV  LVNHEWHVLP  QESVFRFSFV  300
PVVDGDFLSD  TPEALINAGD  FHGLQVLVGV  VKDEGSYFLV  YGAPGFSKDN  350
ESLISRAEFL  AGVRVGVPQV  SDLAAEAVVL  HYTDWLHPED  PARLREALSD  400
VVGDHNVVCP  VAQLAGRLAA  QGARVYAYVF  EHRASTLSWP  LWMGVPHGYE  450
IEFIFGIPLD  PSRNYTAEEK  IFAQRLMRYW  ANFARTGDPN  EPRDPKAPQW  500
PPYTAGAQQY  VSLDLRPLEV  RRGLRAQACA  FWNRFLPKLL  SATASEAPST  550
CPGFTHGEAA  PRPGLPLPLL  LLHQLLLLFL  SHLRRL                 585
```

YT encodes a leader sequence of 31 amino acids
  Antigen mutation is numbered by counting Met as 1.
  C-terminal 29 amino acids are cleaved from the RBC GPI-linked form.

Carrier molecule

GPI-linked glycoprotein that probably exists as a dimer in the RBC membrane.

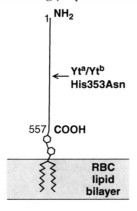

M_r (SDS-PAGE)	160 000 (72 000 as monomer under reducing conditions)
CHO: N-glycan	Three sites
CHO: O-glycan	Present
Cysteine residues	8
Copies per RBC	7000–10 000 (or 3000–5000 dimers)

Molecular basis of antigens

Antigen	Amino acid change	Exon	Nt change
Yt^a/Yt^b	His353Asn	2	1057 C>A

Function

AChE terminates nerve impulse transmission. AChE is in many tissues in various forms as a result of alternative splicing and post-translational modification.

Function in RBC unknown.

Disease association

PNH III RBCs are deficient in AChE. Levels are reduced in myelodysplasias associated with chromosome 7 abnormalities and in some cases of SLE.

Phenotypes (% occurrence)

	Most populations
Yt(a+b−)	91.9
Yt(a+b+)	7.8
Yt(a−b+)	0.3

Null: Inherited Yt(a−b−) phenotype not found.
Unusual: One example of transient Yt(a−b−) RBCs reported[2].

References
[1] Bartels, C.F. et al. (1993) Am. J. Hum. Genet. 52, 928–936.
[2] Rao, N. et al. (1993) Blood 81, 815–819.

Ytᵃ ANTIGEN

Terminology

ISBT symbol (number)	YT1 (011.001)
Other names	Cartwright
History	In 1956 when the antibody to this high prevalence antigen was found, most letters in the patient's name (Cartwright) had been taken by other antigens. The authors thought "why not 'T'?", then "why T" or "Yt"

Occurrence

Most populations: >99.8%; Israeli Jews 98.6%; Israeli Arabs 97.6%; Israeli Druse 97.4%.

Antithetical antigen

Yt^b (**YT2**)

Expression

Cord RBCs	Weak
Altered	Weak or absent from PNH III RBCs

Molecular basis associated with Yt^a antigen[1]

Amino acid	His 353
Nucleotide	C at bp 1057 in exon 2

Nucleotide at position 1431C>T in exon 3 also differentiates Yt^a from Yt^b but does not alter the encoded amino acid. A second silent mutation in exon 5 does not correlate with the Yt polymorphism.

Effect of enzymes/chemicals on Yt^a antigen on intact RBCs

Ficin/papain	Sensitive (variable)
Trypsin	Resistant
α-Chymotrypsin	Sensitive
Pronase	Sensitive
Sialidase	Resistant
DTT 200 mM/50 mM	Sensitive/weakened
Acid	Resistant

In vitro characteristics of alloanti-Yt^a

Immunoglobulin class	IgG
Optimal technique	IAT
Complement binding	Some

Clinical significance of alloanti-Yt^a

Transfusion reaction	No to moderate/delayed
HDN	No

Comments

A report of an apparent alloanti-Yt^a in a Yt(a+) person suggests the possibility of the heterogeneity of Yt^a.[2]

References

[1] Bartels, C.F. et al. (1993) Am. J. Hum. Genet. 52, 928–936.
[2] Mazzi, G. et al. (1994) Vox Sang. 66, 130–132.

Yt^b ANTIGEN

Terminology

ISBT symbol (number)	YT2 (011.001)
History	Identified in 1964 and named when its antithetical relationship to Yt^a was recognized

Occurrence

Europeans 8%; Israeli Jews 21.3%; Israeli Arabs 23.5%; Israeli Druse 26%. Not found in Japanese

Antithetical antigen

Yt^a (**YT1**)

Expression

Cord RBCs	Expressed
Altered	Weak or absent from PNH III RBCs

Molecular basis associated with Yt^b antigen[1]

Amino acid	Asn 353
Nucleotide	A at bp 1057 in exon 2

Nucleotide at position 1431C>T in exon 3 also differentiates Yt^a from Yt^b but does not alter the encoded amino acid. A second silent mutation in exon 5 does not correlate with the Yt polymorphism.

Effect of enzymes/chemicals on Yt^b antigen on intact RBCs

Ficin/papain	Sensitive (variable)
Trypsin	Resistant
α-Chymotrypsin	Sensitive
Pronase	Sensitive
Sialidase	Resistant
DTT 200 mM/50 mM	Sensitive/weakened
Acid	Presumed resistant

In vitro characteristics of alloanti-Yt^b

Immunoglobulin class	IgG
Optimal technique	IAT
Complement binding	No

Clinical significance of alloanti-Yt^b

Transfusion reaction	No
HDN	No

Comments

Anti-Yt^b is rare and usually occurs in sera with other antibodies. The second example of anti-Yt^b was made by a patient with PNH.

Reference
1 Bartels, C.F. et al. (1993) Am. J. Hum. Genet. 52, 928–936.

Number of antigens

2

Terminology

ISBT symbol	XG
ISBT number	012
History	The Xg^a system was established in 1962 when it was found that Xg^a antigen expression was uniquely controlled by the X chromosome

Expression

Other blood cells	Xg^a: expression may be restricted to RBCs CD99: lymphocytes (27 000 sites), platelets (4000 sites)[1]
Tissues	CD99: fibroblasts, fetal liver, lymph nodes, spleen, thymus, pancreatic islet cells, ovarian granulosa cells, sertoli cells, fetal adrenal, adult bone marrow[2]. Most abundant expression is in the most immature stages of the B cell, T cell and granulocyte lineages.

Gene

XG

Chromosome	Xp22.3
Name	*XG (PBDX)*[2]
Organization	10 small exons distributed over approximately 60 kbp of gDNA. Exon 1 to exon 3 are present in the pseudoautosomal region of the X and the Y chromosomes. Exon 4 to exon 10 are only on the X chromosome.[2]
Product	Xg^a glycoprotein

CD99

Chromosome	Xp22.2 and Yp11.2
Name	*MIC2 (CD99)*
Organization	10 exons distributed over 52 kbp of gDNA
Product	CD99

Gene map

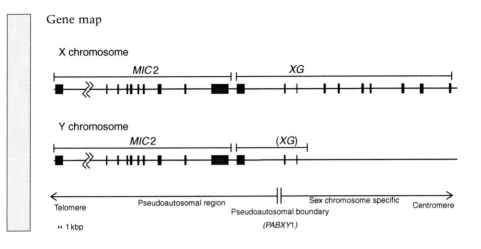

Database accession numbers

GenBank   S73261 (Xg); X16996 (CD99)
www.bioc.aecom.yu.edu/bgmut/index.htm

Xgᵃ amino acid sequence

```
MESWWGLPCL  AFLCFLMHAR  GQRDFDLADA  LDDPEPTKKP  NSDIYPKPKP   50
PYYPQPENPD  SGGNIYPRPK  PRPQPQPGNS  GNSGGYFNDV  DRDDGRYPPR  100
PRPRPPAGGG  GGGYSSYGNS  DNTHGGDHHS  TYGNPEGNMV  AKIVSPIVSV  150
VVVTLLGAAA  SYFKLNNRRN  CFRTHEPENV                          180
```

XG encodes a putative leader sequence of 21 amino acids

CD99 amino acid sequence

```
MARGAALALL  LFGLLGVLVA  APDGGFDLSD  ALPDNENKKP  TAIPKKPSAG   50
DDFDLGDAVV  DGENDDPRPP  NPPKPMPNPN  PNHPSSSGSF  SDADLADGVS  100
GGEGKGGSDG  GGSHRKEGEE  ADAPGVIPGI  VGAVVVAVAG  AISSFIAYQK  150
KKLCFKENAE  QGEVDMESHR  NANAEPAVQR  TLLEK                   185
```

MIC2 encodes a leader sequence of 22 amino acids

339

Carrier molecule

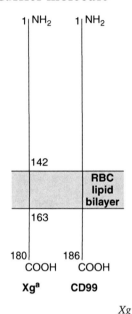

	Xg	*CD99*
M_r (SDS-PAGE)	22 000–29 000	32 500
CHO: N-glycan	0 site	0 site
CHO: O-glycan	11 potential sites[2]	11 potential sites[3]
Cysteine residues	3	1
Copies per RBC	9000 (polyclonal anti-Xg^a)	200 to 2000[3]
	18–450 (monoclonal anti-Xg^a)	

Function

CD99 (12E7 antigen), is an adhesion molecule[2]

Disease association

XG is linked to genes responsible for ichthyosis (*STS*), ocular albimism (*OAI*) and retinoschisis (*RS*).

High levels of CD99 in Ewing's sarcoma, some neuroectodermal tumors, lymphoblastic lymphoma and acute lymphoblastic leukemia[2].

Phenotypes (% occurrence)

	Male	*Female*
Xg(a+)	65.6	88.7
Xg(a−)	34.4	11.3

Phenotypic relationship of Xgᵃ and CD99 antigens

Male	Xg(a+)	CD99 High
	Xg(a−)	CD99 High
	Xg(a−)	CD99 Low
Female	Xg(a+)	CD99 High
	Xg(a+ᵂ)	CD99 High
	Xg(a−)	CD99 Low

Comments

First blood group system to be assigned to the X chromosome. Family studies with anti-Xgᵃ helped to define the mechanism responsible for various sex-chromosome aneuploides.

Xgᵃ and CD99 escape X-chromosome inactivation.

XG gene transcripts were detected in thymus, bone marrow and fetal liver, and in several non erythroid tissues: heart, placenta, skeletal muscle, prostate, thyroid, spinal cord, trachea[4].

References
[1] Latron, F. et al. (1987) Biochem. J. 247, 757–764.
[2] Tippett, P. and Ellis N.A. (1998) Transf. Med. Rev. 12, 233–257.
[3] Fouchet, C. et al. (2000) Immunogenetics 51, 688–694.
[4] Fouchet, C. et al. (2000) Blood 95, 1819–1826.

Xgᵃ ANTIGEN

Terminology

ISBT symbol (number)	XG1 (012.001)
History	Discovered in 1962 when serum from multiply transfused Mr. And detected an antigen with a higher prevalence in females than in males; encoded by a locus on the X chromosome. Named after the X chromosome and "g" from "Grand Rapids", where the patient was treated

Occurrence

Females	89%
Males	66%

Expression

Cord RBCs	Weak
Altered	Weak expression on RBCs from adult females heterozygous for Xg^a. Weak expression on RBCs from adult males is rare

Effect of enzymes/chemicals on Xg^a antigen on intact RBCs

Ficin/papain	Sensitive
Trypsin	Sensitive
α-Chymotrypsin	Sensitive
Pronase	Sensitive
Sialidase	Resistant
DTT 200 mM	Resistant
Acid	Not known

In vitro characteristics of alloanti-Xg^a

Immunoglobulin class	IgG more common than IgM
Optimal technique	RT; IAT; capillary
Complement binding	Some

Clinical significance of alloanti-Xg^a

Transfusion reaction	No
HDN	No

Autoanti-Xg^a

One example reported.

Comments

Some anti-Xg^a are naturally-occurring. Rarely occurs with other alloantibodies.
Xg^a is a poor immunogen.
Xg^a has a phenotypic relationship with CD99, see system pages.
Xg^a escapes X-chromosome inactivation.
Xg^a is found on RBCs from gibbons but not on RBCs from chimpanzees, gorillas, orangutans, baboons, celebes black apes, various monkeys, mice, dogs and coelacanth.

CD99 ANTIGEN

Terminology

ISBT symbol (number)	XG2 (012.002)
Other names	12E7, MIC2, E2, HuLy-m6, FMC29, HEC
History	Became part of the Xg blood group system in 2000 because *MIC2* and *XG* are adjacent, homologous genes and two CD99-negative people were found with the alloantibody

Occurrence

All populations: >99%; the only two CD99-negative probands that have been described are Japanese[1]

Expression

Cord RBCs	Weak

Effect of enzymes/chemicals on CD99 antigen on intact RBCs

Ficin/papain	Sensitive
Trypsin	Sensitive
α-Chymotrypsin	Sensitive
Pronase	Sensitive
Sialidase	Usually resistant
DTT 200 mM	Resistant
Acid	Not known

In vitro characteristics of alloanti-CD99

Immunoglobulin class	IgG
Optimal technique	IAT
Complement binding	No data

Clinical significance of alloanti-CD99

Transfusion reaction	No data
HDN	No data

Comments

CD99 escapes X-chromosome inactivation
CD99 has a phenotypic relationship to Xg^a see system pages

Reference

[1] Uchikawa, M. et al. (1995) Transfusion 35 (Suppl.), 23S (abstract).

343

Number of antigens   4

Terminology

ISBT symbol	SC
ISBT number	013
Other name	Sc
History	Established in 1974; named after the family of the first maker of anti-Sc1

Expression

Other blood cells	Weakly expressed in leukocytes, thymus, lymph nodes, spleen[1]
Tissues	Fetal liver, bone marrow in adults

Gene

Chromosome	1p34
Name	*SC (ERMAP)*
Organization	11 exons spanning 19 kbp of gDNA
Product	Sc glycoprotein [Human erythroid membrane associated protein (HERMAP)]

Gene map

SC1/SC2 (169G>A) encode Sc1/Sc2 (Gly57Arg)

Database accession numbers

GenBank   AY049028, AF311284, AF311285, AJ505027-50
www.bioc.aecom.yu.edu/bgmut/index.htm

Amino acid sequence[1,2]

```
MEMASSAGSW  LSGCLIPLVF  LRLSVHVSGH  AGDAGKFHVA  LLGGTAELLC   50
PLSLWPGTVP  KEVRWLRSPF  PQRSQAVHIF  RDGKDQDEDL  MPEYKGRTVL  100
VRDAQEGSVT  LQILDVRLED  QGSYRCLIQV  GNLSKEDTVI  LQVAAPSVGS  150
LSPSAVALAV  ILPVLVLLIM  VCLCLIWKQR  RAKEKLLYEH  VTEVDNLLSD  200
HAKEKGKLHK  AVKKLRSELK  LKRAAANSGW  RRARLHFVAV  TLDPDTAHPK  250
```

```
LILSEDQRCV  RLGDRRQPVP  DNPQRFDFVV  SILGSEYFTT  GCHYWEVYVG  300
DKTKWILGVC  SESVSRKGKV  TASPANGHWL  LRQSRGNEYE  ALTSPQTSFR  350
LKEPPRCVGI  FLDYEAGVIS  FYNVTNKSHI  FTFTHNFSGP  LRPFFEPCLH  400
DGGKNTAPLV  ICSELHKSEE  SIVPRPEGKG  HANGDVSLKV  NSSLLPPKAP  450
ELKDIILSLP  PDLGPALQEL  KAPSF                               475
```

Antigen mutation is numbered by counting Met as 1.
SC encodes a leader sequence of 29 amino acids.

Carrier molecule[1,3]

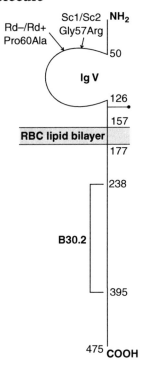

M_r (SDS-PAGE)	60 000–68 000[4]
CHO: N-glycan	Four sites
Cysteine residues	11
Copies per RBC	Not determined

Molecular basis of antigens

Antigen	Amino acid change	Exon	Nt change	Restriction enzyme
Sc1/Sc2	Gly57Arg	3	169G>A	*Sma* I (+/−)
Rd−/Rd+	Pro60Ala	3	178C>G	

Function

Human ERMAP is an erythroid transmembrane adhesion/receptor protein.

Disease association

Not known.

Phenotypes (% occurrence)

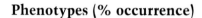

	Caucasians	Blacks
Sc:1,−2	99	100
Sc:1,2	1	0
Sc:−1,2	Very rare	0
Sc:1,−2,Rd+	Very rare	Very rare
Sc:1,2,Rd+	Very rare	0

Null: Sc:−1,−2,−3.

Molecular basis of Sc$_{null}$ phenotypes

54C>T, 76C>T in exon 2 (His26Tyr); 307 delGA in exon 3; frameshift; truncated protein of 113 amino acids.

Comments

The antibodies of 3 Sc:1,−2 individuals did not react with Sc:−1,−2 RBCs, but the antibodies were mutually incompatible[5].

The extracellular IgV domain of HERMAP is homologous with the butyrophilin family of milk proteins, autoantigens, and avian blood group antigens[1].

The intracellular B30.2 domain is highly homologous with a similar domain in a diverse group of proteins, including butyrophilin, pyrin and MID1[1].

References

[1] Su, Y.Y. et al. (2001) Blood Cells Mol. Dis. 27, 938–949.
[2] Xu, H. et al. (2001) Genomics 76, 2–4.
[3] Wagner, F.F. et al. (2003) Blood 101, 752–757.
[4] Spring, F.A. et al. (1990) Vox Sang. 58, 122–125.
[5] Devine, P. et al. (1988) Transfusion 28, 346–349.

Sc1 ANTIGEN

Terminology

ISBT symbol (number)	SC1 (013.001)
Other names	Sm
History	Identified in 1962; name changed from Sm to Sc1 in 1974 when the Scianna system was established. Named for the first maker of anti-Sc1

Occurrence

All populations: 99%.

Antithetical antigen

Sc2 (**SC2**)

Expression

Cord RBCs	Expressed

Molecular basis associated with Sc1 antigen[1]

Amino acid	Gly 57
Nucleotide	169G in exon 3
Restriction enzyme	*Sma* I site present

Effect of enzymes/chemicals on Sc1 antigen on intact RBCs

Ficin/papain	Resistant
Trypsin	Resistant
α-Chymotrypsin	Resistant
Pronase	Weakened
Sialidase	Resistant
DTT 200 mM/50 mM	Sensitive/resistant
Acid	Presumed resistant

In vitro characteristics of alloanti-Sc1

Immunoglobulin class	IgG
Optimal technique	IAT
Complement binding	No

347

Clinical significance of alloanti-Sc1

Transfusion reaction   No
HDN   Positive DAT but no clinical HDN

Autoanti-Sc1

Yes[2]. Some examples are reactive in tests using serum but not plasma[3].

References
1. Wagner, F.F. et al. (2003) Blood 101, 752–757.
2. Owen, I. et al. (1992) Transfusion 32, 173–176.
3. Tregellas, W.M. et al. (1979) Transfusion 19, 650 (abstract).

Sc2 ANTIGEN

Terminology

ISBT symbol (number)   SC2 (013.002)
Other names   Bu[a], Bullee
History   Identified in 1962 and named Bu[a]; renamed Sc2 in 1974 when it was shown to be antithetical to Sm (now Sc1)

Occurrence

1% in Northern Europeans; more common in Mennonites.

Antithetical antigen

Sc1 (SC1)

Expression

Cord RBCs   Expressed

Molecular basis associated with Sc2 antigen[1]

Amino acid   Arg 57
Nucleotide   169A in exon 3
Restriction enzyme   *Sma* I site lost

Effect of enzymes/chemicals on Sc2 antigen on intact RBCs

Ficin/papain	Resistant
Trypsin	Resistant
α-Chymotrypsin	Resistant
Pronase	Sensitive
Sialidase	Resistant
DTT 200 mM/50 mM	Sensitive/resistant
Acid	Presumed resistant

In vitro characteristics of alloanti-Sc2

Immunoglobulin class	IgG
Optimal technique	IAT
Complement binding	No

Clinical significance of alloanti-Sc2

Transfusion reaction	No
HDN	Positive DAT but no clinical HDN

Comment

Sc2 antigen has variable expression among different people.

Reference
[1] Wagner, F.F. et al. (2003) Blood 101, 752–757.

Sc3 ANTIGEN

Terminology

ISBT symbol (number)	SC3 (013.003)
History	Named in 1980 when a person with SC:−1,−2 RBCs made an antibody to a high incidence antigen

Occurrence

All populations	100%

Expression

Cord RBCs   Presumed expressed

Molecular basis associated with Sc3 antigen

For molecular basis of the SC:$-1,-2,-3$ (the null phenotype) see system pages.

Effect of enzymes/chemicals on Sc3 antigen on intact RBCs

Ficin/papain   Resistant ($\uparrow$)
Trypsin   Resistant
α-Chymotrypsin   Resistant
Pronase   Sensitive
Sialidase   Resistant
DTT 200 mM/50 mM   Sensitive/resistant
Acid   Presumed resistant

In vitro characteristics of alloanti-Sc3

Immunoglobulin class   IgG
Optimal technique   IAT
Complement binding   No

Clinical significance of alloanti-Sc3

Transfusion reaction   No to mild/delayed
HDN   Mild

Autoanti-Sc3

Autoanti-Sc3-like antibody in 2 patients with suppressed Sc antigens (1 patient with lymphoma; 1 with Hodgkins disease)[1].

Comments

Of three known probands, the original one was from the Marshall Islands in the South Pacific and the third was from Papua, New Guinea[2]. Six other SC:$-1,-2$ people were found in testing 29 people from her village[2,3].

References
1 Peloquin, P. et al. (1989) Transfusion 29 (Suppl.), 49S (abstract).
2 McCreary, J. et al. (1975) Transfusion 13, 350 (abstract).
3 Woodfield, D.G. et al. (1986) 19th congress of the International Society of Blood Transfusion 651 (abstract).

Sc4 ANTIGEN

Terminology

ISBT symbol (number)	Rd (013.004)
Other names	Radin, Rd[a], 700.015
History	Named after the first family in which the antibody caused HDN. Became part of the SC system when the associated polymorphism in human ERMAP was identified

Occurrence

All populations	<0.01%
Danes	0.5%
Jews, Canadians, African blacks	0.1%

Expression

Cord RBCs	Expressed

Molecular basis associated with Rd antigen[1]

Amino acid	Ala 60
Nucleotide	178G in exon 3
Rd-negative	Pro 60 and C at bp178

Effect of enzymes/chemicals on Rd antigen on intact RBCs

Ficin/papain	Resistant
Trypsin	Variable
α-Chymotrypsin	Variable
Pronase	Sensitive
Sialidase	Resistant
DTT 200 mM/50 mM	Sensitive/resistant
Acid	Presumed resistant

In vitro characteristics of alloanti-Rd

Immunoglobulin class	IgG
Optimal technique	IAT
Complement binding	No

Clinical significance of alloanti-Rd

Transfusion reaction   No
HDN   Mild to severe

Reference
[1] Wagner, F.F. et al. (2003) Blood 101, 752–757.

Number of antigens   5

Terminology

ISBT symbol	DO
ISBT number	014
History	Named after the producer of the first anti-Do^a; identified in 1965

Expression

Other blood cells	Lymphocytes
Tissues	Primarily in adult bone marrow and fetal liver; also in spleen, lymph nodes, intestine, ovary, testes, and fetal heart.

Gene[1–3]

Chromosome	12p13.2–p12.1
Name	*DO (ART4)*
Organization	Three exons distributed over 14 kbp of gDNA
Product	Do glycoprotein

Gene map

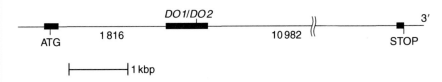

DO1/DO2 (378C>T; 624T>C; 793A>G) encode Do^a/Do^b (126Tyr, 208Leu, Asn265Asp)

Database accession numbers

GenBank   AF290204; XM_017877; X95826
www.bioc.aecom.yu.edu/bgmut/index.htm

Amino acid sequence[1]

Leader sequence: Amino acids 1–44 or, more likely, 22–44 if initiation occurs at the second AUG[4]

```
MGPLINRCKK ILLPTTVPPA TMRIWLLGGL LPFLLLLSGL QSPTEGSEVA  50
IKIDFDFAPG SFDDQYQGCS KQVMEKLTQG DYFTKDIEAQ KNYFRMWQKA 100
HLAWLNQGKV LPQNMTTTHA VAILFYTLNS NVHSDFTRAM ASVARTPQQY 150
ERSFHFKYLH YYLTSAIQLL RKDSIMENGT LCYEVHYRTK DVHFNAYTGA 200
TIRFGQFLST SLLKEEAQEF GNQTLFTIFT CLGAPVQYFS LKKEVLIPPY 250
ELFKVINMSY HPRGNWLQLR STGNLSTYNC QLLK
                                      ASSKKC IPDPIAIASL 300
SFLTSVIIFS KSRV                                        314
```

GPI-anchor motif: Amino acids 298–314 or, more likely 285–314[4].
Antigen mutation is numbered by counting Met as 1.

Carrier molecule

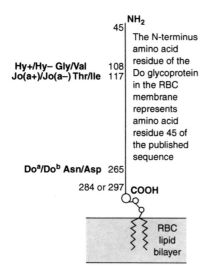

	45	NH$_2$
Hy+/Hy− Gly/Val	108	The N-terminus amino acid residue of the Do glycoprotein
Jo(a+)/Jo(a−) Thr/Ile	117	in the RBC membrane represents amino acid residue 45 of the published sequence
Do^a/Do^b Asn/Asp	265	
	284 or 297	COOH

RBC lipid bilayer

M_r (SDS-PAGE)	47 000–58 000
CHO: N-glycan	5 potential sites
Cysteine residues	4 or 5 in membrane-bound protein

Molecular basis of antigens[4]

Antigen	Amino acid change	Exon	Nt change	Restriction enzyme
Do^a/Do^b	Tyr126Tyr	2	378C>T	DraIII (+/−)
	Leu208Leu	2	624T>C	MnlI (−/+)
	Asn265Asp	2	793 A>G	BSe RI (−/+)
Hy+/Hy−	Gly108Val	2	323G>T	BsaI (+/−)
Jo(a+)/Jo(a−)	Thr117Ile	2	350C>T	XcmI (+/−)

Function

Its function in RBCs is not known. Homologous to ADP-ribosyltransferases (ART) 1, 2 and 3; identical to ART4. [1-3,5]

Disease association

Absent from PNH III RBCs.

Phenotypes (% occurrence)

Phenotype	Do^a	Do^b	Gy^a	Hy	Jo^a	Caucasians	Blacks
Do(a+b−)	+	0	+	+	+	18	11
Do(a+b+)	+	+	+	+	+	49	44
Do(a−b+)	0	+	+	+	+	33	45
Gy(a−)	0	0	0	0	0	Rare	Rare[6]
Hy−	0	wk	wk	0	0/wk[7]	0	Rare
Jo(a−)	wk	0/wk	+	wk	0	0	Rare

Null: Gregory negative [Gy(a−)].

Molecular basis of Do$_{null}$ [Gy(a−)] phenotype[4]

IVS1 −2 a>g: Skipping of exon 2 on *DOB* allele
IVS1 +2 t>c: Skipping of exon 2 on *DOB* allele
Exon 2 442C>T; Gln148Stop on a variant *DO* allele (*GY5*)
Deletion of eight nucleotides (nt 343–350) in exon 2; frameshift; premature stop codon on *DOA* allele[8].

Molecular basis of Hy− and Jo(a−) phenotypes[4]

Hy−
323T, 378C, 624C, 793G in exon 2; 898G in exon 3 (*HY1*): 108Val, 265Asp, 300Val
323T, 378C, 624C, 793G in exon 2; 898C in exon 3 (*HY2*): 108Val, 265Asp, 300Leu
Jo(a−)
350T, 378T, 624T, 793A in exon 2 (*JO*): 117Ile, 265Asn

Comments

Epitope recognized by monoclonal anti-Do (MIMA 52) is restricted to the greater apes[9]. Other monoclonal anti-Do agglutinate RBCs from greater and lesser apes.

References

1 Gubin, A.N. et al. (2000) Blood 96, 2621–2627.
2 Koch-Nolte, F. et al. (1997) Genomics 39, 370–376.
3 Koch-Nolte, F. et al. (1999) Genomics 55, 130.
4 Reid, M.E. (2003) Transfusion 43, 107–114.
5 Grahnert, A. et al. (2002) Biochem. J. 362, 717–723.
6 Smart, E.A. et al. (2000) Vox Sang. 78 (Suppl. 1), P015 (abstract).
7 Scofield, T. et al. (2001) Transfusion 41 (Suppl.), 24S (abstract).
8 Bailly, P. et al. (2001) Transfus. Clin. Biol. 8 (Suppl. 1), 167s (abstract).
9 Rios, M. et al. (2002) Transfusion 42, 52–58.

Doᵃ ANTIGEN

Terminology

ISBT symbol (number)	DO1 (014.001)
Other name	Dombrock
History	Named after the proband who made anti-Doᵃ; reported in 1965

Occurrence

Caucasians	67%
Blacks	55%
Japanese	24%
Thais	14%

Antithetical antigen

Doᵇ (**DO2**)

Expression

Cord RBCs	Expressed
Altered	Absent from PNH III RBCs; absent from Hy− RBCs, and weak on Jo(a−) RBCs

Molecular basis associated with Doᵃ antigen[1]

Amino acid	Asp 265
Nucleotide	C at bp 378 (silent 126Tyr); T at bp 624 (silent 208Leu); A at bp 793; all in exon 2

Restriction enzyme 378 *Dra* III present; 624 *Mnl* I absent; 793 *BSe* RI absent

Effect of enzymes/chemicals on Do^a antigen on intact RBCs

Ficin/papain	Resistant
Trypsin	Sensitive
α-Chymotrypsin	Weakened
Pronase	Sensitive (weakened)
Sialidase	Resistant
DTT 200 mM/50 mM	Sensitive/resistant
Acid	Resistant

In vitro characteristics of alloanti-Do^a

Immunoglobulin class	IgG
Optimal technique	IAT; PEG; enzyme IAT
Complement binding	No

Clinical significance of alloanti-Do^a

Transfusion reaction	Delayed and acute/hemolytic
HDN	Positive DAT but no clinical HDN

Comments

Anti-Do^a is notorious for disappearing *in vivo*.
Poor immunogen; rarely found as a single specificity.

Reference
[1] Gubin, A.N. et al. (2000) Blood 96, 2621–2627.

Do^b ANTIGEN

Terminology

ISBT symbol (number)	DO2 (014.002)
History	Named when it was recognized to be antithetical to Do^a, reported in 1973

357

Occurrence

Caucasians	82%
Blacks	89%

Antithetical antigen

Do^a (**DO1**)

Expression

Cord RBCs	Expressed
Altered	Absent from PNH III RBCs. Weak on Hy−, and absent or weak on Jo(a−) RBCs

Molecular basis associated with Do^b antigen[1]

Amino acid	Asn 265
Nucleotide	T at bp 378 (silent 126Tyr); C at bp 624 (silent 208Leu); G at bp 793; all in exon 2
Restriction enzyme	378 *Dra* III absent; 624 *Mnl* I present; 793 *BSe* RI present

Effect of enzymes/chemicals on Do^b antigen on intact RBCs

Ficin/papain	Resistant
Trypsin	Sensitive
α-Chymotrypsin	Weakened
Pronase	Sensitive (weakened)
Sialidase	Resistant
DTT 200 mM/50 mM	Sensitive/resistant
Acid	Resistant

In vitro characteristics of alloanti-Do^b

Immunoglobulin class	IgG
Optimal technique	IAT; PEG; enzyme IAT
Complement binding	No

Clinical significance of alloanti-Do^b

Transfusion reaction	Acute and delayed
HDN	Positive DAT but no clinical HDN

Comments

Poor immunogen; rarely found as a single specificity.

Reference
[1] Gubin, A.N. et al. (2000) Blood 96, 2621–2627.

Gyᵃ ANTIGEN

Terminology

ISBT symbol (number)	DO3 (014.003)
Other names	Gregory; GY1; 206.001; 900.005
History	Named in 1967 after the last name of the first producer of the antibody. Placed in the Dombrock system in 1992 when it was recognized that Gy(a−) was the null phenotype of Do[1]

Occurrence

Most populations	100%
Eastern European (Romany) ancestry, Japanese	>99%
Blacks	one proband[2]

Expression

Cord RBCs	Weak
Altered	Absent from PNH III RBCs; weak on Hy− RBCs

Molecular basis associated with Gyᵃ antigen

Unknown. Molecular bases underlying the Gy(a−) phenotype (see table on System pages) result in an absence of Do glycoprotein from the membrane.

Effect of enzymes/chemicals on Gyᵃ antigen on intact RBCs

Ficin/papain	Resistant (↑)
Trypsin	Sensitive
α-Chymotrypsin	Weakened

Pronase	Sensitive (weakened)
Sialidase	Resistant
DTT 200 mM/50 mM	Sensitive/resistant
Acid	Resistant

In vitro characteristics of alloanti-Gy^a

Immunoglobulin class	IgG
Optimal technique	IAT
Complement binding	No

Clinical significance of alloanti-Gy^a

Transfusion reaction	No to moderate/delayed
HDN	Positive DAT, but no clinical HDN

Comments

Gy^a is highly immunogenic.

References
[1] Banks, J.A. et al. (1995). Vox Sang. 68, 177–182.
[2] Smart, E.A. et al. (2000). Vox Sang. 78 (Suppl. 1), P015 (abstract).

Hy ANTIGEN

Terminology

ISBT symbol (number)	DO4 (014.004)
Other names	Holley; GY2; 206.002; 900.011
History	Reported in 1967 and named after the proband who made anti-Hy. Joined the Dombrock system in 1995.

Occurrence

Most populations	100%
Blacks	>99%

Expression

Cord RBCs	Weak
Altered	Absent from PNH III RBCs; weak on Jo(a−) RBCs

Molecular basis associated with Hy antigen[1]

Amino acid	Gly 108
Nucleotide	G at bp 323 in exon 2
Restriction enzyme	*Bsa* I site is ablated in Hy− phenotype
Hy−	Val 108 and Asp 265; T at bp 323 and G at bp 793 in exon 2 (*HY2* allele); *HY1* allele also has 898C > G in exon 3 (Leu300Val)

Effect of enzymes/chemicals on Hy antigen on intact RBCs

Ficin/papain	Resistant (↑)
Trypsin	Sensitive
α-Chymotrypsin	Weakened
Pronase	Sensitive (weakened)
Sialidase	Resistant
DTT 200 mM/50 mM	Sensitive/resistant
Acid	Resistant

In vitro characteristics of alloanti-Hy

Immunoglobulin class	IgG
Optimal technique	IAT
Complement binding	No

Clinical significance of alloanti-Hy

Transfusion reaction	No to moderate/delayed
HDN	Positive DAT, but no clinical HDN

Comments

Anti-Hy do not react well by PEG-IAT.

Reference
[1] Reid, M.E. (2003) Transfusion 43, 107–114.

361

Joᵃ ANTIGEN

Terminology

ISBT symbol (number)	DO5 (014.005)
Other names	Joseph; 901.004; 900.010
History	Reported in 1972 and named after the proband who was reported to have made anti-Joᵃ. Joined the Dombrock system in 1992. The original and second probands were later shown to be Hy–! The third proband had a *JO* allele and made anti-Joᵃ. This explains some of the confusion in differentiating Hy and Joᵃ antigens and antibodies

Occurrence

Most populations	100%
Blacks	>99%

Expression

Cord RBCs	Weak
Altered	Absent or weak on Hy− and PNH III RBCs

Molecular basis associated with Joᵃ antigen[1]

Amino acid	Thr 117
Nucleotide	C at bp 350 in exon 2
Restriction enzyme	*Xcm* I site ablated in Jo(a−) phenotype
Jo(a−)	Ile 117, Asn 265; T at bp 350 and A at bp 793 in exon 2

Effect of enzymes/chemicals on Joᵃ antigen on intact RBCs

Ficin/papain	Resistant (↑)
Trypsin	Sensitive
α-Chymotrypsin	Weakened
Pronase	Sensitive (weakened)
Sialidase	Resistant
DTT 200 mM/50 mM	Variable
Acid	Weakened

In vitro characteristics of alloanti-Jo[a]

Immunoglobulin class   IgG
Optimal technique   IAT
Complement binding   No

Clinical significance of alloanti-Jo[a]

Transfusion reaction   No to moderate/delayed
HDN   No

Comment

Anti-Jo[a] do not react well by PEG-IAT.

Reference

[1] Reid, M.E. (2003). Transfusion 43, 107–114.

Number of antigens   3

Terminology

ISBT symbol	CO
ISBT number	015
History	Named in 1967 for the first of the three original producers of anti-Co^a. Should have been named Calton, but the handwriting on the tube was misread

Expression

Tissues: Apical surface of proximal tubules, basolateral membranes subpopulation of collecting ducts in cortex, descending tubules in medulla, liver bile ducts, gall bladder, eye epithelium, cornea, lens, choroid plexus, hepatobiliary epithelia, capillary endothelium[1].

Gene

Chromosome	7p14
Name	*CO [AQP1 (Aquaporin-1)]*
Organization	Four exons distributed over 11.6 kbp of gDNA
Product	Channel-forming integral protein (CHIP); Aquaporin 1 (AQP1); CHIP-1; CHIP28

Gene map

CO 1/CO 2 (134C>T) encode Co^a/Co^b (Ala45Val)

Database accession numbers

GenBank   M77829
www.bioc.aecom.yu.edu/bgmut/index.htm

Amino acid sequence[1]

```
MASEFKKKLF WRAVVAEFLA TTLFVFISIG SALGFKYPVG NNQTAVQDNV  50
KVSLAFGLSI ATLAQSVGHI SGAHLNPAVT LGLLLSCQIS IFRALMYIIA 100
QCVGAIVATA ILSGITSSLT GNSLGRNDLA DGVNSGQGLG IEIIGTLQLV 150
LCVLATTDRR RRDLGGSAPL AIGLSVALGH LLAIDYTGCG INPARSFGSA 200
VITHNFSNHW IFWVGPFIGG ALAVLIYDFI LAPRSSDLTD RVKVWTSGQV 250
EEYDLDADDI NSRVEMKPK                                   269
```
Antigen mutation is numbered by counting Met as 1.

Carrier molecule[1]

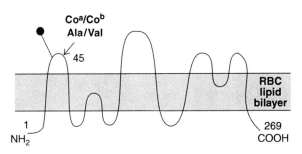

M_r (SDS-PAGE)	28 000 unglycosylated; 40 000–60 000 glycosylated form
CHO: N-glycan	Polylactosaminoglycan that carries ABH determinants at residue 42
Cysteine residues	4
Copies per RBC	120 000–160 000 molecules arranged in tetramers

Molecular basis of antigens[2]

Antigen	Amino acid change	Exon	Nt change	Restriction enzyme
Co^a/Co^b	Ala45Val	1	134C>T	*Pfl*M I (−/+)

Function

Water transport. AQP1 accounts for 80% of water re-absorption in kidneys and is a determinant of vascular permeability in the lung[3].

Disease association

Monosomy 7, weak expression associated with certain chromosome 7 rearrangements, leukemia.

Phenotypes (% occurrence)

	Most populations
Co(a+b−)	90
Co(a−b+)	0.5
Co(a+b+)	9.5
Co(a−b+)	<0.01

Null: Co(a−b−).

Molecular basis of Co(a−b−) phenotype

Exon 1 del[4]
307–308insT, leading to frameshift[4]
113C>T in exon 1, Pro38Leu, which causes a decreased amount of CHIP[4]
614C>A in exon 3, Asn192Lys, which destroys NPA motif, essential for function[5]
232delG in exon 1 leading to frameshift[6]
601delC in exon 3 leading to stop in the next codon[7]

Molecular basis of Co(a−b+^{w})

140A>G in exon 1; Gln47Arg with Val 45[8].

Comments

In RBCs, CHIP-1 exists as a dimer and accounts for 2.4% of the total membrane protein[9,10].

References
1 Preston, G.M. and Agre, P. (1991) Proc. Natl. Acad. Sci. USA 88, 11110–11114.
2 Smith, B.L. et al. (1994) J. Clin. Invest. 94, 1043–1049.
3 King, L.S. et al. (2002) Proc. Natl. Acad. Sci. USA 99, 1059–1063.
4 Preston, G.M. et al. (1994) Science 265, 1585–1587.
5 Chretien, S. and Catron, J.P. (1999) Blood 93, 4021–4023.
6 Joshi, S.R. et al. (2001) Transfusion 41, 1273–1278.
7 Nance, S. et al. (2002) Transfusion 42 (Suppl.), 105S (abstract).
8 Wagner, F.F. and Flegel, W.A. (2002) Transfusion 42 (Suppl.), 24S–25S (abstract).
9 Agre, P. et al. (2002) J. Physiol. London 542, 3–16.
10 Kozono, D. et al. (2002) J. Clin. Invest. 109, 1395–1399.

Coᵃ ANTIGEN

Terminology

ISBT symbol (number)	CO1 (015.001)
Other name	Colton
History	Named in 1967 after the first antibody producer. Should have been named Calton, but the handwriting on the tube was misread

Occurrence

All populations	99.9%

Antithetical antigen

Coᵇ (CO2)

Expression

Cord RBCs	Expressed

Molecular basis associated with Coᵃ antigen[1]

Amino acid	Ala 45
Nucleotide	C at bp 134 in exon 1
Restriction enzyme	*PfIM* I site absent

Effect of enzymes/chemicals on Coᵃ antigen on intact RBCs

Ficin/papain	Resistant
Trypsin	Resistant
α-Chymotrypsin	Resistant
Pronase	Resistant
Sialidase	Resistant
DTT 200 mM	Resistant
Acid	Resistant

In vitro characteristics of alloanti-Coᵃ

Immunoglobulin class	IgG (Rare IgM reported[2])
Optimal technique	IAT
Complement binding	Some

Clinical significance of alloanti-Co^a

Transfusion reaction	No to moderate/delayed; immediate/hemolytic[3]
HDN	Mild to severe[4] (rare)

Autoanti-Co^a

One example

References
[1] Smith, B.L. et al. (1994) J. Clin. Invest. 94, 1043–1049.
[2] Kurtz, S.R. et al. (1982) Vox Sang. 43, 28–30.
[3] Covin, R.B. et al. (2001) Immunohematology 17, 45–49.
[4] Simpson, W.K.H. (1973) S. Afr. Med. J. 47, 1302–1304.

Co^b ANTIGEN

Terminology

ISBT symbol (number)	CO2 (015.002)
History	Named in 1970 when it was shown to be antithetical to Co^a

Occurrence

All populations: 10%

Antithetical antigen

Co^a (CO1)

Expression

Cord RBCs	Expressed
Altered	Co(a−) RBCs with weak expression of Co^b exist. See system pages.

Molecular basis associated with Co^b antigen[1]

Amino Acid	Val 45
Nucleotide	T at bp 134 in exon 1
Restriction enzyme	*PfIM* I digestion site present

Effect of enzymes/chemicals on Co^b antigen on intact RBCs

Ficin/papain	Resistant (↑)
Trypsin	Resistant
α-Chymotrypsin	Resistant
Pronase	Resistant
Sialidase	Resistant
DTT 200 mM	Resistant
Acid	Resistant

In vitro characteristics of alloanti-Co^b

Immunoglobulin class	IgG
Optimal technique	IAT
Complement binding	Rare

Clinical significance of alloanti-Co^b

Transfusion reaction	No to moderate/delayed/hemolytic
HDN	Mild

Comments

Poor immunogen, rarely found as a single specificity.

Reference
[1] Smith, B.L. et al. (1994) J. Clin. Invest. 94, 1043–1049.

Co3 ANTIGEN

Terminology

ISBT symbol (number)	CO3 (015.003)
Other names	Co^{ab}
History	Reported in 1974, when an antibody to a common antigen (called anti-Co^{a}Co^b) was found in a patient whose RBCs typed Co(a−b−)

Occurrence

All populations	100%

Expression

Cord RBCs   Expressed

Molecular basis associated with lack of Co3 antigen

Refer to system pages.

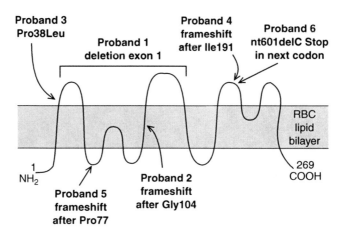

Location of Co3 antigen is not known, but it is absent from RBCs that lack, or have small amounts, of CHIP-1.

Effect of enzymes/chemicals on Co3 antigen on intact RBCs

Ficin/papain	Resistant
Trypsin	Resistant
α-Chymotrypsin	Resistant
Pronase	Resistant
Sialidase	Resistant
DTT 200 mM	Resistant
Acid	Resistant

In vitro characteristics of alloanti-Co3

Immunoglobulin class	IgG
Optimal technique	IAT
Complement binding	Yes

Clinical significance of alloanti-Co3

Transfusion reaction	Not known
HDN	Severe

Autoanti-Co3

One example described as mimicking anti-Co3 made by non-Hodgkins lymphoma patient.

Comments

Poor immunogen, rarely found as a single specificity. RBCs from a baby with congenital dyserythropoietic anemia (CDA) were Co(a−b−), In(a−b−), AnWj− and had a weak expression of LW[1,2].

References
[1] Parsons, S.F. et al. (1994) Blood 83, 860–868.
[2] Agre, P. et al. (1994) J. Clin. Invest. 94, 1050–1058.

Number of antigens 3

Terminology

ISBT symbol	LW
ISBT number	016
CD number	CD242
Other name	ICAM-4
History	Anti-LW, or 'anti-Rh' as it was called, was produced in 1940. However, the phenotypic relationship between LW and the RhD antigen delayed recognition that LW was an independent blood group system until 1963 when it was named to honor Landsteiner and Wiener who made anti-LW in rabbits and guinea pigs after immunizing them with blood from *Macacus rhesus*. In 1982, it became a three-antigen system. The LW_1, LW_2, LW_3 and LW_4 terminology was changed when it was realized that anti-Ne^a (now called anti-LW^b) detects an antigen antithetical to that recognized by anti-LW made by LW_3 people (now called anti-LW^a)[1]

Phenotypes

Obsolete	Obsolete	Current
LW+, D+	LW_1	LW(a+b−) or LW(a+b+)
LW+, D−	LW_2	LW(a+b−) or LW(a+b+)
LW−, D+ or D−	LW_3	LW(a−b+)
LW−, D+ or D−	LW_4	LW(a−b−)
LW−, Rh_{null}	LW_0	LW(a−b−)

Expression

Tissues	May be found in placenta

Gene[2]

Chromosome	19p13.3
Name	*LW*
Organization	3 exons distributed over 2.6 kbp of gDNA
Product	LW glycoprotein

Gene map

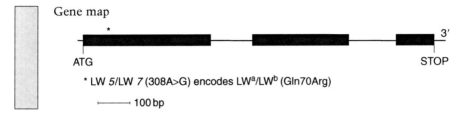

ATG

* LW *5*/LW *7* (308A>G) encodes LW^a/LW^b (Gln70Arg)

STOP

$\longmapsto$ 100 bp

Database accession numbers

GenBank   X93093
www.bioc.aecom.yu.edu/bgmut/index.htm

Amino acid sequence

```
                      MGSLFPLSLL FFLAAAYPGV GSALGRRTKR  -01
AQSPKGSPLA PSGTSVPFWV RMSPEFVAVQ PGKSVQLNCS NSCPQPQNSS   50
LRTPLRQGKT LRGPGWVSYQ LLDVRAWSSL AHCLVTCAGK TRWATSRITA  100
YKPPHSVILE PPVLKGRKYT LRCHVTQVFP VGYLVVTLRH GSRVIYSESL  150
ERFTGLDLAN VTLTYEFAAG PRDFWQPVIC HARLNLDGLV VRNSSAPITL  200
MLAWSPAPTA LASGSIAALV GILLTVGAAY LCKCLAMKSQ A           241
```
Antigen mutation(s) are numbered by counting Arg as 1.
LW encodes a leader sequence of 30 amino acids.

Carrier molecule[3]

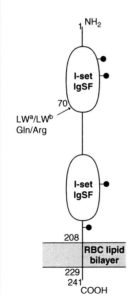

M_r (SDS-PAGE)	37 000–43 000
CHO: N-glycan	Four potential sites at residue 38, 48, 160, 193
CHO: O-glycan	Present
Cysteine residues	Three pairs at residues 39/83, 43/87 and 123/180
Copies per RBC[4]	D+ 4400 (Adult) 5150 (cord)
	D− 2835 (Adult) 3620 (cord)

Molecular basis of antigens[5]

Antigen	Amino acid change	Exon	Nt change	Restriction enzyme
LW[a]/LW[b]	Gln70Arg	1	308A>G	Pvu>II (+/−)

Function

The LW glycoprotein is an intercellular adhesion molecule (ICAM-4) and a ligand for integrins. LW has 30% sequence identity with other ICAMs. ICAM-4 binds to CD11/CD18 ($\alpha_1\beta_2$) integrin and LFA-1 leukocyte integrins $\alpha_4\beta_1$, $\alpha_v\beta_1$, $\alpha_v\beta_5$ and maybe $\alpha_v\beta_3$[6–8]. Possible marker for lymphocyte maturation or differentiation. May assist in stabilizing erythroblastic islands during erythropoiesis. May be involved in removal of senescent RBCs[3].

Disease association

LW antigens may be depressed during pregnancy and some diseases, for example, Hodgkin's disease, lymphoma, leukemia and sarcoma[3]. Autoanti-LW is common in patients with warm AIHA.

Expression of ICAM-4 is elevated on sickle RBCs and interaction between ICAM-4 and vascular endothelial cells may be involved in microvascular occlusions during painful crises of SCD.

Phenotypes (% occurrence)

	Most populations	Finns
LW(a+b−)	97%	93.9%
LW(a+b+)	3%	6.0%
LW(a−b+)	Rare	0.1%

Null: LW(a−b−). Rh$_{null}$ RBCs type LW(a−b−) although *LW* is normal.

There is a phenotypic relationship between LW and D antigens: In adults, D− RBCs have lower expression of LW antigens than D+ RBCs (ratio 1:1.5). In cord RBCs LW is strongly expressed in D− and D+ RBCs.

Molecular basis of phenotypes[2]

Phenotype	Basis
LW(a−b−)	346delACCTGCGCAG in exon 1 (codons 86–89); frameshift; Stop

Comments

LW antigens require intramolecular disulfide bonds and the presence of divalent cations, notably Mg^{++} for expression[9].

Since Rh_{null} RBCs were the only cells that failed to elicit an antibody response in animals, they are presumed to be the only true LW(a−b−).

LW antigens have been detected on RBCs of chimpanzee, gorilla, orangutan, baboon and various species of monkey but not on RBCs of rabbit, mouse, rat, sheep, goat, horse and cattle.

References
1 Sistonen, P. and Tippett, P. (1982) Vox Sang. 42, 252–255.
2 Hermand, P. et al. (1996) Blood 87, 2962–2967.
3 Parsons, S.F. et al. (1999) Baillieres Clin. Haematol. 12, 729–745.
4 Mallinson, G. et al. (1986) Biochem. J. 234, 649–652.
5 Hermand, P. et al. (1995) Blood 86, 1590–1594.
6 Bailly, P. et al. (1995) Eur. J. Immunol. 25, 3316–3320.
7 Spring, F.A. et al. (2001) Blood 98, 458–466.
8 Zennadi, R. et al. (2002). Blood 100 (Suppl., Part 1), 118a (abstract).
9 Bloy, C. et al. (1990) J. Biol. Chem. 265, 21482–21487.

LW[a] ANTIGEN

Terminology

ISBT symbol (number)	LW5 (016.005)
Other names	LW, LW1 in D+, LW2 in D−
History	Named LW[a] in 1982 when the antithetical relationship to Ne[a] (LW[b]) was recognized. LW1 to LW4 were made obsolete because they had been used to designate phenotypes

Occurrence

All populations   100%

Antithetical antigen

LW^b (**LW7**)

Expression

Cord RBCs	Well expressed on D-positive and D-negative
Altered	Weak on D−RBCs from adults
	Weak or absent on RBCs stored in EDTA

Molecular basis associated with LW^a antigen[1]

Amino acid	Gln 70
Nucleotide	A at bp 308 in exon 1
Restriction enzyme	*Pvu* II site

Effect of enzymes/chemicals on LW^a antigen on intact RBCs

Ficin/papain	Resistant
Trypsin	Resistant
α-Chymotrypsin	May be weakened
Pronase	Sensitive
Sialidase	Resistant
DTT 200 mM/50 mM	Sensitive/sensitive
Acid	Resistant

In vitro characteristics of alloanti-LW^a

Immunoglobulin class	IgG (usually); IgM
Optimal technique	RT or IAT
Complement binding	No

Clinical significance of alloanti-LW^a

Transfusion reaction	No to mild/delayed (Rare; D−, LW^a+ RBCs survive well)
HDN	No to mild (very rare)

Autoanti-LW^a

Autoanti-LW^a with suppression of LW antigens has been reported. Common in serum of patients with warm AIHA.

Comment

Antigen expression requires Mg^{++}.

Reference
[1] Hermand, P. et al. (1995) Blood 86, 1590–1594.

LW^{ab} ANTIGEN

Terminology

ISBT symbol (number)	LW6 (016.006)
Other names	Bigelow, Big, LW
History	LW$_4$ was renamed LW^{ab} when the LW blood group system was established

Occurrence

All populations: 100%.

Expression

Cord RBCs	Well expressed on D-positive and D-negative
Altered	Weak on D– RBCs from adults
	Weak or absent on RBCs stored in EDTA

Molecular basis associated with LW^{ab} antigen

Not known. See system pages for molecular basis of LW(a−b−) phenotype.

Effect of enzymes/chemicals on LW^{ab} antigen on intact RBCs

Ficin/papain	Resistant
Trypsin	Resistant

α-Chymotrypsin	May be weakened
Pronase	Sensitive
Sialidase	Resistant
DTT 200 mM/50 mM	Sensitive/sensitive
Acid	Resistant

In vitro characteristics of alloanti-LW^{ab}

Immunoglobulin class	IgG; IgM
Optimal technique	37°C; IAT
Complement binding	No

Clinical significance of alloanti-LW^{ab}

| Transfusion reaction | No data |
| HDN | Mild |

Autoanti-LW^{ab}

Autoanti-LW^{ab} with suppression of LW antigens occurs[1]. Common in serum of patients with warm AIHA.

Comments

Only one alloanti-LW^{ab} has been described in an LW(a−b−) person with an LW(a−b−) brother.

When LW antigens are suppressed, the anti-LW^{ab} may mimic an alloantibody and is a more common specificity than autoanti-LW^a.

Antigen expression required Mg^{++}.

Reference
[1] Storry, J.R. (1992) Immunohematology 8, 87–93.

LW^b ANTIGEN

Terminology

ISBT symbol (number)	LW7 (016.007)
Other names	Ne^a, LW3
History	Name changed from Ne^a when the antithetical relationship to LW^a was recognized in 1982

Occurrence

Estonians 8%, Finns 6%, Latvians and Lithuanians 5%, Poles and Russians 2% and other Europeans less than 1%[1].

Antithetical antigen

LW[a] (**LW5**)

Expression

Cord RBCs	Well expressed on D-positive and D-negative
Altered	Weak on D− RBCs from adults
	Weak or absent on RBCs stored in EDTA

Molecular basis associated with LW[b] antigen[2]

Amino acid	Arg 70
Nucleotide	G at bp 308 in exon 1
Restriction enzyme	*Pvu* II site last

Effect of enzymes/chemicals on LW[b] antigen on intact RBCs

Ficin/papain	Resistant (↑↑)
Trypsin	Resistant (↑↑)
α-Chymotrypsin	May be weakened
Pronase	Sensitive
Sialidase	Resistant
DTT 200 mM/50 mM	Sensitive/sensitive
Acid	Resistant

In vitro characteristics of alloanti-LW[b]

Immunoglobulin class	IgG; IgM
Optimal technique	37°C; IAT
Complement binding	No

Clinical significance of alloanti-LW[b]

Transfusion reaction	No to mild
HDN	No to mild

Comment

Antigen expression requires Mg^{++}.

References
[1] Sistonen, P. et al. (1999) Hum Hered. 49, 154–158.
[2] Hermand, P. et al. (1995) Blood 86, 1590–1594.

Number of antigens	9

Terminology

ISBT symbol	CH/RG
ISBT number	017
History	Named after the first antibody producers, Chido and Rodgers. Anti-Ch was reported in 1967 and when anti-Rg was described in 1976, there were obvious similarities between them. Ch and Rg appeared to be RBC antigens and were given blood group system status. However, the antigens were later located on the fourth component of complement (C4), which becomes bound to RBCs from the plasma

Expression

Soluble form	In plasma or serum
Altered	GPA-deficient RBCs have a weak expression of Ch and Rg antigens

Gene

Chromosome	6p21.3
Name	*CH (C4B); RG(C4A)*
Organization	*C4A* 41 exons distributed over 22 kpb of gDNA
	C4B 41 exons distributed over 22 or 16 kbp of gDNA after loss of a 6.8 kbp intron
Product	C4A complement component (Rg)
	C4B complement component (Ch)

Database accession numbers

GenBank   K02403; M59815; M59816; U24578
http://www.bioc.aecom.yu.edu/bgmut/index.htm

Carrier molecule

C4A and C4B are glycoproteins which are adsorbed onto the RBC membrane. C4A binds preferentially to protein and C4B to carbohydrate. C4A migrates more quickly in electrophoresis than C4B. C4A and C4B are 99% identical in their amino acid sequences.

Ch and Rg antigens are located in the C4d region of C4B or C4A, respectively. C4d is a tryptic fragment of C4.

The different antigens are usually identified in plasma by hemagglutination inhibition studies.

Molecular basis of antigens[1]

Allotype	Ch/Rg Type	Phenotype	Amino acid residue							
			1054	1101	1102	1105	1106	1157	1188	1191
C4A*3	Ch−Rg+	Ch:−1,−2,−3, −4, −5,−6 Rg:1,2	D	P	C	L	D	N	V	L
C4A*1	Ch+Rg−	Ch:1,−2,3,−4, 5,6 Rg:-1,-2	G	P	C	L	D	S	A	R
C4A*3	Ch−Rg+ WH+	Ch:−1,−2,−3, −4−5,6 Rg: 1,−42	D	P	C	L	D	S	V	L
C4B*3	Ch+Rg−	Ch:1,2,3,4,5, 6 Rg:−1,−2	G	L	S	I	H	S	A	R
C4B*1	Ch+Rg−	Ch:1,2,-3,4,5, −6 Rg:−1,−2	G	L	S	I	H	N	A	R
C4B*2	Ch+Rg−	Ch:1,−2,3,4,−5, 6 Rg:−1,−2	D	L	S	I	H	S	A	R
C4B*5	Ch+Rg+ WH+	Ch:−1,−2,−3,4, −5, 6 Rg:1,−2	D	L	S	I	H	S	V	L

Function

There are functional differences between C4A and C4B allotypes: C4A is more effective than C4B at solubilizing immune complexes and inhibiting immune precipitation. C4B binds more effectively to the RBC surface (through sialic acid) and thus is more effective at promoting hemolysis. A single amino acid substitution at position 1106 (aspartic acid for histidine) converts the functional activity of C4B to C4A[2], whereas the substitution of cysteine for serine at position 1102 affects hemolytic activity and IgG binding.

Disease association

Inherited low levels of C4 may be a predisposing factor for diseases such as insulin-dependent diabetes and autoimmune chronic active hepatitis. Specific C4 allotypes and null genes have been associated with numerous autoimmune disorders including Graves' disease and rheumatoid arthritis (for list, see reference 3). Lack of C4B (Ch−) gives increased susceptibility to bacterial meningitis in children. Rg− individuals (lack of C4A) have a much greater susceptibility for SLE.

Phenotypes (% occurrence)

Chido antigens	Most populations	Japanese	Rodgers antigens	Most populations	Japanese
CH/RG:1,2,3	88.2	75	CH/RG:11,12	95	100
CH/RG:1,−2,3	4.9	24	CH/RG:11,−12	3	0
CH/RG:1,2,−3	3.1	0	CH/RG:−11,−12	2	0
CH/RG:−1,−2,−3	3.8	1			
CH/RG:−1,2,−3	rare	0			
CH/RG:1,−2,−3	rare	0			

Null: C4-deficient RBCs.

Comments

Antigens of this system are stable in stored serum or plasma. Phenotypes and antibodies of this system are most accurately defined by hemagglutination inhibition tests.

RBCs coated with C4 (+C3) by use of low ionic strength 10% sucrose solution give enhanced reactivity with anti-Ch and anti-Rg. Sialidase-treated RBCs do not take up C4.

C4 molecule (adapted from Daniels[4])

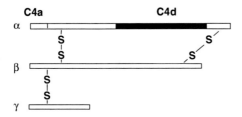

References

[1] Yu, C.Y. et al. (1988) Immunogenetics 27, 399–405.

[2] Carroll, M.C. et al. (1990) Proc. Natl. Acad. Sci. USA 87, 6868–6872.

[3] Moulds, J.M. (1994) In: Immunobiology of Transfusion Medicine (Garratty, G. ed) Marcel Dekker, Inc., New York, pp. 273–297.

[4] Daniels, G. (1995) In: Molecular Basis of Human Blood Group Antigens (Cartron, J.-P. and Rouger, P. eds) Plenum Press, New York, pp. 397–419.

Ch1 ANTIGEN

Terminology

ISBT symbol (number)	CH/RG1 (017.001)
Other names	Ch; Chª; Chido
History	Named in 1967 after Mrs. Chido who made "anti-Chido" (considered a nebulous antibody)

Occurrence

Most populations	96%
Japanese	99%

Expression

Cord RBCs	Absent or weak
Altered	Weak on some dominant Lu(a−b−) and on GPA-deficient RBCs

Molecular basis associated with Ch1 antigen

Requires alanine at residue 1188 and arginine at 1191 of C4[1,2]. See system pages.

Effect of enzymes/chemicals on Ch1 antigen on intact RBCs

Ficin/papain	Sensitive
Trypsin	Sensitive
α-Chymotrypsin	Sensitive
Pronase	Sensitive
Sialidase	Resistant
DTT 200 mM	Resistant
Acid	Resistant

In vitro characteristics of alloanti-Ch1

Immunoglobulin class	IgG (mostly IgG2 and IgG4)
Optimal technique	IAT
Neutralization	Antigen-positive serum or plasma
Complement binding	No

Clinical significance of alloanti-Ch1

Transfusion reaction   No; anaphylactic reactions from plasma products and platelets

HDN   No

Comments

Soluble plasma antigen in donor blood may neutralize patient's antibody. Anti-Ch1 reacts strongly with C4-coated RBCs.

Virtually all anti-Ch contain anti-Ch1.

References
1. Giles, C.M. (1988) Exp. Clin. Immunogenet. 5, 99–114.
2. Daniels, G. (1995) In: Molecular Basis of Human Blood Group Antigens (Cartron, J.-P. and Rouger, P. eds) Plenum Press, New York, pp. 397–419.

Ch2 ANTIGEN

Terminology

ISBT symbol (number)   CH/RG2 (017.002)

History   Defined in 1985 when plasma inhibition studies revealed that there are at least 6 Chido antigens (Ch1 to Ch6) of high prevalence

Occurrence

Most populations   >90%

Japanese   75%

Molecular basis associated with Ch2 antigen

Antigen expression requires presence of Ch4 and Ch5, i.e., glycine at 1054, leucine at 1101, serine at 1102, isoleucine at 1105 and histidine at 1106[1,2].

See system pages.

Comments

Anti-Ch2 + anti-Ch4 was detected in a Ch:1,−2,3,−4,5,6 Rg:1,2 person[3].

Anti-Ch2 + anti-Ch5 was detected in a Ch:1,−2,3,4,−5,6 Rg:1,2 person[4].

References

1. Giles, C.M. (1988) Exp. Clin. Immunogenet. 5, 99–114.
2. Daniels, G. (1995) In: Molecular Basis of Human Blood Group Antigens (Cartron, J.-P. and Rouger, P. eds) Plenum Press, New York, pp. 397–419.
3. Fisher, B. et al. (1993) Transf. Med. 3 (Suppl. 1), 84 (abstract).
4. Giles, C.M. et al. (1987) Vox Sang. 52, 129–133.

Ch3 ANTIGEN

Terminology

ISBT symbol (number)	CH/RG3 (017.003)
History	See Ch2 antigen

Occurrence

Caucasians	93%
Japanese	>99%

Molecular basis associated with Ch3 antigen

Antigen expression requires presence of Ch1 and Ch6, i.e., serine at residue 1157, alanine at 1188 and arginine at 1191[1,2]. See system pages.

References

1. Giles, C.M. (1988). Exp. Clin. Immunogenet. 5, 99–114.
2. Daniels, G. (1995). In: Molecular Basis of Human Blood Group Antigens (Cartron, J.-P. and Rouger, P. eds) Plenum Press, New York, pp. 397–419.

Ch4 ANTIGEN

Terminology

ISBT symbol (number)	CH/RG4 (017.004)
History	See Ch2 antigen

Occurrence

All populations   >99%.

Molecular basis associated with Ch4 antigen

Antigen expression requires presence of leucine 1101, serine at 1102, isoleucine at 1105 and histidine at 1106[1,2]. See system pages. Detected on all C4B allotypes.

References
1 Giles, C.M. (1988). Exp. Clin. Immunogenet. 5, 99–114.
2 Daniels, G. (1995). In: Molecular Basis of Human Blood Group Antigens (Cartron, J.-P. and Rouger, P. eds) Plenum Press, New York, pp. 397–419.

Ch5 ANTIGEN

Terminology
ISBT symbol (number)   CH/RG5 (017.005)
History   See Ch2 antigen

Occurrence
All populations   >99%

Molecular basis associated with Ch5 antigen

Antigen expression requires glycine at 1054[1,2]. See system pages.

References
1 Giles, C.M. (1988). Exp. Clin. Immunogenet. 5, 99–114.
2 Daniels, G. (1995). In: Molecular Basis of Human Blood Group Antigens (Cartron, J.-P. and Rouger, P. eds) Plenum Press, New York, pp. 397–419.

Ch6 ANTIGEN

Terminology
ISBT symbol (number)   CH/RG6 (017.006)
History   See Ch2 antigen

Occurrence
All populations   >99%

Molecular basis associated with Ch6 antigen

Antigen expression requires serine at residue 1157 of C4[1,2]. See system pages.

Comments

Rare specificity, two examples reported.

References
1 Giles, C.M. (1988) Exp. Clin. Immunogenet. 5, 99–114.
2 Daniels, G. (1995) In: Molecular Basis of Human Blood Group Antigens (Cartron, J.-P. and Rouger, P. eds) Plenum Press, New York, pp. 397–419.

WH ANTIGEN

Terminology

ISBT symbol (number)	CH/RG7 (017.007)
History	Named after the person who was thought to carry a hybrid of C4A and C4B

Occurrence

Caucasians   15%

Molecular basis associated with WH antigen

Associated with Ch:6, Rg:1,−2 phenotype. Antigen expression requires serine at residue 1157, valine at 1188 and leucine at 1191[1-3]. See system pages.

In one individual (WH), a single amino acid substitution encoded by the C4A*3 gene at codon 1157 gives rise to Asp in the wild type being replaced by Ser in WH type[2,3].

Comments

Rare specificity, two examples reported[4].

References
1 Giles, C.M. (1988) Exp. Clin. Immunogenet. 5, 99–114.
2 Daniels, G. (1995) In: Molecular Basis of Human Blood Group Antigens (Cartron, J.-P. and Rouger, P. eds) Plenum Press, New York, pp. 397–419.
3 Moulds, J.M. et al. (1995) Transfusion 35 (Suppl.), 53S (abstract).
4 Giles, C.M. and Jones J.W. (1987) Immunogenetics 26, 392–394.

Rg1 ANTIGEN

Terminology

ISBT symbol (number)	CH/RG11 (017.011)
Other names	Rodgers; Rg; Rg^a
History	"Generic" anti-Rg reported in 1976 and named after antibody maker. All anti-Rg contain anti-Rg1 (strongest component) and anti-Rg2

Occurrence

All populations	>98%

Expression

Cord RBCs	Absent or weak
Altered	Weak on Dominant Lu(a−b−) RBCs

Molecular basis associated with Rg1 antigen

Antigen expression requires valine at residue 1188 and leucine at 1191[1,2]. See system pages

Effect of enzymes/chemicals on Rg1 antigen on intact RBCs

Ficin/papain	Sensitive
Trypsin	Sensitive
α-Chymotrypsin	Sensitive
Pronase	Sensitive
Sialidase	Resistant
DTT 200 mM	Resistant
Acid	Resistant

In vitro characteristics of alloanti-Rg1

Immunoglobulin class	IgG
Optimal technique	IAT
Complement binding	No
Neutralization	Antigen-positive serum or plasma

Clinical significance of alloanti-Rg1

Transfusion reaction   No; anaphylactic reactions from plasma products and platelets

HDN   No

References

1 Giles, C.M. (1988). Exp. Clin. Immunogenet. 5, 99–114.
2 Daniels, G. (1995). In: Molecular Basis of Human Blood Group Antigens (Cartron, J.-P. and Rouger, P. eds) Plenum Press, New York, pp. 397–419.

Rg2 ANTIGEN

Terminology

ISBT symbol (number)   CH/RG12 (017.012)

History   See Rg1 antigen

Molecular basis associated with Rg2 antigen

Antigen expression requires asparagine at residue 1157, valine at 1188 and leucine at 1191[1,2]. See system pages.

Comments

All anti-Rg contain anti-Rg1 (strongest component) and anti-Rg2.

References

1 Giles, C.M. (1988). Exp. Clin. Immunogenet. 5, 99–114.
2 Daniels, G. (1995). In: Molecular Basis of Human Blood Group Antigens (Cartron, J.-P. and Rouger, P. eds) Plenum Press, New York, pp. 397–419.

Number of antigens   1

Terminology

ISBT symbol	H
ISBT number	018
Other name	O
History	In 1948, Morgan and Watkins suggested changing the terms "anti-O" and "O substance" to "anti-H" and "H substance" as this would differentiate it as a <u>hetero</u>genetic, basic or primary substance common to the great majority of red cells irrespective of their ABO

Expression

Soluble form	Saliva and all fluids (in secretors) except CSF
Other blood cells	Lymphocytes, platelets
Tissues	Broad tissue distribution (see ABO section)

Gene

Chromosome	19q13.3
Name	H (FUT1)
Organization	Four exons distributed over 8 kbp of gDNA
Product	2-α-fucosyltransferase (α2Fuc-T; 2-α-L-fucosyltransferase; α1,2-fucosyltransferase)

Gene map

The homologous gene (FUT2; Se) encoding 2-α-L-fucosyltransferase is 35 kbp closer to the centromere at 19q13.3. This transferase adds fucose to galactose on type 1 chains (secretory tissues).

Database accession numbers

GenBank   Z69587
http://www.bioc.aecom.yu.edu/bgmut/index.htm

Amino acid sequence of α-2-L-fucosyltransferase

```
MWLRSHRQLC  LAFLLVCVLS  VIFFLHIHQD  SFPHGLGLSI  LCPDRRLVTP   50
PVAIFCLPGT  AMGPNASSSC  PQHPASLSGT  WTVYPNGRFG  NQMGQYATLL  100
ALAQLNGRRA  FILPAMHAAL  APVFRITLPV  LAPEVDSRTP  WRELQLHDWM  150
SEEYADLRDP  FLKLSGFPCS  WTFFHHLREQ  IRREFTLHDH  LREEAQSVLG  200
QLRLGRTGDR  PRTFVGVHVR  RGDYLQVMPQ  RWKGVVGDSA  YLRQAMDWFR  250
ARHEAPVFVV  TSNGMEWCKE  NIDTSQGDVT  FAGDGQEATP  WKDFALLTQC  300
NHTIMTIGTF  GFWAAYLAGG  DTVYLANFTL  PDSEFLKIFK  PEAAFLPEWV  350
GINADLSPLW  TLAKP                                          365
```

Carrier molecule[1,2]

H antigen is not the primary gene product. The *FUT1* product, a fucosyl-transferase, attaches a fucose to galactose on type 2 carbohydrate chains attached to proteins or lipids. The immunodominant fucose is the H antigen, which is the precursor of A and B antigens (see **ABO** blood group system).

Copies per RBC adult   group O   1 700 000   group A, B, AB   70 000
                newborn   group O   325 000

Function

Fucosylated glycans that are the products of *FUT1* and *FUT2* may serve as ligands in cell adhesion or as receptors for certain microorganisms.

Disease association

Increased expression with hematopoetic stress.
Weakened expression in acute leukemia and carcinomatous tissue cells.
Children with leukocyte adhesion deficiency (LADII) have mental retardation and severe recurrent infections with a high white blood cell count and their RBCs are H–.

Phenotypes (% occurrence)

	Caucasians	Blacks	Asians	Mexican
Group O	45	49	43	55

Null: O_h (Bombay).
Unusual: Para-Bombay.
Most RBCs have some H antigen: $O > A_2 > B > A_2B > A_1 > A_1B > Para-Bombay$.

Type	H antigen on RBCs	H antigen in secretion	Predicted genotype	Antibody
Common				
Secretor	Yes	Yes	*HH* or *Hh; SeSe* or *Sese*	
Non-secretor	Yes	No	*HH* or *Hh; sese*	
H-deficient				
Bombay	No	No	*hh; sese*	Anti-H
Para-Bombay	Weak	No	*(H); sese*	Anti-H
Para-Bombay	Weak	Yes	*(H); SeSe* or *Sese*	Anti-HI
H_m (dominant)	Weak	Yes	*HH* or *Hh; SeSe* or *Sese*	None
LADII (CDGII)	No	No	*HH; SeSe* or *sese*	Anti-H

Molecular basis of H-deficient phenotypes due to mutations in *FUT1*

Mutations in the *FUT1* gene or the GDP-fucose-transporter gene give rise to H-deficient phenotypes.

Inactive alleles of *FUT1* fail to express the H epitopes on RBCs. People having these (h) alleles (in the homozygous state) have Bombay or Para-Bombay phenotypes. In Bombay people, mutant alleles of both *FUT1* and *FUT2* fail to express the corresponding fucosyltransferases and these people lack ABH antigens on RBCs and in secretions. In Para-Bombay people *FUT2* is expressed but mutant alleles of *FUT1* usually do not result in active enzyme. There are two types of Para-Bombay people: those who lack ABO antigens on RBCs but possess them in secretions, and those who possess very few ABH antigens on RBCs but may or may not possess them in secretions.

Bombay (non-secretors)

Nucleotide in FUT1	Amino acid	Ethnic origin
349C>T[†]	His117Thr	Reunion[3]
461A>G*	Tyr154Cys	Europe[4]
462C>A	Tyr154Stop	Japan[5,6]
513G>C	Trp171Cys	Europe[4]
695G>A	Trp232Stop	Japan[5]
725T>G**	Leu242Arg	India[7,3]
776T>A	Val259Glu	Europe[8]
785G>A; 786C>A	Ser262Lys	Europe[8]
801G>C	Trp267Cys	Europe[9]
801G>T	Trp267Cys	Europe[9]

Nucleotide in FUT1	Amino acid	Ethnic origin
944C>T	Ala315Val	Europe[8]
948C>G	Tyr316Stop	USA[10]
969–970 delCT	323fs	Europe[4]
1047G>C	Trp349Cys	Europe[4]
Para-Bombay (No H RBC antigen; secretors)		
491T>A	Leu164His	USA[10]
826C>T	Gln276Stop	USA[10]
Para-Bombay (Weak H RBC antigen expression)		
442G>T	Asp148Tyr	Japan[5]
460T>C; 1042G>A	Tyr154His; Glu348Lys	Japan[5,6]
721T>C	Tyr241His	Japan[5]
990delG	330fs	Japan[5,9]
969–970delCT	323fs	European[4]
Para-Bombay – others		
460T>C	Tyr154His	Taiwan[11]
522C>A	Phe174Leu	China[12]
547–548delAG	182fs	Taiwan; China[11,12]
658C>T	Arg220Cys	Taiwan[11,13]
659G>A	Arg220His	Taiwan[13]
832G>A	Asp278Asn	Not given[9]
880–881delTT	294fs	Taiwan, China[9,11,12]
904–906insAAC	His302-Thr303insAsn	Japan[14]

*Silent mutations also present: 474A>G; 954T>A.
** Travels with total deletion of *FUT2*.
† Travels with *FUT2* 428G>A.

For more alleles and details, see http://www.bioc.aecom.yu.edu/bgmut/index.htm

H-deficient RBC phenotypes due to mutations in GDP-fucose transporter gene[15,16]

Mutations in this gene result in no H expression; thus the RBCs have the Bombay, Le(a–b–) phenotype and WBCs lack sialyl Le^X. Gives rise to LADII (CDGII).

439C>T	Arg147Lys	Turkish
923C>G	Thr308Arg	Arab
588delG	Ser195fs; 34 novel amino acids; stop	Brazilian

Comments

In secretions, the H-specified transferase is encoded by the *FUT2* (*Se*) gene. O_h (Bombay) and Para-Bombay are the result of point mutations in the 2-α-fucosyltransferase gene[1]. O_h is associated with Tyr 316 substituted by a stop codon. Several mutations give rise to the Para-Bombay phenotype[9,10].

References
1 Oriol, R. (1995) In: Molecular Basis of Human Blood Group Antigens (Cartron, J.-P. and Rouger, P. eds) Plenum Press, New York, pp. 37–73.
2 Lowe, J.B. (1995) In: Molecular Basis of Human Blood Group Antigens (Cartron, J.-P. and Rouger, P. eds) Plenum Press, New York, pp. 75–115.
3 Fernandez-Mateos, P. et al. (1998) Vox Sang. 75, 37–46.
4 Wagner, F.F. and Flegel, W.A. (1997) Transfusion 37, 284–290.
5 Kaneko, M. et al. (1997) Blood 90, 839–849.
6 Wang, B. et al. (1997) Vox Sang. 72, 31–35.
7 Koda, Y. et al. (1997) Biochem. Biophys. Res. Commun. 238, 21–25.
8 Flegel, W.A. and Wagner, F.F. (2000) Vox Sang. 78, 109–115.
9 Johnson, P.H. et al., (1994) Vox Sang 67 (Suppl. 2), 25 (abstract).
10 Kelly, R.J. et al. (1994) Proc. Natl. Acad. Sci. USA 91, 5843–5847.
11 Yu, L.-C. et al. (1997) Vox Sang. 72, 36–40.
12 Chee, K. et al. (2000) Transfusion 40 (Suppl.), 118S (abstract).
13 Sun, C.F. et al. (2000) Ann Clin. Lab. Sci. 30, 387–390.
14 Ogasawara, K. et al. (2000) Vox Sang. 78 (Suppl. 1), P004 (abstract).
15 Lühn K., et al. (2001) Nature Genet. 28, 69–72.
16 Hidalgo, A. et al. (2003) Blood 101, 1705–1712.

H ANTIGEN

Terminology

ISBT symbol (number)	H1 (018.001)
History	see H Blood Group System pages

Occurrence

All populations: 99.9%

H-deficient people (Bombay and Para-Bombay) 1 in 8000 in Taiwan, 2 in 300000 in Japan, 1 in 10000 in India; 1 per million in Europe.

Expression

Adult RBCs	(Strongest) $O > A_2 > B > A_2B > A_1 > A_1B$
Cord RBCs	Weak
Altered	Weak on some Para-Bombay

Molecular basis associated with H antigen[1]

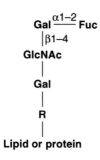

Effect of enzymes/chemicals on H antigen on intact RBCs

Ficin/papain	Resistant (↑↑)
Trypsin	Resistant (↑↑)
α-Chymotrypsin	Resistant (↑↑)
Pronase	Resistant (↑↑)
Sialidase	Resistant
DTT 200 mM	Resistant
Acid	Resistant

In vitro characteristics of alloanti-H

Immunoglobulin class	IgM more common than IgG
Optimal technique	RT or 4°C
Neutralization	Saliva, all body fluids except CSF (secretors)
Complement binding	Some

Clinical significance of alloanti-H in Bombay (O_h) and Para-Bombay people

Transfusion reaction	No to severe; Immediate/delayed/hemolytic
HDN	Possible in O_h mothers

Autoanti-H

Yes, usually cold reactive.

Comments

With the exception of O_h people (whose serum contains anti-A, -B and -H), anti-IH is more common than anti-H. Anti-IH is commonly found in the serum of pregnant group A_1 women. Type-specific blood will be cross-match compatible.

Treatment of serum with a thiol reagent inactivates IgM anti-H and anti-IH, thereby facilitating identification of IgG alloantibodies.

Reference

[1] Lowe, J.B. (1995) In: Molecular Basis of Human Blood Group Antigens (Cartron, J.-P. and Rouger, P. eds) Plenum Press, New York, pp. 75–115.

Number of antigens   1

Terminology

ISBT symbol	XK
ISBT number	019
History	Named in 1990 when the Kx antigen was assigned system status. XK was used as the ISBT symbol after the gene name

Expression

Tissues   Fetal liver, adult skeletal muscle, brain, pancreas, heart, low levels in adult liver, kidney, spleen

Gene

Chromosome	Xp21.1
Name	*XK*
Organization	Three exons; sizes of introns have not been determined
Product	XK protein (Kx protein)

Gene map

Database accession numbers

GenBank   Z32684 (nucleotide 'A' of 'ATG' is at bp 83)
www.bioc.aecom.yu.edu/bgmut/index.htm.

Amino acid sequence[1]

(Note: the amino acid residues 204 and 205 listed here are the corrected data in GenBank accession number Z32684, which has not yet appeared in published form.)

```
MKFPASVLAS  VFLFVAETTA  ALSLSSTYRS  GGDRMWQALT  LLFSLLPCAL   50
VQLTLLFVHR  DLSRDRPLVL  LLHLLQLGPL  FRCFEVFCIY  FQSGNNEEPY  100
VSITKKRQMP  KNGLSEEIEK  EVGQAEGKLI  THRSAFSRAS  VIQAFLGSAP  150
QLTLQLYISV  MQQDVTVGRS  LLMTISLLSI  VYGALRCNIL  AIKIKYDEYE  200
VKVKPLAYVC  IFLWRSFEIA  TRVVVLVLFT  SVLKTWVVVI  ILINFFSFFL  250
YPWILFWCSG  SPFPENIEKA  LSRVGTTIVL  CFLTLLYTGI  NMFCWSAVQL  300
KIDSPDLISK  SHNWYQLLVY  YMIRFIENAI  LLLLWYLFKT  DIYMYVCAPL  350
LVLQLLIGYC  TAILFMLVFY  QFFHPCKKLF  SSSVSEGFQR  WLRCFCWACR  400
QQKPCEPIGK  EDLQSSRDRD  ETPSSSKTSP  EPGQFLNAED  LCSA        444
```

Carrier molecule[1]

In the RBC membrane, XK protein is covalently linked at Cys 72 to Cys 347 of the Kell glycoprotein

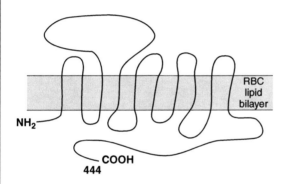

M_r (SDS-PAGE)	37 000
Glycosylation	None
Cysteine residues	16
Copies per RBC	1 000

Function

Not known, but XK has structural characteristics of a membrane transport protein, and a homolog, ced-8, is involved in regulating cell death in *C. elegans*[2]. Involved in maintenance of normal cell membrane integrity.

Disease association

Absence of XK protein is associated with acanthocytosis and the McLeod syndrome, which manifests a compensated hemolytic anemia, elevated serum creatinine kinase and neuromuscular disorders including chorea, areflexia, skeletal muscle atrophy and cardiomyopathy[3,4].

Some males with the McLeod phenotype have X-linked CGD. See web site: www.nefo.med.uni-muenchen.de/~adanek/McLeod.html.

Phenotypes

Null   McLeod
Unusual   Kx antigen has an increased expression on RBCs
                        that lack or have a reduced expression of Kell anti-
                        gens. See tables in Kell blood group system section.

Molecular bases of McLeod phenotype[4,5]

Unless otherwise stated found in one proband.

Deletion of XK gene or exon
XK gene deletion (several probands)
Promoter and exon 1 deletion
Exon 1 deletion (few probands)
Exon 2 deletion
Intron 2 and exon 3 deletion

Deletion of nucleotide(s) leading to frameshift and premature
stop codon in XK
351delA in exon 2
350delT in exon 2   Tyr90fs
768–769delTT in exon 3   Phe229fs + Pro264Stop
853delG in exon 3   Trp257fs + Ile267Stop
938–942delCTCTA in exon 3   Leu286fs + Lys301Stop
1020–1033del in exon 3   Asn313fs + Tyr336Stop
1095delT in exon 3[6]   Phe338fs + Ile408Stop

Insertion of nucleotide leading to frameshift and premature stop codon in XK
533–534insC in exon 2   Pro150 fs

Splice site mutation in XK
IVS1 +1 g>c (intron 1 in 5' splice site, g>c) (few probands)
IVS2 +1 g>a (intron 2 in 5' splice site, g>a) (few probands)
IVS2 +5 g>a (intron 2 in 5' splice site, g>a)*
IVS2 −1 g>a (intron 2 g>a in 3' splice site, g>a)

Nonsense mutation in XK
189G>A in exon 1, Trp36Stop
479C>T in exon 2, Arg133Stop
545C>T in exon 2, Gln155Stop
789G>A in exon 3, Trp236Stop
977C>T in exon 3, Gln299Stop
1023G>A in exon 3, Trp314Stop

Missense mutations in XK
746C>G in exon 3, Arg222Gly
962T>C in exon 3, Cys294Arg

*Very weak expression of Kx antigen (resembling a McLeod phenotype)
together with extreme depression of Kell system antigens on the RBCs of a

German proband were caused by the simultaneous presence of a single base change in the donor splice site of *XK*, and homozygosity for *Kp*^{a} at the *KEL* locus[7].

References

[1] Ho, M. et al. (1994) Cell 77, 869–880.
[2] Stanfield, G.M. and Horvitz, H.R. (2000) Mol. Cell 5, 423–433.
[3] Lee, S. et al. (2000) Semin. Hematol. 37, 113–121.
[4] Danek, A. et al. (2001) Ann. Neurol. 50, 755–764.
[5] Russo, D.C.W. et al. (2002) Transfusion 42, 287–293.
[6] Hanaoka, N. et al. (1999) J. Neurol. Sci. 165, 6–9.
[7] Daniels, G.L. et al. (1996) Blood 88, 4045–4050.

Kx ANTIGEN

Terminology

ISBT symbol (number)	XK1 (019.001)
Other names	006.015; K15
History	Named in 1975 when Kx was shown to be associated with the Kell blood group system but controlled by a gene on the X chromosome

Occurrence

All populations	100%

Expression

Cord RBCs	Expressed
Altered	Weak on RBCs of common Kell phenotype
	Expression of Kx antigen is enhanced on RBCs with reduced expression of Kell [K_0, K_{mod}, thiol-treated RBCs, Kp(a+b−)] even though levels of XK protein may be reduced[1].

Molecular basis associated with Kx antigen

Not known. For molecular basis associated with a lack of Kx antigen, see table in system pages.

Effect of enzymes/chemicals on Kx antigen on intact RBCs

Ficin/papain	Resistant
Trypsin	Resistant
α-Chymotrypsin	Resistant
Pronase	Resistant
Sialidase	Resistant
DTT 200 mM	Resistant (↑)
Acid	Not known

In vitro characteristics of alloanti-Kx

Immunoglobulin class	IgG
Optimal technique	IAT
Complement binding	No

Clinical significance of alloanti-Kx

Transfusion reaction	Mild/delayed
HDN	Not applicable

Autoanti-Kx

One example reported in a man with common Kell phenotype.

Comments

Anti-Km is made by non-CGD McLeod males; both McLeod and K_0 blood will be compatible.

Anti-Kx + anti-Km (sometimes called anti-KL) is made by males with the McLeod phenotype and CGD; only McLeod blood will be compatible.

Anti-Kx has been made by one non-CGD McLeod male[2].

Anti-Kx can be prepared by adsorption of anti-Kx + anti-Km (anti-KL) onto and elution from K_0 RBCs.

XK is subject to X-chromosome inactivation and female carriers have a mixed population of normal and acanthocytic RBCs. The range has been as much as 99% and as few as 1% McLeod RBCs.

References
1 Lee, S. et al. (2000) Semin. Hematol. 37, 113–121.
2 Russo, D.C. et al. (2000) Transfusion 40, 1371–1375.

Number of antigens   7

Terminology

ISBT symbol	GE
ISBT number	020
CD number	CD236
Other name	ISBT Collection 201
History	Named in 1960 after one of three mothers who were found at the same time and whose serum contained the antibody defining Ge; became a system in 1990

Expression

Other blood cells	Erythroblasts
Tissues	Fetal liver, renal endothelium

Gene

Chromosome	2q14–q21
Name	*GE (GYPC)*
Organization	Four exons distributed over 13.5 kbp of gDNA
Product	Glycophorin C (GPC) and glycophorin D (GPD)

Gene map

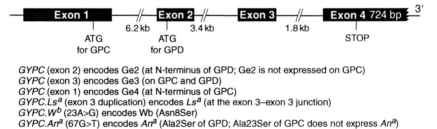

GYPC (exon 2) encodes Ge2 (at N-terminus of GPD; Ge2 is not expressed on GPC)
GYPC (exon 3) encodes Ge3 (on GPC and GPD)
GYPC (exon 1) encodes Ge4 (at N-terminus of GPC)
GYPC.Ls^a (exon 3 duplication) encodes *Ls^a* (at the exon 3–exon 3 junction)
GYPC.W^b (23A>G) encodes Wb (Asn8Ser)
GYPC.An^a (67G>T) encodes *An^a* (Ala2Ser of GPD; Ala23Ser of GPC does not express *An^a*)
GYPC.Dh^a (40C>T) encodes *Dh^a* (Leu14Phe)

Database accession numbers

GenBank   M36284
www.bioc.aecom.yu.edu/bgmut/index.htm

Amino acid sequence[1]

Glycophorin C:

```
MWSTRSPNST AWPLSLEPDP GMASASTTMH TTTIAEPDPG MSGWPDGRME    50
TSTPTIMDIV VIAGVIAAVA IVLVSLLFVM LRYMYRHKGT YHTNEAKGTE   100
FAESADAALQ GDPALQDAGD SSRKEYFI                           128
```

Glycophorin D:

```
            MASASTTMH TTTIAEPDPG MSGWPDGRME    29
TSTPTIMDIV VIAGVIAAVA IVLVSLLFVM LRYMYRHKGT YHTNEAKGTE    79
FAESADAALQ GDPALQDAGD SSRKEYFI                           107
```

Antigen mutation is numbered by counting Met as 1.

Carrier molecule

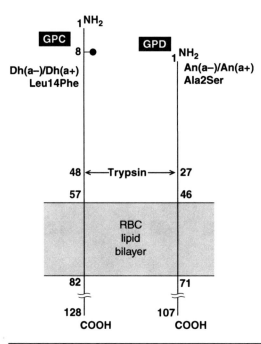

	GPC	*GPD*
M_r (SDS-PAGE):	40 000	30 000
CHO: N-glycan:	1 site	0 site
O-glycan:	13 sites	8 sites
Copies per RBC:	135 000	50 000

Molecular basis of antigens

Antigen	Amino acid change	Exon	Nt change
Wb−/Wb+	Asn8Ser of GPC	1	23A>G
An(a−)/An(a+)	Ala2Ser of GPD	2	67G>T
Dh(a−)/Dh(a+)	Leu14Phe of GPC	1	40C>T

Function

Maintenance of RBC membrane integrity via interaction with protein 4.1.
Contributes to the negatively charged glycocalyx.

Disease association

GPC and GPD are markedly reduced in protein 4.1-deficient RBCs and as
such can be associated with hereditary elliptocytosis. RBC receptors for
influenza A and influenza B.

Phenotypes (% occurrence)

	Most populations	Melanesians
Ge:2,3,4 (Ge+)	>99.9%	50–90%
Ge-negative		
Ge:−2,3,4 (Yus type)	Rare	Not found
Ge:−2,−3,4 (Gerbich type)	Rare	10–50%
Ge:−2,−3,−4 (Leach type)	Rare	Not found

Null: Leach (PL and LN types) (Ge:−2,−3,−4).
Unusual: Gerbich (Ge:−2,−3,4), Yus (Ge:−2,3,4).

Molecular basis of phenotypes

Phenotype	Basis
Ge:−2,3,4	Exon 2 deleted
Ge:−2,−3,4	Exon 3 deleted
Ge:−2,−3,−4 (PL)	Exon 3 and exon 4 deleted
Ge:−2,−3,−4 (LN)	Nt 134 deleted in exon 5, codon 45 (Pro45Arg), frameshift, a new amino acid sequence until residue 55 where the new codon is a stop codon

405

Comments

The LN type has a different RFLP pattern after treatment with *Msp* 1 restriction enzyme.

The majority of RBC samples with Leach or Gerbich phenotypes have a weak expression of Kell blood group system antigens. Some anti-Vel fail to react with Ge:−2,−3,4 RBCs.

Gerbich antigens are weak on protein 4.1-deficient RBCs due to reduced levels of GPC and GPD in these membranes.

GPC variants have been described that have amino acids encoded by duplicated exon 2 or duplicated exon 3.

A woman with an apparent Ge:2,3,4 phenotype who made anti-Ge2 had *GYPC.Ge* (deletion of exon 3) and *GYPC* with 173A>T in exon 3 encoding Asp58Val in GPC and Asp37Val in GPD.

Reference

[1] Colin, Y. et al. (1986) J. Biol. Chem. 261, 229–233.

Ge2 ANTIGEN

Terminology

ISBT symbol (number)	GE2 (020.002)
Other names	Ge; 201.002
History	Antigen lacking from all Ge-negative phenotypes. Originally defined by the "Yus-type" antibody found in 1961; later referred to as anti-Ge1,2 and now as anti-Ge2

Occurrence

All populations	100%

Expression

Cord RBCs	Expressed
Altered	Weak on protein 4.1-deficient RBCs
	Absent from Yus, Gerbich and Leach phenotype RBCs

Molecular basis associated with Ge2 antigen[1]

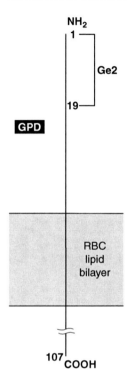

Ge2 as determined with alloanti-Ge2 is not expressed on GPC

Effect of enzymes/chemicals on Ge2 antigen on intact RBCs

Ficin/papain   Sensitive
Trypsin   Sensitive
α-Chymotrypsin   Weakened
Pronase   Sensitive
Sialidase   Variable
DTT 200 mM   Variable
Acid   Resistant

In vitro characteristics of alloanti-Ge2

Immunoglobulin class   Usually IgG
Optimal technique   IAT
Complement binding   Yes; some hemolytic

Clinical significance of alloanti-Ge2

Transfusion reaction   No to moderate/immediate/delayed
HDN   Positive DAT but no clinical HDN

Autoanti-Ge2

Yes; detects a determinant on GPC.

Comments

Alloanti-Ge2 can be made by individuals with Yus, Gerbich or Leach phenotypes, detects an antigen on GPD, and may be naturally-occurring.
The reciprocal gene to *GYPC.Yus* encodes two copies of amino acids encoded
  by exon 2.

Reference

[1] Reid, M.E. and Spring, F.A. (1994) Transf. Med. 4, 139–149.

Ge3 ANTIGEN

Terminology

ISBT symbol (number)   GE3 (020.003)
Other names   Ge; 201.003
History   Antigen originally defined by the "Ge-type"
         serum (identified in 1960). The defining
         antibody was termed anti-Ge1,2,3 and later
         renamed to anti-Ge3.

Occurrence

Most populations   99.9%
Melanesians   50%

Expression

Cord RBCs   Expressed
Altered   Weak on protein 4.1-deficient RBCs
         Absent from Gerbich and Leach pheno-
         type RBCs

Molecular basis associated with Ge3 antigen[1]

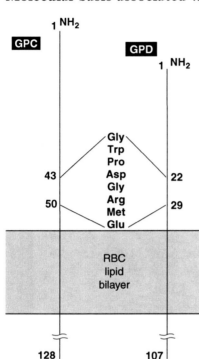

The Ge3 antigen amino acid sequence is encoded by exon 3 of *GYPC*.

Effect of enzymes/chemicals on Ge3 antigen on intact RBCs

Ficin/papain	Resistant
Trypsin	Variable
α-Chymotrypsin	Resistant
Pronase	Sensitive
Sialidase	Resistant
DTT 200 mM	Resistant
Acid	Resistant

In vitro characteristics of alloanti-Ge3

Immunoglobulin class	IgM less common than IgG
Optimal technique	IAT
Complement binding	Yes; some hemolytic

Clinical significance of alloanti-Ge3

Transfusion reaction	No to moderate, immediate or delayed
HDN	Positive DAT to severe[2]

Autoanti-Ge3

Yes.

Comments

Alloanti-Ge3 and autoanti-Ge3 detect the antigen on both GPC and GPD. Alloanti-Ge3 can be made by individuals with either Gerbich or Leach phenotypes.

References
[1] Reid, M.E. and Spring, F.A. (1994) Transf. Med. 4, 139–149.
[2] Arndt, P. et al. (2002). Transfusion 42 (Suppl.), 19S (abstract).

Ge4 ANTIGEN

Terminology

ISBT symbol (number)	GE4 (020.004)
Other names	201.004
History	Ge4 was given the next number when an antibody was found that agglutinated Ge-positive and Ge-negative (both Yus and Gerbich type) RBCs but not RBCs with the Leach (Ge$_{null}$) phenotype.

Occurrence

All populations	100%

Expression

Cord RBCs	Expressed
Altered	Weak on protein 4.1-deficient RBCs
	Absent from Leach phenotype RBCs

Molecular basis associated with Ge4 antigen[1]

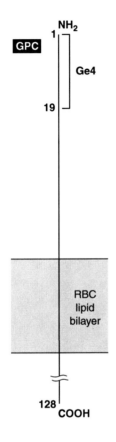

Effect of enzymes/chemicals on Ge4 antigen on intact RBCs

Ficin/papain	Sensitive
Trypsin	Sensitive
α-Chymotrypsin	Resistant
Pronase	Sensitive
Sialidase	Sensitive
DTT 200 mM	Resistant
Acid	Resistant

In vitro characteristics of alloanti-Ge4

Immunoglobulin class	IgG
Optimal technique	IAT
Complement binding	No

Clinical significance of alloanti-Ge4

No information because only one alloanti-Ge4 has been described.

Comments

Ge4 is also expressed on the N-terminal domain of GPC.Yus, GPC.Gerbich, GPC.Wb, GPC.Ls^a, GPC.An^a and GPC.Dh^a.

Reference
[1] Reid, M.E. and Spring, F.A. (1994) Transf. Med. 4, 139–149.

Wb ANTIGEN

Terminology

ISBT symbol (number)	GE5 (020.005)
Other names	Webb; 201.005; 700.009
History	Found in 1963 and named after the donor whose group O RBCs were agglutinated by a high-titer ABO typing serum. Shown to be on a variant form of GPC in 1986.

Occurrence

Less than 0.01%.
May be less rare in Wales and Australia.

Expression

Cord RBCs	Presumed expressed

Molecular basis associated with Wb antigen[1]

Amino acid   Ser 8 of GPC
This substitution results in a loss of the N-glycan and possibly a gain of an O-glycan[2]. Thus, GPC.Wb has an M_r of approximately 2700 less than GPC.

Nucleotide	G at bp 23 in exon 1
Wb−	Asn 8 and A at nt 23

Effect of enzymes/chemicals on Wb antigen on intact RBCs

Ficin/papain	Sensitive
Trypsin	Sensitive
α-Chymotrypsin	Resistant
Pronase	Sensitive
Sialidase	Sensitive
DTT 200 mM	Resistant
Acid	Resistant

In vitro characteristics of alloanti-Wb

Immunoglobulin class	IgM and IgG
Optimal technique	RT; IAT
Complement binding	No

Clinical significance of alloanti-Wb

Transfusion reaction	No
HDN	No

Comment

Anti-Wb are usually naturally-occurring[3].

References
[1] Reid, M.E. and Spring, F.A. (1994) Transf. Med. 4, 139–149.
[2] Reid, M.E. et al. (1987) Biochem. J. 244, 123–128.
[3] Bloomfield, L. et al. (1986) Hum. Hered. 36, 352–356.

Ls^a ANTIGEN

Terminology

ISBT symbol (number)	GE6 (020.006)
Other names	Lewis II; Rl^a (Rosenlund); 700.007; 700.024; 201.006
History	Anti-Ls^a identified in an anti-B typing serum in 1963. Originally called Lewis II after the antigen-positive donor but later renamed Ls^a to avoid confusion with the established Lewis antigens. Associated with Ge in 1990

Occurrence

Less than 0.01% in most populations; 2% of Blacks; 1.6% of Finns.

Expression

Cord RBCs   Presumed expressed
Altered   Increased on RBCs with three copies of
                              amino acids encoded by exon 3

Molecular basis associated with Ls^a antigen[1,2]

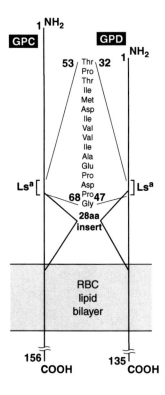

Ls^a antigen is located within an amino acid sequence encoded by nucleotides at the junction of the duplicated exon 3 to exon 3.

Effect of enzymes/chemicals on Ls^a antigen on intact RBCs

Ficin/papain   Sensitive
Trypsin   Sensitive

α-Chymotrypsin	Resistant
Pronase	Sensitive
Sialidase	Resistant
DTT 200 mM	Resistant
Acid	Presumed resistant

In vitro characteristics of alloanti-Ls^a

Immunoglobulin class	IgM and IgG
Optimal technique	RT; IAT
Complement binding	No

Clinical significance of alloanti-Ls^a

Transfusion reaction	Not known
HDN	No

Comment

Anti-Ls^a is naturally-occurring.

References

[1] Reid, M.E. and Spring, F.A. (1994) Transf. Med. 4, 139–149.
[2] Reid, M.E. et al. (1994) Transfusion 34, 966–969.

An^a ANTIGEN

Terminology

ISBT symbol (number)	GE7 (020.007)
Other names	Ahonen; 700.020
History	Identified in 1972 and named after the donor (Ahonen) whose RBCs were agglutinated by a patient's serum. Joined Ge in 1990 when the antigen was located on a variant of GPD

Occurrence

Most populations	0.01%
Finns	0.2%

Expression

Cord RBCs	Presumed expressed

Molecular basis associated with An^a antigen[1,2]

Amino acid   Ser 2 of GPD
The altered GPC (Ser 23) does not express An^a.
Nucleotide   T at bp 67 in exon 2 of *GYPC*
An(a−)   GPC has Ala 23 and GPD has Ala 2 and G at
                          nt 67

Effect of enzymes/chemicals on An^a antigen on intact RBCs

Ficin/papain   Sensitive
Trypsin   Sensitive
α-Chymotrypsin   Weakened
Pronase   Sensitive
Sialidase   Sensitive
DTT 200 mM   Resistant
Acid   Presumed resistant

In vitro characteristics of alloanti-An^a

Immunoglobulin class   IgM and IgG
Optimal technique   RT; IAT
Complement binding   No

Clinical significance of alloanti-An^a

Transfusion reaction   Not known
HDN   No

Comment

Anti-An^a may be naturally-occurring.

References
1 Daniels, G. et al. (1993) Blood 82, 3198–3203.
2 Reid, M.E. and Spring, F.A. (1994) Transf. Med. 4, 139–149.

Dh^a ANTIGEN

Terminology

ISBT symbol (number)	GE8 (020.008)
Other names	Duch; 700.031
History	Identified in 1968 during pretransfusion testing and named after the antigen-positive Danish blood donor. Joined Ge in 1990 when Dh^a was located on a variant of GPC

Occurrence

Less than 0.01%.

Expression

Cord RBCs	Presumed expressed

Molecular basis associated with Dh^a antigen[1,2]

Amino acid	Phe 14 of GPC
Nucleotide	T at bp 40 in exon 1
Dh(a−)	Leu 14 and C at nt 40

Effect of enzymes/chemicals on Dh^a antigen on intact RBCs

Ficin/papain	Sensitive
Trypsin	Sensitive
α-Chymotrypsin	Resistant
Pronase	Sensitive
Sialidase	Sensitive
DTT 200 mM	Resistant
Acid	Presumed resistant

In vitro characteristics of alloanti-Dh^a

Immunoglobulin class	IgM and IgG
Optimal technique	RT and IAT
Complement binding	No

417

Clinical significance of alloanti-Dh^a

Transfusion reaction   Not known
HDN   No

Comment

Anti-Dh^a may be naturally-occurring.

References

[1] King, M.J. et al. (1992) Vox Sang. 63, 56–58.
[2] Reid, M.E. and Spring, F.A. (1994) Transf. Med. 4, 139–149.

Number of antigens 11

Terminology

ISBT symbol	CROM
ISBT number	021
CD Number	CD55
Other name	Collection 202; DAF
History	Named after the first antigen in this system, Cr[a]

Expression

Soluble form	Low levels in plasma, serum and urine
Other blood cells	Leukocytes; platelets
Tissues	Apical surfaces of trophoblasts in placenta

Gene

Chromosome	1q32[1]
Name	CROM (DAF)
Organization	11 exons distributed over 40 kbp of gDNA
Product	Decay accelerating factor (DAF; CD55)

Gene map

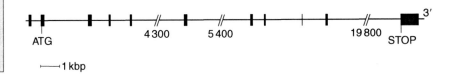

Database accession numbers

GenBank   M31516; M30142; 35156
www.bioc.aecom.yu.edu/bgmut/index.htm.

Amino acid sequence[1]

```
           MTVA RPSVPAALPL LGELPRLLLL VLLCLPAVWG   -1
DCGLPPDVPN AQPALEGRTS FPEDTVITYK CEESFVKIPG EKDSVICLKG   50
SQWSDIEEFC NRSCEVPTRL NSASLKQPYI TQNYFPVGTV VEYECRPGYR  100
REPSLSPKLT CLQNLKWSTA VEFCKKKSCP NPGEIRNGQI DVPGGILFGA  150
TISFSCNTGY KLFGSTSSFC LISGSSVQWS DPLPECREIY CPAPPQIDNG  200
IIQGERDHYG YRQSVTYACN KGFTMIGEHS IYCTVNNDEG EWSGPPPECR  250
GKSLTSKVPP TVQKPTTVNV PTTEVSPTSQ KTTTKTTTPN AQATRSTPVS  300
RTTKHFHETT PNKGSGTTSG TTRLLSGHTC FTLTGLLGTL VTMGLLT     347
```

The 28 carboxyl terminal amino acids are cleaved prior to attachment of DAF to its GPI-linkage. Antigen mutations are numbered by counting Asp as 1.

Carrier molecule

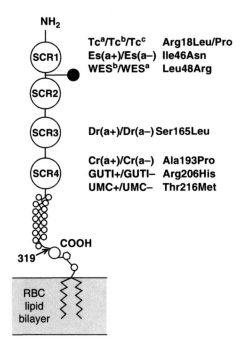

Tc^a/Tc^b/Tc^c	Arg18Leu/Pro
Es(a+)/Es(a–)	Ile46Asn
WES^b/WES^a	Leu48Arg
Dr(a+)/Dr(a–)	Ser165Leu
Cr(a+)/Cr(a–)	Ala193Pro
GUTI+/GUTI–	Arg206His
UMC+/UMC–	Thr216Met

M_r (SDS-PAGE)	Reduced: 64 000–73 000
	Non-reduced: 60 000–70 000
CHO: N-glycan	1 site
CHO: O-glycan	15 sites (32 potential)
Cysteine residues	14
Copies per RBC	20 000

Molecular basis of antigens[2–4]

Antigen	Amino acid substitution	Exon	Nucleotide mutation	Restriction enzyme
Cr(a+)/Cr(a−)	Ala193Pro	6	679G>C	
Tc^a/Tc^b	Arg18Leu	2	155G>T	Rsa I (+/−)
				Stu I (−/+)
Tc^a/Tc^c	Arg18Pro	2	155G>C	Rsa I (+/−)
				Stu I (−/−)
Dr(a+)/Dr(a−)	Ser165Leu	5	596C>T	Taq I (+/−)
Es(a+)/Es(a−)	Ile46Asn	2	239T>A	Sau3 A I (+/−)
IFC+/IFC−	Trp53 Stop	2	261G>A	Bcl I (−/+)
	Ser54Stop	2	263C>A	Mbo II (+/−)
WES^b/WES^a	Leu48Arg	2	245T>G	Afl II (+/−)
UMC+/UMC−	Thr216Met	6	749C>T	
GUTI+/GUTI−	Arg206His	6	719G>A	Mae II (+/−)

Function

Complement regulation; inhibits assembly and accelerates decay of C3 and C5 convertases.

Disease association

Five of six known individuals with the Inab phenotype have intestinal disorders. PNH III RBCs are deficient in DAF. A patient with splenic infarcts had an acquired and transient form of the Inab phenotype, in whom the CD55 deficiency was limited to RBCs; he made anti-ICF[5].

Phenotypes

Null   Inab (IFC−)
Unusual   Dr(a−) RBCs weakly express inherited Cromer antigens

Molecular basis of phenotypes

Inab (Cr_{null})
261G>A in exon 2; Trp53Stop (proband Inaba)[6,7]
263C>A in exon 2; Ser 54; forms a cryptic splice site; 26 bp deletion; frameshift; Stop (proband HA)[6].

Dr(a−)
596C>T in exon 5; Ser165Leu. This transition results in two cDNA species, one encoding full length DAF with the single amino acid change and the other, which is more abundant that uses the novel branch point, which leads to use of a downstream cryptic acceptor splice site, a 44 bp deletion, and a frameshift in exon 5 (proband KZ)[6].

Comments

Antibodies in the Cromer blood group system do not cause HDN. DAF is strongly expressed on the apical surface of placental trophoplasts[8] and will absorb antibodies in the Cromer system.

References
1   Lublin, D.M. and Atkinson J.P. (1989) Ann. Rev. Immunol. 7, 35–58.
2   Lublin, D.M. et al. (2000) Transfusion 40, 208–213.
3   Storry, J.R. et al. (2003) Transfusion 43, 340–344.
4   Storry, J.R. and Reid, M.E. (2002) Immunohematology 18, 95–103.
5   Matthes, T. et al. (2002) Transfusion 42, 1448–1457.
6   Lublin, D.M. et al. (1994) Blood 84, 1276–1282.
7   Wang, L. et al. (1998) Blood 91, 680–684.
8   Holmes, C.H. et al. (1990) J. Immunol. 144, 3099–3105.

Cr^a ANTIGEN

Terminology

ISBT symbol (number)	CROM1 (021.001)
Other names	Go^b; 202.001; 900.013
History	Named in 1975 after Mrs. Cromer, a Black antenatal patient who made the antibody. Originally thought to be antithetical to Go^a

Occurrence

All populations	100%
Blacks	>99%
Hispanics	One Cr(a−) found

Expression

Cord RBCs	Expressed
Altered	Weak on Dr(a−) and negative on PNH III RBCs

Molecular basis associated with Cr^a antigen[1]

Amino acid	Ala 193 in SCR 4
Nucleotide	G at bp 679 in exon 6
Cr(a−)	Pro 193 and C at bp 679

Effect of enzymes/chemicals on Cr^a antigen on intact RBCs

Ficin/papain	Resistant
Trypsin	Resistant
α-Chymotrypsin	Sensitive
Pronase	Sensitive
Sialidase	Resistant
DTT 200 mM/50 mM	Weakened/resistant
Acid	Resistant

In vitro characteristics of alloanti-Cr^a

Immunoglobulin class	IgG
Optimal technique	IAT
Neutralization	With concentrated plasma/serum/urine
Complement binding	No

Clinical significance of alloanti-Cr^a

Transfusion reaction	No to moderate
HDN	No

Comments

The abundance of DAF on apical surface of trophoblasts in placenta may absorb maternal antibodies to antigens in the Cromer system; thereby explaining why HDN is unlikely.

Reference
1   Lublin, D.M. et al. (2000) Transfusion 40, 208–213.

Tcᵃ ANTIGEN

Terminology

ISBT symbol (number)	CROM2 (021.002)
Other names	202.002; 900.020
History	Named in 1980 and placed in the Cromer system when the antibody was shown to be non-reactive with Inab RBCs. The first two examples of the antibody were GT and DLC, hence Tcᵃ

Occurrence

All populations	100%
Blacks	>99%

Antithetical antigens

Tcᵇ (**CROM3**); Tcᶜ (**CROM4**)

Expression

Cord RBCs	Expressed
Altered	Weak on Dr(a−) and negative on PNH III RBCs

Molecular basis associated with Tcᵃ antigen[1]

Amino acid	Arg 18 in SCR 1
Nucleotide	G at bp 155 in exon 2
Restriction enzyme	Gains *Rsa* I site

Effect of enzymes/chemicals on Tcᵃ antigen on intact RBCs

Ficin/papain	Resistant
Trypsin	Resistant
α-Chymotrypsin	Sensitive
Pronase	Sensitive
Sialidase	Resistant
DTT 200 mM/50 mM	Weakened/resistant
Acid	Resistant

In vitro characteristics of alloanti-Tc^a

Immunoglobulin class	IgG
Optimal technique	IAT
Neutralization	With concentrated serum/plasma/urine
Complement binding	No

Clinical significance of alloanti-Tc^a

Transfusion reaction	No to severe[2]
HDN	No

Comments

Only three examples of anti-Tc^a have been reported. All Tc(a−) Blacks are Tc(b+).

The abundance of DAF on apical surface of trophoblasts in placenta may absorb maternal antibodies to antigens in the Cromer system; thereby explaining why HDN is unlikely.

References

[1] Lublin, D.M. et al. (2000) Transfusion 40, 208–213.
[2] Kowalski, M.A. et al. (1999) Transfusion 39, 948–950.

Tc^b ANTIGEN

Terminology

ISBT symbol (number)	CROM3 (021.003)
Other names	202.003; 700.035
History	Original antibody found in a serum containing anti-Go^a; named in 1985 when it was recognized to be antithetical to Tc^a

Occurrence

Caucasians	0%
Blacks	6%

Antithetical antigens

Tc^a (**CROM2**); Tc^c (**CROM4**)

Expression

Cord RBCs	Expressed

Molecular basis associated with Tc^b antigen[1]

Amino acid	Leu 18 in SCR 1
Nucleotide	T at bp 155 in exon 2
Restriction enzyme	Gains a *Stu* I site; ablates *Rsa* I site

Effect of enzymes/chemicals on Tc^b antigen on intact RBCs

Ficin/papain	Resistant
Trypsin	Resistant
α-Chymotrypsin	Sensitive
Pronase	Sensitive
Sialidase	Resistant
DTT 200 mM/50 mM	Weakened/resistant
Acid	Resistant

In vitro characteristics of alloanti-Tc^b

Immunoglobulin class	IgG
Optimal technique	IAT
Complement binding	No

Clinical significance of alloanti-Tc^b

No data.

Reference

[1] Lublin, D.M. et al. (2000) Transfusion 40, 208–213.

Tc^c ANTIGEN

Terminology

ISBT symbol (number)	CROM4 (021.004)
Other names	202.004; 700.036
History	Described in 1982 and named when it was recognized to be antithetical to Tc^a

Occurrence

Caucasians	Less than 0.01%; two Tc(a−b−c+) families
Blacks	0%

Antithetical antigens

Tc^a (**CROM2**); Tc^b (**CROM3**)

Expression

Cord RBCs	Expressed

Molecular basis associated with Tc^c antigen[1]

Amino acid	Pro 18 in SCR 1
Nucleotide	C at bp 155 in exon 2
Restriction enzyme	Ablates a *Rsa* I site

Effect of enzymes/chemicals on Tc^c antigen on intact RBCs

Ficin/papain	Presumed resistant
Trypsin	Presumed resistant
α-Chymotrypsin	Presumed sensitive
Pronase	Presumed sensitive
Sialidase	Presumed resistant
DTT 200 mM/50 mM	Presumed weakened/resistant
Acid	Presumed resistant

427

In vitro characteristics of alloanti-Tc^c

Immunoglobulin class	IgG
Optimal technique	IAT
Complement binding	No

Clinical significance of alloanti-Tc^c

Transfusion reaction	No to mild
HDN	No

Comments

A female with the rare Tc(a−b−c+) phenotype made an antibody that appears to be an inseparable anti-Tc^{a}Tc^b.

Reference
[1] Lublin, D.M. et al. (2000) Transfusion 40, 208–213.

Dr^a ANTIGEN

Terminology

ISBT symbol (number)	CROM5 (021.005)
Other names	202.005; 900.021
History	Reported in 1984 and named after the Israeli proband, Mrs. Drori

Occurrence

All populations: 100%
Dr(a−) have been found in Bukharan Jews and Japanese.

Expression

Cord RBCs	Expressed
Altered	Absent from PNH III RBCs

Molecular basis associated with Dr^a antigen[1]

Amino acid   Ser 165 in SCR 3
Nucleotide   C at nt 596 in exon 5
Restriction enzyme   *Taq* I site present

Dr(a−) has Leu 165; the *DAF* gene has T at bp 596, which introduces a branch point that leads to use of a downstream cryptic acceptor splice site, deletion of 44 bp and a frame shift.

Effect of enzymes/chemicals on Dr^a antigen on intact RBCs

Ficin/papain   Resistant
Trypsin   Resistant
α-Chymotrypsin   Sensitive
Pronase   Sensitive
Sialidase   Resistant
DTT 200 mM/50 mM   Weakened/resistant
Acid   Presumed resistant

In vitro characteristics of alloanti-Dr^a

Immunoglobulin class   IgG
Optimal technique   IAT
Neutralization   With concentrated serum/plasma/urine
Complement binding   No

Clinical significance of alloanti-Dr^a

Transfusion reaction   No to mild
HDN   No

Comments

All inherited Cromer antigens are expressed weakly on Dr(a−) RBCs due to markedly reduced copy number of DAF[1].

Dr^a is the receptor for uropathogenic *E. coli*[2].

The abundance of DAF on the apical surface of placental trophoblasts may result in the absorption of maternal antibodies to antigens in the Cromer system, thereby explaining why HDN is unlikely.

References
[1] Lublin, D.M. et al. (1994) Blood 84, 1276–1282.
[2] Hasan, R.J. et al. (2002) Infect. Immun. 70, 4485–4493.

Es^a ANTIGEN

Terminology

ISBT symbol (number) CROM6 (021.006)
Other names 202.006; 900.022
History Named in 1984 after Mrs. Escandoz, whose Mexican parents were cousins

Occurrence

All populations: 100%
Three Es(a−) probands are known: 1 Mexican; 1 South American[1]; and 1 Black[2]

Expression

Cord RBCs Expressed
Altered Weak on Dr(a−), WES(a+b−), and negative on PNH III RBCs

Molecular basis associated with Es^a antigen[3]

Amino acid Ile 46 in SCR 1
Nucleotide T at bp 239 in exon 2
Es(a−) form of DAF has Asn 46 and the *DAF* gene has A at bp 239, which ablates a *Sau*3 AI site.

Effect of enzymes/chemicals on Es^a antigen on intact RBCs

Ficin/papain Resistant
Trypsin Resistant
α-Chymotrypsin Sensitive
Pronase Sensitive
Sialidase Resistant
DTT 200 mM/50 mM Weakened/resistant
Acid Presumed resistant

In vitro characteristics of alloanti-Es^a

Immunoglobulin class IgG
Optimal technique IAT
Complement binding No

Clinical significance of alloanti-Es^a

One report of a mild transfusion reaction.

Comments

Es(a−) RBCs have a weak expression of WES^b

The abundance of DAF on the apical surface of placental trophoblasts may result in the absorption of maternal antibodies to antigens in the Cromer system, thereby explaining why HDN is unlikely.

References

[1] Hustinx, H., Poole, J. and Storry, J. (2003) Personal communication.
[2] Reid, M.E. et al. (1996) Immunohematology 12, 112–114.
[3] Lublin, D.M. et al. (2000) Transfusion 40, 208–213.

IFC ANTIGEN

Terminology

ISBT symbol (number)	CROM7 (021.007)
Other names	202.007
History	Anti-IFC is made by people with the Inab phenotype. Named in 1986 after the first three probands

Occurrence

All populations	100%

Expression

Cord RBCs	Expressed
Altered	Weak on Dr(a−) and absent from PNH III RBCs

Molecular basis associated with IFC antigen

Not determined. Refer to system pages for molecular basis associated with an absence of IFC (the Inab phenotype).

Effect of enzymes/chemicals on IFC antigen on intact RBCs

Ficin/papain	Resistant
Trypsin	Resistant
α-Chymotrypsin	Sensitive
Pronase	Sensitive
Sialidase	Resistant
DTT 200 mM/50 mM	Weakened/resistant
Acid	Presumed resistant

In vitro characteristics of alloanti-IFC

Immunoglobulin class	IgG
Optimal technique	IAT
Neutralization	With concentrated serum/plasma/urine
Complement binding	No

Clinical significance of alloanti-IFC

Transfusion reaction	No to mild
HDN	No

Comments

The only phenotype that lacks IFC is the Inab phenotype because the RBCs do not express DAF[1].

A patient with an acquired and transient form of the Inab phenotype who made anti-IFC and had splenic infarctions has been reported[2].

The abundance of DAF on the apical surface of placental trophoblasts may result in the absorption of maternal antibodies to antigens in the Cromer system, thereby explaining why HDN is unlikely.

References
[1] Lublin, D.M. et al. (1994) Blood 84, 1276–1282.
[2] Matthes, T. et al. (2002) Transfusion 42, 1448–1457.

WESᵃ ANTIGEN

Terminology

ISBT symbol (number)	CROM8 (021.008)
Other names	202.008; 700.042
History	Named after the first antibody producer, SW, in 1987

Occurrence

Most populations	Less than 0.01%
Blacks (America)	0.48%
Blacks (N. London)	2.04%
Finns	0.56%

One probable *WESᵃ* homozygote identified.

Antithetical antigen

WESᵇ (**CROM9**)

Expression

Cord RBCs	Expressed

Molecular basis associated with WESᵃ antigen[1]

Amino acid	Arg 48 in SCR 1
Nucleotide	G at bp 245 in exon 2
Restriction enzyme	Loss of *Afl*l II and *Mse* I sites

Effect of enzymes/chemicals on WESᵃ antigen on intact RBCs

Ficin/papain	Resistant
Trypsin	Resistant
α-Chymotrypsin	Sensitive
Pronase	Sensitive
Sialidase	Resistant
DTT 200 mM/50 mM	Weakened/resistant
Acid	Presumed resistant

In vitro characteristics of alloanti-WES[a]

Immunoglobulin class	IgG
Optimal technique	IAT
Neutralization	With concentrated serum/plasma/urine
Complement binding	No

Clinical significance of alloanti-WES[a]

Transfusion reaction	No to mild
HDN	No

Comment

WES(a+b−) RBCs have a weak expression of Es[a].

Reference
[1] Lublin, D.M. et al. (2000) Transfusion 40, 208–213.

WES[b] ANTIGEN

Terminology

ISBT symbol (number)	CROM9 (021.009)
Other names	202.004; 900.033
History	Named in 1987 when it was recognized to be antithetical to WES[a]

Occurrence

All populations: 100%
Only two WES(a+b−) probands (one Finn, one Black).

Antithetical antigen

WES[b] (CROM8)

Expression

Cord RBCs	Expressed
Altered	Weak on Dr(a−) and Es(a−), and negative on PNH III RBCs

Molecular basis associated with WES^b antigen[1]

Amino acid	Leu 48 in SCR 1
Nucleotide	T at bp 245 in exon 2
Restriction enzyme	Has *Afl* II and *Mse* I sites

Effect of enzymes/chemicals on WES^b antigen on intact RBCs

Ficin/papain	Resistant
Trypsin	Resistant
α-Chymotrypsin	Sensitive
Pronase	Sensitive
Sialidase	Resistant
DTT 200 mM/50 mM	Weakened/resistant
Acid	Presumed resistant

In vitro characteristics of alloanti-WES^b

Immunoglobulin class	IgG
Optimal technique	IAT
Neutralization	Concentrated serum/plasma/urine
Complement binding	No

Clinical significance of alloanti-WES^b

Only one example of anti-WES^b, produced as a result of pregnancy, is described. The baby's RBCs had a positive DAT but there were no clinical signs of HDN.

Comments

WES(a+b−) RBCs have a weak expression of Es^a antigen.
The abundance of DAF on the apical surface of placental trophoblasts may result in the absorption of maternal antibodies to antigens in the Cromer system, thereby explaining why HDN is unlikely.

Reference
[1] Lublin, D.M. et al. (2000) Transfusion 40, 208–213.

UMC ANTIGEN

Terminology

ISBT symbol (number)	CROM10 (021.010)
Other names	202.010
History	Named in 1989, from the proband's name

Occurrence

All populations: 100%.
Only one UMC– proband and her UMC– brother (Japanese) have been described

Expression

Cord RBCs	Expressed
Altered	Weak on Dr(a−) and absent from PNH III RBCs

Molecular basis associated with UMC antigen[1]

Amino acid	Thr 216 in SCR 4
Nucleotide	C at bp 749 in exon 6

UMC– form of DAF has Met 216 and the *DAF* gene has T at bp 479.

Effect of enzymes/chemicals on UMC antigen on intact RBCs

Ficin/papain	Resistant
Trypsin	Resistant
α-Chymotrypsin	Sensitive
Pronase	Sensitive
Sialidase	Resistant
DTT 200 mM/50 mM	Weakened/resistant
Acid	Presumed resistant

In vitro characteristics of alloanti-UMC

Immunoglobulin class	IgG
Optimal technique	IAT
Neutralization	With concentrated serum/plasma/urine
Complement binding	No

Clinical significance of alloanti-UMC

Transfusion reaction   No data
HDN   The proband had three children with no signs or symptoms of HDN

Comments

The abundance of DAF on the apical surface of placental trophoblasts may result in the absorption of maternal antibodies to antigens in the Cromer system, thereby explaining why HDN is unlikely.

Reference
[1] Lublin, D.M. et al. (2000) Transfusion 40, 208–213.

GUTI ANTIGEN

Terminology

ISBT symbol (number)   CROM11 (021.011)
History   Named in 2002 after the first producer of the antibody

Occurrence

All populations: 100%
Only one GUTI– proband (Chilean) and his sister; 15% of Mapuche Indians are heterozygotes.

Expression

Cord RBCs   Expressed
Altered   Weak on Dr(a−) and negative on PNH III RBCs

Molecular basis associated with GUTI antigen[1]

Amino acid   Arg 206 in SCR 4
Nucleotide   G at bp 719 in exon 6

The GUTI– form of DAF has His 206 and the gene has A at bp 719; ablates *Mae* II site.

Effect of enzymes/chemicals on GUTI antigen on intact RBCs

Ficin/papain	Resistant
Trypsin	Resistant
α-Chymotrypsin	Sensitive
Pronase	Sensitive
Sialidase	Resistant
DTT 200 mM/50 mM	Weakened/resistant
Acid	Resistant

In vitro characteristics of alloanti-GUTI

Immunoglobulin class	IgG
Optimal technique	IAT
Complement binding	No

Clinical significance of alloanti-GUTI

No data.

Comments

The abundance of DAF on the apical surface of placental trophoblasts may result in the absorption of maternal antibodies to antigens in the Cromer system, thereby explaining why HDN is unlikely.

Reference
[1] Storry, J.R. et al. (2003) GUTI: A new antigen in the Cromer blood group system Transfusion 43, 340–344.

Number of antigens   8

Terminology

ISBT symbol	KN
ISBT number	022
CD number	CD35
Other names	ISBT collection 205, complement receptor 1 (CR1)
History	Reported in 1970 and named in honor of the first patient who made anti-Kn^a. Knops was established as a system in 1992 when the antigens were located to CR1

Expression

Soluble form	Present in low levels in plasma
Other blood cells	Granulocytes, B cells, a subset of T cells, monocytes, macrophages, neutrophils, eosinophils
Tissues	Glomerular podocytes, follicular dendritic cells in spleen and lymph nodes

Gene

Chromosome	1q32
Name	*KN (CR1)*
Organization:	Distributed over 130 to 160 kbp of gDNA: CR1*1 has 39 exons; CR1*2 has 47 exons; CR1*3 has 30 exons and CR1*4 has 31 exons[1,2]
Product	Complement receptor type 1 (CR1; CD35)

Gene map

Database accession numbers

GenBank   Y00816
www.bioc.aecom.yu.edu/bgmut/index.htm

Amino acid sequence of CR1*1[3]

```
         M GASSPRSPEP VGPPAPGLPF CCGGSLLAVV VLLALPVAWG  -1
QCNAPEWLPF ARPTNLTDEF EFPIGTYLNY ECRPGYSGRP FSIICLKNSV  50
WTGAKDRCRR KSCRNPPDPV NGMVHVIKGI QFGSQIKYSC TKGYRLIGSS 100
SATCIISGDT VIWDNETPIC DRIPCGLPPT ITNGDFISTN RENFHYGSVV 150
TYRCNPGSGG RKVFELVGEP SIYCTSNDDQ VGIWSGPAPQ CIIPNKCTPP 200
NVENGILVSD NRSLFSLNEV VEFRCQPGFV MKGPRRVKCQ ALNKWEPELP 250
SCSRVCQPPP DVLHAERTQR DKDNFSPGQE VFYSCEPGYD LRGAASMRCT 300
PQGDWSPAAP TCEVKSCDDF MGQLLNGRVL FPVNLQLGAK VDFVCDEGFQ 350
LKGSSASYCV LAGMESLWNS SVPVCEQIFC PSPPVIPNGR HTGKPLEVFP 400
FGKAVNYTCD PHPDRGTSFD LIGESTIRCT SDPQGNGVWS SPAPRCGILG 450
HCQAPDHFLF AKLKTQTNAS DFPIGTSLKY ECRPEYYGRP FSITCLDNLV 500
WSSPKDVCKR KSCKTPPDPV NGMVHVITDI QVGSRINYSC TTGHRLIGHS 550
SAECILSGNA AHWSTKPPIC QRIPCGLPPT IANGDFISTN RENFHYGSVV 600
TYRCNPGSGG RKVFELVGEP SIYCTSNDDQ VGIWSGPAPQ CIIPNKCTPP 650
NVENGILVSD NRSLFSLNEV VEFRCQPGFV MKGPRRVKCQ ALNKWEPELP 700
SCSRVCQPPP DVLHAERTQR DKDNFSPGQE VFYSCEPGYD LRGAASMRCT 750
PQGDWSPAAP TCEVKSCDDF MGQLLNGRVL FPVNLQLGAK VDFVCDEGFQ 800
LKGSSASYCV LAGMESLWNS SVPVCEQIFC PSPPVIPNGR HTGKPLEVFP 850
FGKAVNYTCD PHPDRGTSFD LIGESTIRCT SDPQGNGVWS SPAPRCGILG 900
HCQAPDHFLF AKLKTQTNAS DFPIGTSLKY ECRPEYYGRP FSITCLDNLV 950
WSSPKDVCKR KSCKTPPDPV NGMVHVITDI QVGSRINYSC TTGHRLIGHS 1000
SAECILSGNT AHWSTKPPIC QRIPCGLPPT IANGDFISTN RENFHYGSVV 1050
TYRCNLGSRG RKVFELVGEP SIYCTSNDDQ VGIWSGPAPQ CIIPNKCTPP 1100
NVENGILVSD NRSLFSLNEV VEFRCQPGFV MKGPRRVKCQ ALNKWEPELP 1150
SCSRVCQPPP EILHGEHTPS HQDNFSPGQE VFYSCEPGYD LRGAASLHCT 1200
PQGDWSPEAP RCAVKSCDDF LGQLPHGRVL FPLNLQLGAK VSFVCDEGFR 1250
LKGSSVSHCV LVGMRSLWNN SVPVCEHIFC PNPPAILNGR HTGTPSGDIP 1300
YGKEISYTCD PHPDRGMTFN LIGESTIRCT SDPHGNGVWS SPAPRCELSV 1350
RAGHCKTPEQ FPFASPTIPI NDFEFPVGTS LNYECRPGYF GKMFSISCLE 1400
NLVWSSVEDN CRRKSCGPPP EPFNGMVHIN TDTQFGSTVN YSCNEGFRLI 1450
GSPSTTCLVS GNNVTWDKKA PICEIISCEP PPTISNGDFY SNNRTSFHNG 1500
TVVTYQCHTG PDGEQLFELV GERSIYCTSK DDQVGVWSSP PPRCISTNKC 1550
TAPEVENAIR VPGNRSFFSL TEIIRFRCQP GFVMVGSHTV QCQTNGRWGP 1600
KLPHCSRVCQ PPPEILHGEH TLSHQDNFSP GQEVFYSCEP SYDLRGAASL 1650
HCTPQGDWSP EAPRCTVKSC DDFLGQLPHG RVLLPLNLQL GAKVSFVCDE 1700
GFRLKGRSAS HCVLAGMKAL WNSSVPVCEQ IFCPNPPAIL NGRHTGTPFG 1750
DIPYGKEISY ACDTHPDRGM TFNLIGESSI RCTSDPQGNG VWSSPAPRCE 1800
LSVPAACPHP PKIQNGHYIG GHVSLYLPGM TISYTCDPGY LLVGKGFIFC 1850
TDQGIWSQLD HYCKEVNCSF PLFMNGISKE LEMKKVYHYG DYVTLKCEDG 1900
YTLEGSPWSQ CQADDRWDPP LAKCTSRAHD ALIVGTLSGT IFFILLIIFL 1950
SWIILKHRKG NNAHENPKEV AIHLHSQGGS SVHPRTLQTN EENSRVLP   1998
```

Signal peptide: 41 amino acid residues.
Antigen mutations numbered by counting Met as 1.

Carrier molecule[3]

The CR1*1 allotype has 30 complement control protein repeats (CCPs) each comprising about 60 amino acids with sequence homology (also called short consensus repeats (SCRs)). Each CCP has four cysteine residues and is maintained in a folded conformation by two disulphide bonds. The other allotypes have a similar structure.

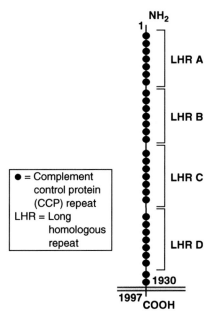

● = Complement
control protein
(CCP) repeat
LHR = Long
homologous
repeat

M_r (SDS-PAGE)	CR1*1 (A allotype) 220 000; CR1*2 (B allotype) 250 000; CR1*3 (C allotype) 190 000; CR1*4 (D allotype) 280 000 under non-reducing conditions.
CHO: N-glycan	25 sites: probably 6–8 occupied
CHO: O-glycan	None
Cysteine residues	Four per CCP
Copies per RBC	20–1500[4]

Molecular basis of antigens[5,6]

Antigen	Amino acid change	Exon	Nt change
Kn^a/Kn^b	Val1561Met	29	4708 G>A
McC^a/McC^b	Lys1590Glu	29	4795 A>G
Sl^a/Vil	Arg1601Gly	29	4828 A>G
Sl3+/Sl3−	Ser1610Thr	29	4855 A>G

Function

CR1 binds C3b and C4b and has an inhibitory effect on complement activation by classical and alternative pathways, protecting RBCs from autohemolysis. Erythrocyte CR1 is important in processing immune complexes by binding them for transport to the liver and spleen for removal from the circulation.

CR1 binds particles coated with C3b and C4b, thereby mediating phagocytosis by neutrophils and monocytes. The presence of CR1 on other blood cells and tissues suggests it has multiple roles in the immune response, for example, activation of B lymphocytes.

Disease association

Knops antigens (CR1 copy number) depressed in SLE, CHAD, PNH, hemolytic anemia, insulin-dependent diabetes mellitus, AIDS, some malignant tumors, any condition with increased clearance of immune complexes. Low levels of CR1 on RBCs may result in deposition of immune complexes on blood vessel walls with subsequent damage to the walls.

CR1 is a ligand for the rosetting of *Plasmodium falciparum* infected RBCs to uninfected RBCs[5]. [See Sl[a] (KN4).]

Almost 75% of HIV-1+ patients have an *in vivo* CR1 cleavage fragment of M_r 160 000, suggesting that RBC CR1 may have a role in HIV infection. This compares with 6.5% of healthy donors and 13.5% of patients with immune complex diseases[7].

Phenotypes (% occurrence)

	Caucasians	Blacks
Kn(a+b−)	94.5	99.9
Kn(a−b+)	1	0
Kn(a+b+)	4.5	0.1
McC(a+)	98	94
Sl(a+)	98	60
Yk(a+)	92	98

Null: Some RBCs (e.g. Helgeson) type as Kn(a−b−), McC(a−), Sl(a−) and Yk(a−) because these RBCs have low copy numbers of CR1 (approximately 10% of normal)[4].

Molecular basis of phenotypes

See table in Sl3 antigen [KN8].

Comments

The CR1 copy number per RBC may be decreased in stored specimens.
Allotypes may have arisen as a result of intragenic unequal crossing-over.

References

[1] Wong, W.W. et al. (1989) J. Exp. Med. 169, 847–863.
[2] Vik, D.P. and Wong, W.W. (1993) J. Immunol. 151, 6214–6224.
[3] Cohen, J.H. et al. (1999) Mol. Immunol. 36, 819–825.
[4] Moulds, J.M. et al. (1992) Vox Sang. 62, 230–235.
[5] Moulds, J.M. et al. (2001) Blood 97, 2879–2885.
[6] Moulds, J.M. et al. (2002) Transfusion 42, 251–256.
[7] Moulds, J.M. et al. (1995) Transfusion 35 (Suppl.), 59S (abstract).

Kn^a ANTIGEN

Terminology

ISBT symbol (number)	KN1 (022.001)
Other names	Knops; COST4; 205.004
History	Named after the Kn(a−) patient who made anti-Kn^a. The three Kn(a−) siblings in the original paper (1970) were later shown to have the Helgeson phenotype

Occurrence

Caucasians	98%
Blacks	99%

Antithetical antigen

Kn^b (KN2)

Expression

Cord RBCs	Weakened
Altered	Weak on dominant Lu(a−b−) RBCs and RBCs of patients with autoimmune diseases

443

Molecular basis associated with Kn[a] antigen[1]

Amino acid   Val 1561 in SCR 24
Nucleotide   G at bp 4708 in exon 29

Effect of enzymes/chemicals on Kn[a] antigen on intact RBCs

Ficin/papain   Weakened (especially ficin)
Trypsin   Sensitive
α-Chymotrypsin   Sensitive
Pronase   Resistant
Sialidase   Resistant
DTT 200 mM/50 mM   Sensitive/resistant
Acid   Resistant

In vitro characteristics of alloanti-Kn[a]

Immunoglobulin class   IgG
Optimal technique   IAT
Complement binding   No

Clinical significance of alloanti-Kn[a]

Transfusion reaction   No
HDN   No

Comments

Anti-Kn[a] frequently found in multispecific sera.

Disease processes causing RBC CR1 deficiency can lead to "false" negative Kn[a] typing.

Variable results in tests on different samples from the same patient have been described[2].

References

[1] Moulds, J.M. (2003) Personal communication.
[2] Rolih, S. (1990) Immunohematology 6, 59–67.

Kn^b ANTIGEN

Terminology

ISBT symbol (number)	KN2 (022.002)
Other names	COST5; 205.005
History	Identified in 1980 when an antibody was found to react with Kn(a−) RBCs

Occurrence

Caucasians, 4.5%; less than 0.01% in Blacks.

Antithetical antigen

Kn^a (**KN1**)

Expression

Cord RBCs	Weak

Molecular basis associated with Kn^b antigen[1]

Amino acid	Met 1561 in SCR 24
Nucleotide	A at bp 4708 in exon 29

Effect of enzymes/chemicals on Kn^b antigen on intact RBCs

Ficin/papain	Weakened (especially ficin)
Trypsin	Sensitive
α-Chymotrypsin	Sensitive
Pronase	Presumed resistant
Sialidase	Presumed resistant
DTT 200 mM/50 mM	Presumed sensitive/resistant
Acid	Presumed resistant

In vitro characteristics of alloanti-Kn^b

Immunoglobulin class	IgG
Optimal technique	IAT
Complement binding	No

Clinical significance of alloanti-Kn^b

No data available. Only one example of anti-Kn^b, in a serum containing anti-Kp^a, has been reported[2].

Comments

Disease processes causing RBC CR1 deficiency can lead to "false" negative Kn^b typing. Variable results in tests on different samples from the same patient have been described[3].

References
[1] Moulds, J.M. (2003) Personal communication.
[2] Mallan, M.T. et al. (1980) Transfusion 20, 630–631.
[3] Rolih, S. (1990) Immunohematology 6, 59–67.

McC^a ANTIGEN

Terminology

ISBT symbol (number)	KN3 (022.003)
Other names	McCoy; COST6; 205.006
History	Identified in 1978 and named after the patient who made the first anti-McC^a. Associated with Kn^a because 53% of McC(a−) RBCs were also Kn(a−)

Occurrence

Caucasians	98%
Blacks	94%

Antithetical antigen

McC^b (KN6)

Expression

Cord RBCs	Weak
Altered	Weak on dominant Lu(a−b−) RBCs and RBCs of patients with autoimmune diseases

Molecular basis associated with McC^a antigen[1]

Amino acid   Lys 1590 in SCR 25
Nucleotide   A at bp 4795 in exon 29

Effect of enzymes/chemicals on McC^a antigen on intact RBCs

Ficin/papain   Weakened (especially ficin)
Trypsin   Sensitive
α-Chymotrypsin   Sensitive
Pronase   Weakened
Sialidase   Resistant
DTT 200 mM/50 mM   Sensitive/resistant
Acid   Resistant

In vitro characteristics of alloanti-McC^a

Immunoglobulin class   IgG
Optimal technique   IAT
Complement binding   No

Clinical significance of alloanti-McC^a

Transfusion reaction   No
HDN   No

Comments

Disease processes causing RBC CR1 deficiency can lead to "false" negative typing. Variable results in tests on different samples from the same patient have been described.

Reference
[1] Moulds, J.M. et al. (2001) Blood 97, 2879–2885.

Sl^a ANTIGEN

Terminology

ISBT symbol (number)	KN4 (022.004)
Other names	Sl1; Swain–Langley; 205.007; COST7; McC^c
History	Reported in 1980 and named after the first two antibody producers: Swain and Langley

Occurrence

Caucasians	98%
Blacks	50–60%; 30% in West Africa

Antithetical antigens

Vil (**Sl2; KN7**)

Expression

Cord RBCs	Weak
Altered	Weak on dominant Lu(a−b−) RBCs and RBCs of patients with autoimmune diseases

Molecular basis associated with Sl^a antigen[1]

Amino acid	Arg 1601 in SCR 25
Nucleotide	A at bp 4828 in exon 29

See Sl3 [KN8].

Effect of enzymes/chemicals on Sl^a antigen on intact RBCs

Ficin/papain	Weakened (especially ficin)
Trypsin	Sensitive
α-Chymotrypsin	Sensitive
Pronase	Resistant
Sialidase	Resistant

| DTT 200 mM/50 mM | Sensitive/resistant |
| Acid | Resistant |

In vitro characteristics of alloanti-Sl^a

Immunoglobulin class	IgG
Optimal technique	IAT
Complement binding	No

Clinical significance of alloanti-Sl^a

| Transfusion reaction | No |
| HDN | No |

Comments

Sl^a has been subdivided, see Sl3 [KN8].

Anti-Sl^a is a common specificity produced by Blacks and initially may be confused with anti-Fy3 because most Fy(a−b−) RBCs are also likely to be Sl(a−).

Disease processes causing RBC CR1 deficiency can lead to "false" negative typing. Variable results in tests on different samples from the same patient have been described.

Reference
[1] Moulds, J.M. et al. (2001) Blood 97, 2879–2885.

Yk^a ANTIGEN

Terminology

ISBT symbol (number)	KN5 (022.005)
Other names	York; COST3; 205.003
History	Briefly described in 1969 and initially thought to be anti-Cs^a because the serum was non-reactive with two Cs(a−) RBC samples. Named in 1975 after the first producer of the antibody, Mrs. York

Occurrence

Caucasians	92%
Blacks	98%

Expression

Cord RBCs	Weak
Altered	Weak on dominant Lu(a−b−) RBCs and RBCs of patients with autoimmune diseases

Effect of enzymes/chemicals on Yk^a antigen on intact RBCs

Ficin/papain	Weakened (especially ficin)
Trypsin	Sensitive
α-Chymotrypsin	Sensitive
Pronase	Resistant
Sialidase	Sensitive
DTT 200 mM/50 mM	Sensitive/resistant
Acid	Resistant

In vitro characteristics of alloanti-Yk^a

Immunoglobulin class	IgG
Optimal technique	IAT
Complement binding	No

Clinical significance of alloanti-Yk^a

Transfusion reaction	No
HDN	No

Comments

Approximately 12% of Caucasian Yk(a−) RBCs and 16% of Black Yk(a−) RBCs are Cs(a−)[1].

Disease processes causing RBC CR1 deficiency can lead to "false" negative typing. Variable results in tests on different samples from the same patient have been described.

Reference
[1] Rolih, S. (1990) Immunohematology 6, 59–67.

McC^b ANTIGEN

Terminology

ISBT symbol (number)	KN6 (022.006)
History	Identified in 1983; antibody recognized an antigen antithetical to McC^a on RBCs of Blacks. Confirmed by molecular analysis and became a Knops system antigen in 2000

Occurrence

Caucasians	0%
Blacks	45%

Antithetical antigen

McC^a (**KN3**)

Expression

Cord RBCs	Weak

Molecular basis associated with McC^b antigen[1]

Amino acid	Glu 1590 in SCR 25
Nucleotide	G at bp 4795 in exon 29

Effect of enzymes/chemicals on McC^b antigen on intact RBCs

Ficin/papain	Variable
Trypsin	Presumed sensitive
α-Chymotrypsin	Presumed sensitive
Pronase	Presumed weakened
Sialidase	Presumed resistant
DTT 200 mM/50 mM	Presumed sensitive/resistant
Acid	Presumed resistant

In vitro characteristics of alloanti-McC^b

Immunoglobulin class	IgG
Optimal technique	IAT
Complement binding	No

Clinical significance of alloanti-McC^b

No data but unlikely to be significant.

Comments

Disease processes causing RBC CR1 deficiency can lead to "false" negative typing. Variable results in tests on different samples from the same patient have been described.

Reference

1   Moulds, J.M. et al. (2001) Blood 97, 2879–2885.

Vil ANTIGEN

Terminology

ISBT symbol (number)	KN7 (022.007)
Other names	Sl2; KN7; Villien; McC^d
History	Reported in 1980 and named after the first patient who made the antibody and before the antithetical relationship to Sl^a was established. Joined the Knops system in 2000 after molecular analysis confirmed the relationship with Sl^a

Occurrence

Caucasians	0%
Blacks	80%

Antithetical antigens

Sl^a (Sl1; KN4)

Expression

Cord RBCs	Weak

Molecular basis associated with Vil antigen[1]

Amino acid	Gly 1601 in SCR 25
Nucleotide	G at bp 4828 in exon 29

See Sl3 [KN8].

Effect of enzymes/chemicals on Vil antigen on intact RBCs

Ficin/papain	Presumed weakened
Trypsin	Presumed sensitive
α-Chymotrypsin	Presumed sensitive
Pronase	Presumed resistant
Sialidase	Presumed resistant
DTT 200 mM/50 mM	Presumed sensitive/resistant
Acid	Presumed resistant

In vitro characteristics of alloanti-Vil

Immunoglobulin class	IgG
Optimal technique	IAT
Complement binding	No

Clinical significance of alloanti-Vil

No data but unlikely to be significant.

Reference
[1] Moulds, J.M. et al. (2001) Blood 97, 2879–2885.

Sl3 ANTIGEN

Terminology

ISBT symbol (number)	KN8 (022.008)
Other names	KMW
History	Subdivision of Sl[a] reported in 2002 when differences were noted in the reactivity of various anti-Sl[a] (used for population studies). The definitive anti-Sl[a] (anti-Sl3) was made by a Caucasian woman (KMW).

Occurrence

All populations: 100%.
Only one Sl:1, −2, −3 person reported[1]. See table below.

Expression

Cord RBCs	Weak

453

Molecular basis associated with Sl3 antigen[1]

Amino Acid	Arg 1601 and Ser 1610 in SCR 25
Nucleotide	A at bp 4828 and A at 4855 in exon 29
Sl:1,−2,−3	G at bp 4855 and Thr1610. See table below.

Effect of enzymes/chemicals on Sl3 antigen on intact RBCs

Ficin/papain	Presumed weakened
Trypsin	Presumed sensitive
α-Chymotrypsin	Presumed sensitive
Pronase	Presumed resistant
Sialidase	Presumed resistant
DTT 200 mM/50 mM	Presumed sensitive/resistant
Acid	Presumed resistant

In vitro characteristics of alloanti-Sl3

Immunoglobulin class	IgG
Optimal technique	IAT
Complement binding	No

Clinical significance of alloanti-Sl3

Transfusion reaction	No data
HDN	No data

Relationship of Sl phenotypes

Phenotype	Amino acid 1601	Amino acid 1610	Ethnic association
Sl:1,−2,3	Arg	Ser	Most common in Whites
Sl:−1,2,−3	Gly	Ser	Common in Blacks
Sl:1,−2,−3	Arg	Thr	Found only in one White (KMW)

Reference
[1] Moulds J.M. et al. (2002) Transfusion 42, 251–256.

Number of antigens   2

Terminology

ISBT symbol	IN
ISBT number	023
CD number	CD44
Other name	ISBT Collection 203
History	Named because the first In(a+) people were from India

Expression

Other blood cells	Neutrophils, lymphocytes, monocytes
Tissues	Brain, breast, colon epithelium, gastric, heart, kidney, liver, lung, placenta, skin, spleen, thymus, fibroblasts

Gene

Chromosome	11p13
Name	IN (CD44)
Organization	At least 19 exons distributed over 50 kbp of gDNA (10 exons are variable). The hemopoietic isoform uses exons 1–5, 15–17 and 19.
Product	CD44

Gene map

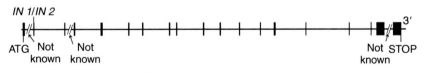

IN 1/IN 2 (252C>G) encode In^a/In^b (Pro46Arg)

⊢—⊣ 1 kbp

Database accession numbers

GenBank   M69215
www.bioc.aecom.yu.edu/bgmut/index.htm.

Amino acid sequence[1]

```
MDKFWWHAAW GLCLVPLSLA QIDLNITCRF AGVFHVEKNG RYSISRTEAA   50
DLCKAFNSTL PTMAQMEKAL SIGFETCRYG FIEGHVVIPR IHPNSICAAN  100
NTGVYILTYN TSQYDTYCFN ASAPPEEDCT SVTDLPNAFD GPITITIVNR  150
DGTRYVQKGE YRTNPEDIYP SNPTDDDVSS GSSSERSSTS GGYIFYTFST  200
VHPIPDEDSP WITDSTDRIP ATRDQDTFHP SGGSHTTHES ESDGHSHGSQ  250
EGGANTTSGP IRTPQIPEWL IILASLLALA LILAVCIAVN SRRRCGQKKK  300
LVINSGNGAV EDRKPSGLNG EASKSQEMVH LVNKESSETP DQFMTADETR  350
NLQNVDMKIG V                                            361
```

Signal peptide: 20 amino acid residues; sometimes cleaved.
Antigen mutation numbered by counting Met as 1.

Carrier molecule[1,2]

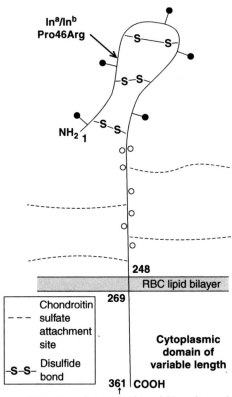

M_r (SDS-PAGE)   Reduced 80 000
CHO: N-glycan   6 sites
CHO: O-glycan   Depends on isoform
Chondroitin sulfate   4 sites
Cysteine residues   Depends on isoform
Copies per RBC   2000–5000 on mature RBCs

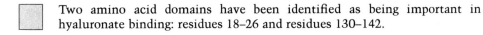

Two amino acid domains have been identified as being important in hyaluronate binding: residues 18–26 and residues 130–142.

Molecular basis of antigens

Antigen	Amino acid change	Exon	Nt change
In^a/In^b	Pro46Arg	2	252C>G

Other CD44 polymorphisms have been described but they are not associated with the In^a/In^b polymorphism.

Function

CD44 is an adhesion molecule in lymphocytes, monocytes and other tumor cells.
CD44 binds to hyaluronate and other components of the extracellular matrix.
CD44 is involved in immune stimulation and signaling between cells.

Disease association

One In(a−b−) individual with CDA was also Co(a−b−)[3].

Joint fluid from patients with inflammatory synovitis has higher than normal levels of soluble CD44[2]. Serum CD44 is elevated in some patients with lymphoma.

Phenotypes (% occurrence)

Phenotype	Caucasians/Blacks	Indians (South Asians)	Iranians/Arabs
In(a+b−)	Rare	Rare	Rare
In(a−b+)	99.9	96	90
In(a+b+)	<0.1	4	10

Comments

CD44 is present in reduced (variable) amounts in dominant type Lu(a−b−) RBCs but is expressed normally in other cells from these people.
Ser-Gly is a potential chondroitin sulfate linkage site. After Thr 202 various sequences can be generated by alternative splicing of at least 10 exons. Different splicing events occur during different stages of hemopoiesis. In mature RBCs, 10 exons are usually encoded. A protein of 361 amino acids is the predominant type in the RBC membrane (hematopoietic isoform CD44).

References

1 Spring, F.A. et al. (1988) Immunology 64, 37–43.
2 Moulds, J.M. (1994) In: Immunobiology of Transfusion Medicine (Garratty, G. ed) Marcel Dekker, Inc., New York, pp. 273–297.
3 Parsons, S.F. et al. (1994) Blood 83, 860–868.

In[a] ANTIGEN

Terminology

ISBT symbol (number)	IN1 (023.001)
Other names	203.001
History	In is an abbreviation of Indian, in honor of the ethnic group in which this antigen was first found

Occurrence

Caucasians	0.1%
Asians and Blacks	0.1%
Indians (South Asians)	4%
Iranians	10.6%
Arabs	11.8%

Antithetical antigen

In[b] (IN2)

Expression

Cord RBCs	Weak
Altered	Weak on RBCs from pregnant women

Molecular basis associated with In[a] antigen[1]

Amino acid	Pro 46
Nucleotide	C at bp 252 (nucleotide numbered as in Stamenkovic et al.[2]) in exon 2

Other polymorphisms exist in CD44 but are apparently not involved in In[a] antigen expression.

Effect of enzymes/chemicals on In^a antigen on intact RBCs

Ficin/papain	Sensitive
Trypsin	Sensitive
α-Chymotrypsin	Sensitive
Pronase	Sensitive
Sialidase	Resistant
DTT 200 mM/50 mM	Sensitive/sensitive
Acid	Resistant

In vitro characteristics of alloanti-In^a

Immunoglobulin class	IgG
Optimal technique	IAT
Complement binding	No

Clinical significance of alloanti-In^a

Transfusion reaction	Decreased cell survival
HDN	Positive DAT, no clinical HDN

References

1 Telen, M.J. et al. (1996) J. Biol. Chem. 271, 7147–7153.
2 Stamenkovic, I. et al. (1989) Cell 56, 1057–1062.

In^b ANTIGEN

Terminology

ISBT symbol (number)	IN2 (023.002)
Other names	203.002; Salis
History	Named when its antithetical relationship to In^a was identified

Occurrence

Caucasians	99%
Indians (South Asians)	96%

Antithetical antigen

In^a (**IN1**)

Expression

Cord RBCs   Weak
Altered   Weak on dominant Lu(a−b−) RBCs. Weak
                          on RBCs from pregnant women

Molecular basis associated with In[b] antigen[1]

Amino acid   Arg 46
Nucleotide   G at bp 252 (nucleotide numbered as in
                          Stamenkovic et al.[2]) in exon 2

Effect of enzymes/chemicals on In[b] antigen on intact RBCs

Ficin/papain   Sensitive
Trypsin   Sensitive
α-Chymotrypsin   Sensitive
Pronase   Sensitive
Sialidase   Resistant
DTT 200 mM/50 mM   Sensitive/sensitive
Acid   Resistant

In vitro characteristics of alloanti-In[b]

Immunoglobulin class   IgG
Optimal technique   IAT
Complement binding   No

Clinical significance of alloanti-In[b]

Transfusion reaction   No to severe/delayed and hemolytic[3]
HDN   Positive DAT, but no clinical HDN[4]

References
1   Telen, M.J. et al. (1996) J. Biol. Chem. 271, 7147–7153.
2   Stamenkovic, I. et al. (1989) Cell 56, 1057–1062.
3   Joshi, S.R. (1992) Vox Sang. 63, 232–233.
4   Longster, G.H. et al. (1981) Clin. Lab. Haemat. 3, 351–356.

OK OK blood group system

Number of antigens   1

Terminology

ISBT symbol	OK
ISBT number	024
CD number	CD147
Other name	EMMPRIN[1], M6 leukocyte activation antigen, basigin
History	The Ok[a] antigen achieved system status, becoming the OK system in 1998 when the antigen was located on CD147

Expression

Other blood cells	White blood cells, platelets
Tissues	Epithelium in kidney cortex and medullary, liver, acinar cells of pancreas, trachea, cervix, testes, colon, skin, smooth muscle, neural cells, forebrain, cerebellum[2-4]

Gene

Chromosome	19p13.3
Name	*OK (EMPRIN)*
Organization	7 exons distributed over 1.8 kbp of gDNA
Product	CD147 glycoprotein (OK glycoprotein)

Gene map

* OK (274G > A) encodes Ok(a+)/Ok(a−) (Glu92Lys)

|—————| 1 kbp

Database accession numbers

GenBank   X64364
www.bioc.aecom.yu.edu/bgmut/index.htm

Amino acid sequence[5]

```
                              M AAALFVLLGF ALLGTHGASG  -1
AAGTVFTTVE DLGSKILLTC SLNDSATEVT GHRWLKGGVV LKEDALPGQK  50
TEFKVDSDDQ WGEYSCVFLP EPMGTANIQL HGPPRVKAVK SSEHINEGET 100
AMLVCKSESV PPVTDWAWYK ITDSEDKALM NGSESRFFVS SSQGRSELHI 150
ENLNMEADPG QYRCNGTSSK GSDQAIITLR VRSHLAALWP FLGIVAEVLV 200
LVTIIFIYEK RRKPEDVLDD DDAGSAPLKS SGQHQNDKGK NVRQRNSS   248
```

OK encodes a leader sequence of 21 amino acids.

The amino acid substitution associated with the Ok(a+)/Ok(a−) phenotypes is at residue 92 counting from the Met as 1.

Carrier molecule[2,3]

Single pass type I membrane glycoprotein with two IgSF domains

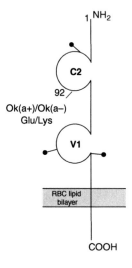

M_r (SDS-PAGE)	35 000–69 000
CHO: *N*-glycan	3
Cysteine residues	4
Copies per RBC	3000

Molecular basis of antigens

Antigen	Amino acid change	Exon	Nt change
Ok(a+)/Ok(a−)	Glu92Lys	4	274G>A

Function

Human CD147 (EMMPRIN – extracellular matrix metalloproteinase inducer) on tumor cells is thought to bind an unknown ligand on fibroblasts, which stimulates their production of collagenase and other extracellular matrix metalloproteinases, thus enhancing tumor cell invasion and metastases[1,5]. The monocarboxylate (lactate) transporters, MCT1 and MCT4, require CD147 for their correct plasma membrane expression and function[6].

Disease association

Expression is increased on granulocytes in rheumatoid and reactive arthritis. May be involved in tumor metastases.

Phenotypes (% occurrence)

	Caucasians	Blacks	Japanese
Ok(a+)	100%	100%	<100%
Ok(a−)	0	0	8 probands

Comments

CD147 is the human homolog of rat OX-47 antigen, mouse basigin, and chicken neurothelin.

References
1. Biswas, C. et al. (1995) Cancer Res. 55, 434–439.
2. Williams, B.P. et al. (1988) Immunogenetics 27, 322–329.
3. Anstee, D.J. and Spring, F.A. (1989) Transf. Med. Rev. 3, 13–23.
4. Spring, F.A. et al. (1997) Eur. J. Immunol. 27, 891–897.
5. Barclay, A.N., et al. (1997) Leucocyte Antigen FactsBook, 2nd Edition, Academic Press, San Diego, CA.
6. Wilson, M.C. et al. (2002) J. Biol. Chem. 277, 3666–3672.

Ok^a ANTIGEN

Terminology

ISBT symbol (number)	OK1 (024.001)
Other names	901.006; 900.016
History	Named in 1979 after the family name of the patient (S.Ko.G.) whose RBCs lacked the antigen and whose plasma contained the antibody

Occurrence

All populations, 100%; all eight Ok(a−) probands are Japanese.

Expression

Cord RBCs	Expressed
Other blood cells	All tested[2,3]
Tissues	All tested[2,3]

Molecular basis associated with Ok^a antigen[1]

Amino acid	Glu 92
Nucleotide	G at bp 274 in exon 4 and a silent mutation of T at bp 384
Ok(a−)	Lys 92 and A at bp 274, and C at bp 384

Effect of enzymes/chemicals on Ok^a antigen on intact RBCs

Ficin/papain	Resistant
Trypsin	Resistant
α-Chymotrypsin	Resistant
Pronase	Resistant
Sialidase	Resistant
DTT 200 mM	Resistant
Acid	Resistant

In vitro characteristics of alloanti-Ok^a

Immunoglobulin class	IgG
Optimal technique	IAT
Complement binding	No

Clinical significance of alloanti-Ok^a

Transfusion reaction	^{51}Cr cell survival studies indicated reduced RBC survival
HDN	No

References

1 Spring, F.A. et al. (1997) Eur. J. Immunol. 27, 891–897.
2 Williams, B.P. et al. (1988) Immunogenetics 27, 322–329.
3 Anstee, D.J. and Spring, F.A. (1989) Transf. Med. Rev. 3, 13–23.

Number of antigens

1

Terminology

ISBT symbol	MER2
ISBT number	025
Other name	901.011; Raph
History	Named RAPH after the first producer of alloanti-MER2; achieved system status in 1998. The antigen had previously been recognized by a monoclonal antibody, MER2, and the only antigen in the system retains this name

Expression

Other blood cells	CD34+ cells. There is a rapid decrease in expression during *ex vivo* erythropoiesis
Tissues	Fibroblasts

Gene

Chromosome	11p15.5
Name	*MER2*

The gene has not been cloned.

Carrier molecule

M_r (SDS-PAGE)   40 000

Disease association

An absence of the antigen in three of four probands with anti-MER2 maybe associated with kidney disease.

MER2 ANTIGEN

Terminology

ISBT symbol	MER2
Other names	Raph; Raf; RAPH1; 901.011

History   The first red cell surface polymorphism to
                                     be defined by a monoclonal antibody
                                     (MER2) was described in 1987[1]

Occurrence

All populations   92%

Expression

Cord RBCs   Expressed
Altered   Weakened (slightly) on RBCs with the
                                     dominant type of Lu(a−b−)

Effect of enzymes/chemicals on MER2 antigen on intact RBCs

Ficin/papain   Resistant
Trypsin   Sensitive
α-Chymotrypsin   Sensitive
Pronase   Sensitive
Sialidase   Resistant
DTT 200 mM   Variable
Acid   Not known

In vitro characteristics of alloanti-MER2

Immunoglobulin class   IgG
Optimal technique   IAT
Complement binding   2 of 3 human anti-MER2

Clinical significance of alloanti-MER2

Transfusion reaction   No
HDN   No information

Comments

Three individuals (two probands) with anti-MER2 (previously called anti-Raph) had renal failure requiring dialysis; two had made the antibody before receiving transfusion. All were Indian (South Asian) Jews. A fourth example of alloanti-MER2 was in a healthy Turkish blood donor who had never been transfused but had been pregnant twice.
   Antigen strength varies between different RBC samples.

Reference
1   Daniels, G.L. et al. (1988) Vox Sang. 55, 161–164.

JMH blood group system

Number of antigens   1

Terminology

ISBT symbol	JMH
ISBT number	026
CD number	CD108
Other name	901.007, Sema7A, H-Sema-L, Semaphorin CD108
History	JMH became a system in 2000 after it was shown that the JMH glycoprotein is CD108[1] and the gene encoding CD108 was cloned[2,3]

Expression

Other blood cells	Weak on lymphocytes, strong on activated lymphocytes and activated macrophages[1]
Tissues	Neurons of central nervous system; respiratory epithelium, placenta, testes and spleen[4]. Low expression in brain and thymus

Gene

Chromosome	15q22.3–q23[2,3]
Name	*JMH (SEMA-L; CD108)*
Organization	At least 13 exons over 9 kbp of gDNA
Product	Semaphorin CD108 (H-Sema-L)

Gene map

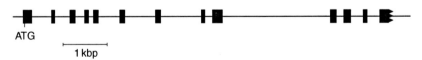

ATG

1 kbp

Database accession numbers

GenBank   AF069493, AF030698

Amino acid sequence[3]

```
MTPPPPGRAA PSAPRARVPG PPARLGLPLR LRLLLLLWAA AASAQGHLRS  50
GPRIFAVWKG HVGQDRVDFG QTEPHTVLFH EPGSSSVWVG GRGKVYLFDF 100
PEGKNASVRT VNIGSTKGSC LDKRDCENYI TLLERRSEGL LACGTNARHP 150
SCWNLVNGTV VPLGEMRGYA PFSPDENSLV LFEGDEVYST IRKQEYNGKI 200
PRFRRIRGES ELYTSDTVMQ NPQFIKATIV HQDQAYDDKI YYFFREDNPD 250
```

```
KNPEAPLNVS  RVAQLCRGDQ  GGESSLSVSK  WNTFLKAMLV  CSDAATNKNF  300
NRLQDVFLLP  DPSGQWRDTR  VYGVFSNPWN  YSAVCVYSLG  DIDKVFRTSS  350
LKGYHSSLPN  PRPGKCLPDQ  QPIPTETFQV  ADRHPEVAQR  VEPMGPLKTP  400
LFHSKYHYQK  VAVHRMQASH  GETFHVLYLT  EPGEQEHSFA  FNIMEIQPFR  450
RAAAIQTMSL  DAERRKLYVS  SQWEVSQVPL  DLCEVYGGGC  HGCLMSRDPY  500
CGWDQGRCIS  IYSSERSVLQ  SINPAEPHKE  CPNPKPDKAP  LQKVSLAPNS  550
RYYLSCPMES  RHATYSWRHK  ENVEQSCEPG  HQSPNCILFI  ENLTAQQYGH  600
YFCEAQEGSY  FREAQHWQLL  PEDGIMAEHL  LGHACALAAS  LWLGVLPTLT  650
LGLLVH                                                     656
```

A signal peptide of 44 amino acids is cleaved after membrane attachment.
A GPI anchor motif of 19 amino acids is cleaved.

Carrier molecule[1]

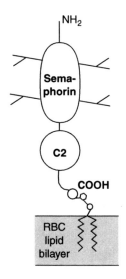

M_r (SDS-PAGE)	75 000–76 000
CHO: N-glycan	5
Cysteine residues	19

Function

Function of CD108 on RBCs is not known. Secreted and membrane-bound semaphorins function as signals, which guide axons in developing nervous tissue. Semaphorins and their receptors may also be involved in control of cellular functions, e.g., in cell-cell repulsion[5]. CD108 contains an Arg-Gly-Asp (267–269) cell attachment motif, which is common in adhesion molecules.

References
1 Mudad, R. et al. (1995) Transfusion 35, 566–570.
2 Yamada, A. et al. (1999) J. Immunol. 162, 4094–4100.
3 Lange, C. et al. (1998) Genomics 51, 340–350.
4 Bobolis, K.A. et al. (1992) Blood 79, 1574–1581.
5 Tamagnone, L. and Comoglio P.M. (2000) Trends Cell Biol. 10, 377–383.

JMH ANTIGEN

Terminology

ISBT symbol (number)	JMH (026.001)
Other names	John Milton Hagen; "Old Boys"; 900.018; JMH1
History	Named after the first antibody producer, John Milton Hagen

Occurrence

All populations	100%

Expression

Cord RBCs	Weak (some variation)
Altered	JMH variants

Molecular basis associated with JMH antigen

Not known; JMH antigen is carried on the GPI-linked CD108 glycoprotein.

Effect of enzymes/chemicals on JMH antigen on intact RBCs

Ficin/papain	Sensitive
Trypsin	Sensitive
α-Chymotrypsin	Sensitive
Pronase	Sensitive
Sialidase	Resistant
DTT 200 mM/50 mM	Sensitive/sensitive
Acid	Resistant

In vitro characteristics of alloanti-JMH

Immunoglobulin class   IgG (predominantly IgG4 in acquired JMH-negative people)

Optimal technique   IAT

Complement binding   No

Clinical significance of alloanti-JMH

Transfusion reaction   No. Decreased survival in a JMH variant[1]

HDN   No

Autoanti-JMH

Yes.

Comments

Autoanti-JMH is often found in elderly persons with an acquired absent or weak expression of JMH; the DAT may be positive. One family has shown dominant inheritance of the JMH-negative phenotype in 3 generations[2]. Alloanti-JMH is present in JMH-positive individuals whose RBCs express variant forms of CD108[1]. These alloantibodies (RM, VG, GP, DW) are not mutually compatible.

References

[1] Mudad, R. et al. (1995) Transfusion 35, 925–930.
[2] Kollmar, M. et al. (1981) Transfusion 21, 612 (abstract).

Number of antigens   1

Terminology

ISBT symbol	I
ISBT number	027
Other name	207; Ii collection
History	The I antigen was placed in a system in 2002 when mutations of the *I* gene encoding the transferase responsible for converting i-active straight chains to I-active branched chains were identified

Expression

Soluble form	Human milk, saliva, amniotic fluid, urine, ovarian cyst fluid (small amounts in serum/plasma)
Other blood cells	Lymphocytes, monocytes, granulocytes, platelets
Tissues	Wide tissue distribution

Gene

Chromosome	6p24
Name	*I (IGnT, GCNT2)*[1]
Organization	Three exons spread over approximately 100 kbp of gDNA. Three forms of exon 1 are differentially spliced to give one of three transcripts: IGnTA, IGnTB, or IGnTC[2,3]
Product	*N*-acetylglucosaminyltransferase (β6GlcNAc-transferase). The branching enzyme for I antigen expression on RBCs is encoded by *IGnTC*. That for expression of I antigen on lens epithelium is encoded by *IGnTB*[2]

Gene map

Database accession numbers

IGnTA, AF458024; IGnTB, AF458025; IGnTC, AF458026
www.bioc.aecom.yu.edu/bgmut/index.htm

Amino acid sequence for IGnTC β6GlcNAc-transferase[2,3]

```
MNFWRYCFFA FTLLSVVIFV RFYSSQLSPP KSYEKLNSSS ERYFRKTACN  50
HALEKMPVFL WENILPSPLR SVPCKDYLTQ NHYITSPLSE EEAAFPLAYV 100
MVIHKDFDTF ERLFRAIYMP QNVYCVHVDE KAPAEYKESV RQLLSCFQNA 150
FIASKTESVV YAGISRLQAD LNCLKDLVAS EVPWKYVINT CGQDFPLKTN 200
REIVQHLKGF KGKNITPGVL PPDHAIKRTK YVHQEHTDKG GFFVKNTNIL 250
KTSPPHQLTI YFGTAYVALT REFVDFVLRD QRAIDLLQWS KDTYSPDEHF 300
WVTLNRVSGV PGSMPNASWT GNLRAIKWSD MEDRHGGCHG HYVHGICIYG 350
NGDLKWLVNS PSLFANKFEL NTYPLTVECL ELRHRERTLN QSETAIQPSW 400
YF                                                    402
```

Carrier molecule

The *IGnTC* gene product adds β6GlcNAc to i-active, linear chains of repeating *N*-acetyllactosamine units, on lipids and proteins on RBCs and to proteins in plasma. See figure in Section III.

Present on proteins with polylactosamine-containing N-glycans (band 3, glucose transporter, etc.)

Copies per RBC   500000

Function

Not known.

Disease association

A decreased expression of I antigen and concomitant increased expression of the reciprocal i antigen are associated with leukemia, Tk polyagglutination, thalassemia, sickle cell disease, HEMPAS, Diamond Blackfan anemia, myeloblastic erythropoiesis, sideroblastic erythropoiesis and any condition that results in stress hematopoiesis.

Anti-I is associated with cold agglutinin hemagglutinin disease (CHAD) and some mycoplasma pneumonia.

Phenotypes associated with I antigen and the reciprocal i antigen

RBCs	Antigens		Occurrence
	I	*i*	
Adult	Strong	Weak	Common
Cord	Weak	Strong	All
i Adult	Trace	Strong	Rare

Molecular basis of I-negative phenotype

Taiwanese with cataracts: no β6GlcNAc-transferase activity[3,4]

Family S	*IGnTA*	1049 G>A in exon 3	Gly350Glu
	IGnTB	1043 G>A in exon 3	Gly348Glu
	IGnTC	1049 G>A in exon 3	Gly350Glu
Family W	*IGnTA*	1049 G>A in exon 3	Gly350Glu; 1154G>A, Arg385His
	IGnTB	1043 G>A in exon 3	Gly348Glu; 1148G>A, Arg383His
	IGnTC	1049 G>A in exon 3	Gly350Glu; 1154G>A, Arg385His
Family C	Deletion of *IGnT* exons 1B, 1C, 2, and 3		

Caucasians without cataracts: markedly reduced β6GlcNAc-transferase activity[2]

4 probands	505G>A in exon 1C Ala169Thr
1 proband	505G>A in exon 1C Ala169Thr; 683G>A, Arg228Gln

Comments

I antigens occur on precursor A-, B- and H-active oligosaccharide chains.

References

1 Bierhuizen, M.F. et al. (1993) Genes Dev. 7, 468–478.
2 Yu, L.C. et al. (2001) Blood. 98, 3840–3845.
3 Yu, L.C. et al. (2003) Blood. 101, 2081–2087.
4 Inaba, N. et al. (2003) Blood. 101, 2870–2876.

I ANTIGEN

Terminology

ISBT symbol (number)	I1 (027.001)
Other names	900.026, 207.001, Individual
History	Reported in 1956; named I to emphasize the high degree of the 'Individuality' of blood samples failing to react with a potent cold agglutinin. Placed in a collection with i antigen in 1990 and a made one antigen system in 2002 when the gene encoding the branching transferase was cloned.

Occurrence

Adults	100%w

Reciprocal antigen

i (See Ii collection [207])

Expression

Cord RBCs	Weaker than on adult RBCs; frequently appear to be I-negative
Altered	Weakened on RBCs produced under hematopoietic stress and on South East Asian ovalocytes

Molecular basis associated with I antigen[1]

Branched type 2 chains: Galβ1–4GlcNAcβ1

$$Gal\beta1–4(GlcNAc\beta1–3Gal\beta1–4)_n\text{-Glc-Cer}$$

Galβ1–4GlcNAcβ1

See System pages for molecular basis associated with I-negative (i adult) phenotype.

Effect of enzymes and chemicals on I antigen on intact RBCs

Ficin/papain	Resistant (↑↑)
Trypsin	Resistant (↑↑)
α-Chymotrypsin	Resistant (↑↑)
Pronase	Resistant (↑↑)
Sialidase	Resistant (↑)
DTT 200 mM	Resistant
Acid	Resistant

In vitro characteristics of anti-I

Immunoglobulin class	IgM (rarely IgG)
Optimal technique	RT or 4°C
Complement binding	Yes; some hemolytic

Clinical significance of anti-I

Transfusion reaction	No (may need to prewarm blood). Increased destruction of I+ RBCs transfused to people with the adult i phenotype and alloanti-I
HDN	No

Autoanti-I

Most people have cold reactive autoanti-I.
A common specificity in CHAD and pregnancy.

Comments

So-called compound antigens have been described: IA, IB, IAB, IH, IP1, ILe[bH]
Alloanti-I is rare because the I– (i adult) phenotype is rare.

Reference
[1] Roelcke, D. (1995) In: Molecular Basis of Human Blood Group Antigens (Cartron, J.-P. and Rouger, P. eds) Plenum Press, New York, pp. 117–152.

GLOB Globoside blood group system

Numer of antigens

1

Terminology

ISBT symbol	GLOB
ISBT number	028
History	The P antigen was removed from the Globoside (GLOB) Collection in 2002 when the molecular basis of globoside deficiency was defined

Expression

Other blood cells	Erythroid precursor cells, lymphocytes, monocytes
Tissues	Endothelium, placenta (trophoblasts and interstitial cells), fibroblasts, fetal liver, fetal heart, kidney, prostate, peripheral nerves[1].

Gene[2,3]

Chromosome	3q25
Name	*GLOB* (β3GalNAcT1)
Organization	Six exons distributed over ~19 kbp
Product	UDP-*N*-acetylgalactosamine (globotriao-sylceramide 3-β-*N*-acetylgalactosaminyl-transferase Gb$_4$Cer/globoside synthase EC2.4.1.79; β3GalNAc-T1; P synthase)

Gene map

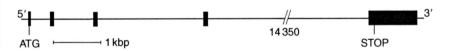

ATG ├───── 1 kbp

14 350

STOP

Database accession numbers

AB050855

Amino acid sequence

```
MASALWTVLP SRMSLRSLKW SLLLLSLLSF FVMWYLSLPH YNVIERVNWM  50
YFYEYEPIYR QDFHFTLREH SNCSHQNPFL VILVTSHPSD VKARQAIRVT 100
WGEKKSWWGY EVLTFFLLGQ EAEKEDKMLA LSLEDEHLLY GDIIRQDFLD 150
TYNNLTLKTI MAFRWVTEFC PNAKYVMKTD TDVFINTGNL VKYLLNLNHS 200
EKFFTGYPLI DNYSYRGFYQ KTHISQEYP FKVFPPYCSG LGYIMSRDLV 250
PRIYEMMGHV KPIKFEDVYV GICLNLLKVN IHIPEDTNLF FLYRIHLDVC 300
QLRRVIAAHG FSSKEIITFW QVMLRNTTCH Y                    331
```

Carrier molecule

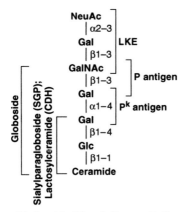

See Globoside Blood Group Collection and Section III
Copies per RBC   14 600 000

Function

This transferase converts P^k (209.002) to P.

Disease association[4]

P is a receptor for Parvovirus B19 and some P-fimbriated *E. coli.*
Anti-P is associated with paroxysmal cold hemoglobinuria (PCH).
   Cytotoxic IgM and IgG3 antibodies directed against P and/or P^k antigens are associated with a higher than normal rate of spontaneous abortion in women with the rare p [Tj(a−)], P_1^k, and P_2^k phenotypes.

Phenotypes

Phenotype	Occurrence	RBC antigens	Antibody
P_1	80%	P, P1, P^k	None
P_2	20%	P, P^k	Anti-P1
P_1^k	Rare	P1, P^k	Anti-P
P_2^k	Rare	P^k	Anti-P
p	Rare	None	Anti-P, P1, P^k (formerly anti-Tj[a])

Molecular basis of P^k phenotype due to mutations in *GLOB* (β3GalNAc-T1)[5]

Exon 5 202C>T resulting in a stop codon following residue 67 (Finnish)
Exon 5 537–538 insA; frameshift from amino acid 180 and a premature stop codon 182 (Arab)

Exon 5 797 A>C Glu266Ala (French)
Exon 5 811 G>A Gly271Arg (English)
See GLOB Blood Group Collection (209) for mutations in α4Gal-T associated with the p phenotype

References

1   Bailly, P. and Bouhours, J.F. (1995) In: Molecular Basis of Human Blood Group Antigens (Cartron, J.-P. and Rouger, P. eds) Plenum Press, New York, pp. 300–329.
2   Amado, M. et al. (1998) J. Biol. Chem. 273, 12770–12778.
3   Okajima, T. et al. (2000) J. Biol. Chem. 275, 40498–40503.
4   Moulds, J.M. et al. (1996) Transfusion 36, 362–374.
5   Hellberg, A., Poole, J. and Olsson, M.L. (2002) J. Biol. Chem. 277, 29455–29459.

P ANTIGEN

Terminology

ISBT symbol (number)	GLOB1 (028.001)
Other names	Globoside; Gb_4Cer; Gb4; 209.001
History	Anti-P recognized in 1955 as a component in sera of p people and in 1959 as the specificity made by P^k people; resulted in renaming the original anti-P as anti-P_1 (now called anti-P1; see P 003)

Occurrence

All populations	>99.9%

Expression

Cord RBCs	Expressed

Molecular basis associated with P antigen[1]

```
GalNAc  ⎤
  |β1–3 ⎥ P antigen
 Gal    ⎦
  |α1–4
 Gal
  |β1–4
 Glc
  |β1–1
Ceramide
```

For molecular basis of P-negative phenotypes, see System pages

Effect of enzymes/chemicals on P antigen on intact RBCs

Ficin/papain	Resistant ($\uparrow\uparrow$)
Trypsin	Resistant ($\uparrow\uparrow$)
α-Chymotrypsin	Resistant ($\uparrow\uparrow$)
Pronase	Resistant ($\uparrow\uparrow$)
Sialidase	Resistant
DTT 200 mM	Resistant
Acid	Resistant

In vitro characteristics of alloanti-P

Immunoglobulin class	IgM and IgG
Optimal technique	RT; 37°C; IAT
Complement binding	Yes; some hemolytic

Clinical significance of alloanti-P

Transfusion reaction	No to severe (rare) because anti-P is rare, (crossmatch would be incompatible)
HDN	No to mild (in P^k mothers with anti-P)

Autoanti-P

Yes, as a biphasic autohemolysin in paroxysmal cold hemoglobinuria (PCH).

Comments

Minority of anti-P made by P^k propositi react very weakly with p RBCs.
Anti-P is associated with paroxysmal cold hemoglobinuria (PCH).
Cytotoxic IgM and IgG3 antibodies directed against P and/or P^k antigens are associated with a higher than normal rate of spontaneous abortion in women with the rare p [Tj(a−)], P_1^k, and P_2^k phenotypes.

Reference
[1] Bailly, P. and Bouhours J.F. (1995) In: Molecular Basis of Human Blood Group Antigens (Cartron, J.-P. and Rouger, P. eds) Plenum Press, New York, pp. 300–329.

GIL blood group system

Number of antigens
1

Terminology
ISBT symbol	GIL
ISBT number	029
History	Named after the last name of the first antigen-negative proband. Became a system in 2002 after the antigen was located on aquaglyceroporin

Expression
Other blood cells	Absent from platelets
Tissues	Kidney medulla and cortex, basolateral membrane of collecting duct cells, small intestine, stomach, colon, spleen, airways, skin, colon, eye[1–3]

Gene[1,4,5]
Chromosome	9p13
Name	*AQP3*
Organization	Six exons spanning approximately 6 kbp of gDNA
Product	Aquaglyceroporin, AQP3; a member of the major intrinsic protein (MIP) family of water channels

Gene map

ATG   STOP

1 kbp

Database accession numbers
GenBank   AB001325 (Partial Sequence)
www.bioc.aecom.yu.edu/bgmut/index.htm

Amino acid sequence

```
MGRQKELVSR  CGEMLHIRYR  LLRQALAECL  GTLILVMFGC  GSVAQVVLSR   50
GTHGGFLTIN  LAFGFAVTLG  ILIAGQVSGA  HLNPAVTFAM  CFLAREPWIK  100
LPIYTLAQTL  GAFLGAGIVF  GLYYDAIWHF  ADNQLFVSGP  NGTAGIFATY  150
PSGHLDMING  FFDQFIGTAS  LIVCVLAIVD  PYNNPVPRGL  EAFTVGLVVL  200
VIGTSMGFNS  GYAVNPARDF  GPRLFTALAG  WGSAVFTTGQ  HWWWVPIVSP  250
LLGSIAGVFV  YQLMIGCHLE  QPPPSNEEEN  VKLAHVKHKE  QIMGRQKELV  300
SRCGEMLHIR  YRLLRQALAE  CLGTLILVMF  GCGSVAQVVL  SR          342
```

Carrier molecule

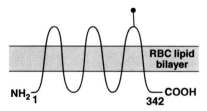

M_r (SDS-PAGE)

CHO: *N*-glycan
Cysteine residues
Copies per RBC

46 000 that reduced to 26 000 after
N-glycosidase F treatment
1
6
25 000

Molecular basis of GIL$_{null}$ phenotype[6]

Intron 5 donor splice site g > a, outsplicing of exon 5; frameshift; Stop.

Function

A water channel that also transports nonionic small molecules such as urea and glycerol. RBCs from a GIL-negative proband had reduced glycerol permeability.

Phenotypes

Null phenotype   GIL-negative

Comments

By Western blotting, RBC membranes from different people have different levels of expression of AQP3. AQP3 is present in the membrane as dimers, trimers and tetramers[7].

References
1   Ishibashi, K. et al. (1994) Proc. Natl. Acad. Sci. USA 91, 6269–6273.
2   Roudier, N. et al. (2002) J. Biol. Chem. 277, 7664–7669.
3   Agre, P. et al. (2002) J. Physiol. London 542, 3–16.
4   Inase, N. et al. (1995) J. Biol. Chem. 270, 17913–17916.
5   Carbrey, J. (2003) Personal Communication.
6   Roudier, N. et al. (2002) J. Biol. Chem. 277, 45854–45859.
7   Ledvinova, J. et al. (1997) Biochim. Biophys Acta 1345, 180–187.

GIL ANTIGEN

Terminology

ISBT symbol (number)	GIL1 (029.001)
Other names	Gill
History	Reported in 1981; name derived from the first antigen-negative proband who made an antibody to a high incidence antigen

Occurrence

All populations, 100%; Five probands (American, French, German)[1]

Expression

Cord RBCs	Slightly weaker than on RBCs from adults

Molecular basis associated with GIL antigen

GIL-negative RBCs lack the aquaglyceroporin (AQP3). See system pages.

Effect of enzymes/chemicals on GIL antigen on intact RBCs

Ficin/papain	Resistant ($\uparrow$)
Trypsin	Resistant ($\uparrow$)
α-Chymotrypsin	Resistant ($\uparrow$)
Pronase	Presumed resistant
Sialidase	Presumed resistant
DTT 200 mM	Resistant
Acid	Resistant

In vitro characteristics of alloanti-Gil

Immunoglobulin class	IgG
Optimal technique	IAT
Complement binding	Yes

Clinical significance of alloanti-Gil

Transfusion reaction	Hemolytic
HDN	Positive DAT but no clinical HDN

Comment

There may be heterogeneity among the five reported anti-GIL[1].

Reference

[1] Daniels, G.L. et al. (1998) Immunohematology 14, 49–52.

Blood group collections

Antigens in these collections have serological, biochemical or genetic connection.

Number	Name
205	Cost
207	Ii
208	Er
209	Globoside
210	Unnamed

Number of antigens   2

Terminology

ISBT symbol   COST
ISBT number   205
Other name   Cost–Stirling
History   This collection of phenotypically associated high prevalence antigens was established in 1988 and named after one of the original patients who made anti-Cs^a (Mrs. Cost). Five of the original antigens from this collection are in the Knops system because they are carried on CR1.

Phenotypes

Null   Some red cells type as Kn(a−b−), McC(a−), Sl(a−), Yk(a−), Cs(a−b−) and have low copy numbers of CR1. However, Cs^a and Cs^b are not carried on CR1[1].

Reference
[1] Moulds, J.M. et al. (1992) Vox Sang. 62, 230–235.

Cs^a ANTIGEN

Terminology

ISBT symbol (number)   COST1 (205.001)
Other names   Cost–Stirling; 900.004
History   Named in 1965 after two of the original patients (Mrs. Cost and Mrs. Stirling) who made anti-Cs^a

Occurrence

Most populations   >98%
Blacks   96%

Antithetical antigen

Cs^b (**COST2**)

Expression

Cord RBCs   Expressed; may be slightly weaker

Effect of enzymes/chemicals on Cs^a antigen on intact RBCs

Ficin/papain   Resistant
Trypsin   Resistant
α-Chymotrypsin   Resistant
Pronase   Resistant
Sialidase   Resistant
DTT 200 mM/50 mM   Variable
Acid   Resistant

In vitro characteristics of alloanti-Cs^a

Immunoglobulin class   IgG
Optimal technique   IAT
Complement binding   No

Clinical significance of alloanti-Cs^a

Transfusion reaction   No
HDN   No

Comments

Cs^a has variable expression on RBCs from different people. RBCs of approximately 12% Caucasians and 15% Blacks with the Yk(a−) phenotype are also Cs(a−)[1].

Reference
[1]   Rolih, S. (1990) Immunohematology 6, 59–67.

Cs^b ANTIGEN

Terminology

ISBT symbol (number)	COST2 (205.002)
History	Identified in 1987 and named when the antigen was shown to be antithetical to Cs^a

Occurrence

All populations	34%

Antithetical antigen

Cs^a (COST1)

Expression

Cord RBCs	Presumed expressed

Effect of enzymes/chemicals on Cs^b antigen on intact RBCs

Ficin/papain	Resistant
Trypsin	Resistant
α-Chymotrypsin	Resistant
Pronase	Resistant
Sialidase	Resistant
DTT 200 mM/50 mM	Variable
Acid	Presumed resistant

In vitro characteristics of alloanti-Cs^b

Immunoglobulin class	IgG
Optimal technique	IAT
Complement binding	No

Clinical significance of alloanti-Cs^b

Only one example of antibody published.

Number of antigens   1

Terminology

ISBT symbol	I
ISBT number	207
Other name	Individual
History	I and i antigens were placed in the Ii Blood Group Collection in 1990. In 2002, the I antigen was promoted to a blood group system, leaving i alone in Blood Group Collection 207

Gene

The gene has not been identified; i antigen is produced by the sequential action of multiple glycosyltransferases.

Carrier molecule

The i antigen is on unbranched carbohydrate chains of repeating N-acetyl-lactosamine units on lipids and proteins on RBCs and proteins in plasma. With the action of the branching enzyme, β6GlcNAc-transferase, these i antigen-carrying chains become part of the precursor chain of the I antigen (see I Blood Group System [027]).

Function

Not known.

Disease association

Enhanced expression of i antigens is associated with leukemia, Tk polyag-glutination, thalassemia, sickle cell disease, HEMPAS, Diamond Blackfan anemia, myeloblastic erythropoiesis, sideroblastic erythropoiesis and any condition that results in stress hematopoiesis.

Anti-i is associated with infectious mononucleosis and other lymphopro-liferative disorders (e.g. Hodgkins disease) and occasionally with CHAD.

Phenotypes associated with i antigen and the reciprocal I antigen

See I Blood Group System [027].

Molecular basis of i adult phenotype

See I Blood Group System [027].

Comments

The i antigen occurs on precursor I-active and A-, B-, H-active oligosaccharide chains.

i ANTIGEN

Terminology

ISBT symbol (number)   I2 (207.002)
Other names   900.027
History   Named in 1960 because of its reciprocal, but not classical antithetical association with the I antigen

Occurrence

All RBCs of adults have trace amounts of i antigen. The adult i phenotype is rare.

Reciprocal antigen

I (See I [027] blood group system).

Expression

Cord RBCs   Strong
Altered   Enhanced on CDA II RBCs and RBCs produced under hematopoietic stress

Molecular basis associated with i antigen[1]

Linear type 2 chains $Gal\beta1–4(GlcNAc\beta1–3Gal\beta1–4)_n$-Glc-Cer
See I Blood Group System (027) pages for molecular basis associated with i adult phenotype.

Effect of enzymes/chemicals on i antigen on intact RBCs

Ficin/papain	Resistant (↑↑)
Trypsin	Resistant (↑↑)
α-Chymotrypsin	Resistant (↑↑)
Pronase	Resistant (↑↑)
Sialidase	Resistant
DTT 200 mM	Resistant
Acid	Resistant

In vitro characteristics of anti-i

Immunoglobulin class	IgM (rarely IgG)
Optimal technique	RT or 4°C
Complement binding	Yes; some hemolytic

Clinical significance of anti-i

Transfusion reaction	No
HDN	Rare

Autoanti-i

Anti-i are considered to be autoantibodies. Transient autoanti-i can occur in infectious mononucleosis and some lymphoproliferative disorders.

Comments

So-called compound antigens have been described: iH, iP1, iHLe[b].

RBCs with the dominant Lu(a−b−) phenotype have a depressed expression of i antigen whereas RBCs with the X-linked form of the Lu(a−b−) phenotype have enhanced expression of i antigen. i antigen is often enhanced in patients with hematopoietic stress.

Horse RBCs have a strong expression of i antigen and can be used as a diagnostic tool of infectious mononucleosis.

Reference
1 Roelcke, D. (1995) In: Molecular Basis of Human Blood Group Antigens (Cartron, J.-P. and Rouger, P. eds) Plenum Press, New York, pp. 117–152.

ER | Er blood group collection

Number of antigens   2

Terminology
ISBT symbol   ER
ISBT number   208
History   Became a blood group collection in 1990

Phenotypes (% occurrence)
Null   Er(a−b−)

Er^a ANTIGEN

Terminology
ISBT symbol (number)   ER1 (208.001)
Other names   Rosebush; Ros; Min; Rod
History   Reported in 1982; named after the first
                                proband to make the antibody

Occurrence
All populations, 100%; three Er(a−) European probands.

Antithetical antigen
Er^b (**Er2**)

Expression
Cord RBCs   Expressed
Altered   RBCs from a Japanese woman and two of
                                her siblings reacted with three of eight
                                anti-Er^a

Effect of enzymes/chemicals on Er^a antigen on intact RBCs
Ficin/papain   Resistant
Trypsin   Resistant
α-Chymotrypsin   Resistant

Pronase	Resistant
Sialidase	Resistant
DTT 200 mM/50 mM	Resistant
Acid	Sensitive

In vitro characteristics of alloanti-Er^a

Immunoglobulin class	IgG
Optimal technique	IAT
Complement binding	No

Clinical significance of alloanti-Er^a

Transfusion reaction	No
HDN	Positive DAT but no clinical HDN

Comment

The mode of inheritance of Er^a is unclear: One Er(a−) proband has two siblings, a mother, two aunts and an uncle all of whom were Er(a−).

Er^b ANTIGEN

Terminology

ISBT symbol (number)	ER2 (208.002)
History	Reported in 1988 when the antibody was shown to recognize the antithetical low prevalence antigen to Er^a

Occurrence

Less than 0.01%.

Antithetical antigen

Er^a (**ER1**)

Expression

Cord RBCs   Expressed

Effect of enzymes/chemicals on Er^b antigen on intact RBCs

Ficin/papain   Resistant
Trypsin   Presumed resistant
α-Chymotrypsin   Presumed resistant
Pronase   Presumed resistant
Sialidase   Presumed resistant
DTT 200 mM/50 mM   Presumed resistant
Acid   Presumed sensitive

In vitro characteristics of alloanti-Er^b

Immunoglobulin class   IgG
Optimal technique   IAT
Complement binding   No

Clinical significance of alloanti-Er^b

Only one example[1].

Reference
[1]   Hamilton, J.R. et al. (1988) Transfusion 28, 268–271.

491

 **GLOB** Globoside blood group collection

Number of antigens   2

Terminology

ISBT symbol (number)	GLOB (209)
Other name	GLOBO
History	P, P^k, and LKE were removed from the P system (see 003) because separate loci control production of P1 antigen (neo-lacto-based antigen) and these antigens. They were gathered into an unnamed collection in 1990 because of their serological and biochemical relationship; named Globoside (GLOB) Collection in 1991. In 2002, P was upgraded to its own system (GLOB 028). The gene encoding the galactosyltransferase responsible for synthesis of P^k was cloned in 2000. However, until the relationship to the P1 polymorphism is clarified, P^k remains as 209.002

Expression

Soluble form	P^k substance is in hydatid cyst fluid and pigeon egg white
Other blood cells	Erythroid precursor cells, lymphocytes, monocytes, platelets
Tissues	Vascular endothelium, placenta (trophoblasts and interstitial cells), fibroblasts, fetal liver, fetal heart, kidney, prostate, smooth muscle, lung, breast, pancreas, epithelium (buccal, vaginal and uro) and weak on peripheral nerves[1]

Gene[2–4]

The sequential action of multiple gene products is required for expression of these antigens. The gene encoding the galactosyltransferase that converts the precursor (CDH) to P^k antigen has been cloned.

Chromosome	22q13.2 (P^k)
Name	P^k ($\alpha 4GAL\text{-}T$)
Organization	Two exons distributed over approximately 2.7 kbp of gDNA
Product	4-α-Galactosyltransferase (α4Gal-T; Gb3 synthase)

Gene map

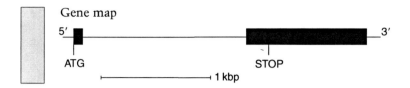

5′ ATG   STOP   3′

|———————————| 1 kbp

Database accession numbers

GenBank   AJ245581

Amino acid sequence

```
MSKPPDLLLR LLRGAPRQRV CTLFIIGFKF TFFVSIMIYW HVVGEPKEKG  50
QLYNLPAEIP CPTLTPPTPP SHGPTPGNIF FLETSDRTNP NFLFMCSVES 100
AARTHPESHV LVLMKGLPGG NASLPRHLGI SLLSCFPNVQ MLPLDLRELF 150
RDTPLADWYA AVQGRWEPYL LPVLSDASRI ALMWKFGGIY LDTDFIVLKN 200
LRNLTNVLGT QSRYVLNGAF LAFERRHEFM ALCMRDFVDH YNGWIWGHQG 250
PQLLTRVFKK WCSIRSLAES RACRGVTTLP PEAFYPIPWQ DWKKYFEDIN 300
PEELPRLLSA TYAVHVWNKK SQGTRFEATS RALLAQLHAR YCPTTHEAMK 350
MYL                                                  353
```

Carrier molecule

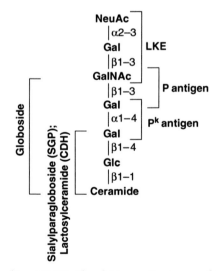

(See GLOB Blood Group System and Section III.)

Disease association

P^k is the physiologic receptor for shiga toxin on renal epithelium, platelets and endothelium. Transcriptional up-regulation of P^k by inflammatory mediators

(IFN, IL1) contributes to shiga toxin toxicity of renal and vascular tissue and is a critical cofactor in the development of E. coli HUS. P^k is involved in signal modulation of α-interferon receptor and CXCR4 (an HIV co-receptor). May play a role in shiga toxin-mediated apoptosis and multi-drug resistance due to P-glycoprotein[5]. These effects may be through lipid rafts.

P^k is a receptor for pyelonephritogenic E. coli and Streptococcus suis. LKE is the preferred receptor for P-fimbriated uropathogenic E. coli.

Cytotoxic IgM and IgG3 antibodies directed against P and/or P^k antigens are associated with a higher than normal rate of spontaneous abortion in women with the rare p [Tj(a−)], P_1^k, and P_2^k phenotypes.

Increased P^k is linked to increased tumor genicity in Burkitt's lymphoma; globo-H is linked to breast cancer; galactosyl-globoside is associated with better prognosis in seminoma; increased LKE and disialo-LKE are associated with metastasis in renal cell carcinoma.

Phenotypes (% occurrence)

See GLOB Blood Group System (028)
Null   p
Unusual P_1^k, and P_2^k (rarer even than p)

Molecular basis of p phenotype due to mutations in α4Gal-T[2,6,7]

Basis	Identification
Exon 2 752 C>T, Pro251Leu; 903 G>C, Pro301Pro	Japanese[2]
Exon 2 109 A>G, Met37Val; 752 C>T, Pro252Leu; 987 G>A,Thr329Thr	Japanese[2]
Exon 2 548T>A, Met183Lys	Swedish[2,6]
Exon 2 548T>A, Met183Lys; 987G>A, Thr329Thr	Swedish[2]
Exon 2 560G>A, Gly187Asp	Swedish[6]
Exon 2 783 G>A, Trp261Stop	Japanese[6]
Exon 2 237–239 del CTT, del Phe81	(Japanese)[7]
Exon 2 1029–1030 ins C, frameshift resulting in a product of 92 additional amino acids	(Japanese)[7]

See GLOB Blood Group System (028) for mutations in βGalNAc-T1.

Comments

It is not understood why, if the p phenotype is the result of mutations in the 4-α-Galactosyltransferase gene, P1 antigen is not produced.

References
1. Spitalnik, P.F. and Spitalnik, S.L. (1995) Transf. Med. Rev. 9, 110–122.
2. Steffensen, R. et al. (2000) J. Biol. Chem. 275, 16723–16729.
3. Keusch, J.J. et al. (2000) J. Biol. Chem. 275, 25308–25314.
4. Kojima, Y. et al. (2000) J. Biol. Chem. 275, 15152–15156.
5. Lala, P. et al. (2000) J. Biol. Chem. 275, 6246–6251.
6. Furukawa, K. et al. (2000) J. Biol. Chem. 275, 37752–37756.
7. Koda, Y. et al. (2002) Transfusion 42, 48–51.

P^k ANTIGEN

Terminology

ISBT symbol (number)	GLOB2 (209.002)
Other names	Trihexosylceramide; Ceramide trihexose (CTH); Globotriaosylceramide (Gb$_3$Cer); Gb3; CD77
History	Named in 1959 when the relationship to P was recognized; the 'k' comes from the last name of the first proband to produce anti-P^k

Occurrence

Strongly expressed on RBCs from 0.01% of the population. All RBCs, except those with the p phenotype, have trace amounts of P^k.

Expression

Cord RBCs	Expressed

Molecular basis associated with P^k antigen[1]

See Globoside Blood Group Collection (209) for figure for P^k antigen and molecular bases of the p phenotype (PP1P^k−) and Section III.

Effect of enzymes/chemicals on P^k antigen on intact RBCs

Ficin/papain	Resistant (↑↑)
Trypsin	Resistant (↑↑)

α-Chymotrypsin   Resistant (↑↑)
Pronase   Resistant (↑↑)
Sialidase   Resistant (↑↑)
DTT 200 mM   Resistant
Acid   Resistant

In vitro characteristics of alloanti-P^k

Immunoglobulin class   IgM and IgG
Optimal technique   RT; 37°C; IAT
Neutralization   Hydatid cyst fluid
Complement binding   Yes; some hemolytic

Clinical significance of alloanti-P^k

No information because anti-P^k does not exist as a single specificity but is found with anti-P and anti-P1 in serum from p people.
    Hydatid cyst fluid inhibits anti-P1 and anti-P^k.

Autoanti-P^k

Yes.

Comments

P^k was thought to be expressed only by P^k phenotype RBCs until it was realized that all RBCs express P^k antigen, albeit weakly. P^k antigen is more strongly expressed on LKE− RBCs than on LKE+ RBCs.
Neuraminidase treatment of RBCs exposes neutral glycosphingolipids (P^k and P antigen) and gangliosides.
Anti-P^k can be separated from some anti-PP1P^k by absorption with P1 RBCs.
Murine monoclonal anti-P^k exists.

Reference
[1] Bailly, P. and Bouhours, J.F. (1995) In: Molecular Basis of Human Blood Group Antigens (Cartron, J.-P. and Rouger, P. eds) Plenum Press, New York, pp. 300–329.

LKE ANTIGEN

Terminology

ISBT symbol (number)	GLOB3 (209.003)
Other names	Luke; SSEA-4; MSGG (monosialo-galactosyl-globoside)
History	In 1986 the name LKE was proposed for the antigen detected by the Luke serum, which was reported in 1965

Occurrence

All populations	98%

Expression

Cord RBCs	Expressed

Molecular basis associated with LKE antigen[1]

See Blood Group Collection pages for figure of LKE antigen and molecular bases of the p phenotype (PP1P^k−), and Section III.

Effect of enzymes/chemicals on LKE antigen on intact RBCs

Ficin/papain	Resistant (↑↑)
Trypsin	Resistant (↑↑)
α-Chymotrypsin	Resistant
Pronase	Resistant
Sialidase	Sensitive
DTT 200 mM	Resistant
Acid	Resistant

In vitro characteristics of alloanti-LKE

Immunoglobulin class	IgM
Optimal technique	RT or lower
Complement binding	Some

Clinical significance of alloanti-LKE

Transfusion reaction   None reported
HDN   No

Comments

Anti-LKE in humans is a rare specificity[2].

The expression of LKE and P^k antigens is inversely related: LKE-negative RBCs express almost twice the P^k expressed by LKE+(strong) RBCs[3].

N-terminal NeuAc is crucial for SSEA-4 determinant; standard methods for sialidase treatment of RBCs do not affect reaction of RBC with anti-LKE (monoclonal antibody).

The presence of *Se* decreases LKE expression; secretors have a 3–4 fold decreased risk of *E. coli* infections.

There are three LKE phenotypes:

LKE+^{S} 80–90%
LKE+^{W} 10–20%
LKE−   1–2%

References

1   Bailly, P. and Bouhours, J.F. (1995) In: Molecular Basis of Human Blood Group Antigens (Cartron, J.-P. and Rouger, P. eds) Plenum Press, New York, pp. 300–329.
2   Bruce, M. et al. (1988) Vox Sang. 55, 237–240.
3   Cooling, L.L. and Kelly, K. (2001) Transfusion 41, 898–907.

Unnamed

Number of antigens   2

Terminology
ISBT symbol   Unnamed
ISBT number   210

Carrier molecule
Glycosphingolipid adsorbed onto RBCs.

Le^c ANTIGEN

Terminology
ISBT symbol (number)   None (210.001)

Occurrence
Most populations   1%

Molecular basis associated with Le^c antigen[1]

```
        Gal
         | β1–3
      GlcNAc
         |
         R
Lacto-N-tetrasylceramide
```

Comments
Anti-Le^c agglutinates Le(a−b−) RBCs from non-secretors.
Anti-Le^c has been made in humans and in goats.

Reference
[1] Mollison, P.L. et al. (1993) Blood Transfusion in Clinical Medicine, 9th Edition, Blackwell, Oxford.

Le^d ANTIGEN

Terminology

ISBT symbol (number)   None (210.002)

Occurrence

Most populations: 6%.

Molecular basis associated with Le^d antigen[1]

$$Gal \xrightarrow{\alpha 1-2} Fuc$$
$$| \beta 1-3$$
$$GlcNAc$$
$$|$$
$$R$$

Type 1 H

Comments

Anti-Le^d agglutinates Le(a−b−) RBCs from non-secretors.
Anti-Le^d has been made in goats.

Reference

[1] Mollison, P.L. et al. (1993) Blood Transfusion in Clinical Medicine, 9th Edition, Blackwell, Oxford.

The 700 series of low incidence antigens

Antigens in this series occur in less than 1% in most populations, have no known alleles and cannot be placed in a Blood Group System or Collection.

Number	Symbol	Name	No. of pro-bands	No. of anti-bodies	Stimulus		Found as		Caused HDN	Enzymes chemicals	Comments
					Immune	Unknown	Multi-specific	Mono-specific			
700.002	By	Batty	Few	Many	×	×	×		+DAT	Papain/ficin/α-chymo-trypsin resistant AET resistant	Original anti-By stimulated by pregnancy; antibody found in AIHA
700.003	Chr^a	Christiansen	2	Few		×		×	No	Trypsin resistant	Found in Danes
700.005	Bi	Biles	Few	Few	×				Probably	Trypsin sensitive	Original anti-Bi stimulated by pregnancy Antibody found in AIHA
700.006	Bx^a	Box	Few	Few		×	×			Papain resistant	
700.017	To^a	Torkildsen	Few	Many		×	×	×		Ficin resistant (lytic)	Some cold reactive, IgM bind complement, others are IgG; Scandinavian M_r approx. 31 600
700.018	Pt^a	Peters	Few	Many		×	×			Trypsin resistant; papain/α-chymo-trypsin/pronase sensitive AET resistant	
700.019	Re^a	Reid	Few	Few	×				Mild +DAT	Papain/ficin/trypsin/α-chymotrypsin resistant	
700.021	Je^a	Jensen	Few	Few		×			No	Papain/ficin sensitive	Danish; IgM
700.028	Li^a	Livesey	Few	Few		×			Mild	Papain/ficin resistant	Maybe part of Lutheran system

Number	Symbol	Name	No. of probands	No. of antibodies	Stimulus		Found as		Caused HDN	Enzymes chemicals	Comments
					Immune	Unknown	Multi-specific	Mono-specific			
700.039		Milne	1	Many		×	×			Papain/ficin resistant [lytic]	IgM
700.040	RASM	Rasmussen	1	1	×			×	+DAT	Papain/ficin/trypsin resistant; DTT, AET sensitive	IgG; not complement binding
700.043	Olª	Oldeide	Few	Few		×	×			Papain resistant	Ol(a+) RBCs have suppressed Rh antigens; Olª is not part of RH; Norwegian; 2 anti-Olª were agglutinins reacting optimally below 37°C
700.044	JFV		2	Few	×			×	+DAT to moderate	Papain resistant (↑) Sialidase resistant DTT, AET resistant	German/Dutch probands
700.045	Kg	Katagiri	1	1	×				Severe	Papain/trypsin resistant	Japanese
700.047	JONES	Jones	2	Few	×			×	Moderate	Papain/ficin/trypsin/α-chymotrypsin/pronase resistant (↑) AET resistant	May be an Rh antigen
700.049	HJK		1	1	×			×	Severe		
700.050	HOFM		1	1	×			×	Mild	Papain resistant (↑) AET, DTT resistant	May be an Rh antigen; associated with weak C antigen; Dutch
700.052	SARA	SARAH	Few	Few		×		×		Papain resistant	Australian; AET resistant
700.054	REIT		Few	1	×				Severe	AET resistant	

Few = 1–5 examples; several = 6–12 examples; many = 13 or more examples; unknown = apparently naturally-occurring.

Antigens in this series occur in greater than 90% of people, but have no known alleles and cannot be placed in a blood group system or collection.

Originally high incidence antigens were in the 900 series. At the 1988 meeting of the ISBT Working Party on Terminology[1], many of the 900 series antigens were transferred to newly established Blood Group Collections, thereby generating many obsolete 900 numbers. Consequently, the 900 Series was replaced by the 901 Series.

Number	Symbol	Name
901.001	Vel	
901.002	Lan	Langereis
901.003	At^a	August
901.005	Jr^a	
901.008	Emm	
901.009	AnWj	Anton
901.012	Sd^a	Sid
901.013		Duclos
901.014	PEL	
901.015	ABTI	
901.016	MAM	

[1] Lewis, M. et al. (1990) Vox Sang. 58, 152–169.

Vel ANTIGEN

Terminology

ISBT symbol (Number)	Vel (901.001)
Other names	Ve^a; 900.001
History	Reported in 1952 and named after the first antigen-negative proband who made anti-Vel

Occurrence

All populations: >99%.

Vel-negative RBCs have been found in 1 in 4000 people and approximately 1 in 1500 in Norwegians and Swedes.

Expression

Cord RBCs	Weak
Adult RBCs	Expression is variable. RBCs with a weak expression of the Vel antigen may be mistyped as Vel$-$[2]

Effect of enzymes/chemicals on Vel antigen on intact RBCs

Ficin/papain	Resistant (↑↑)
Trypsin	Resistant (↑↑)
α-Chymotrypsin	Resistant (↑↑)
Pronase	Resistant (↑↑)
Sialidase	Resistant
DTT 200 mM	Resistant
Acid	Resistant

In vitro characteristics of alloanti-Vel

Immunoglobulin class	IgM and IgG
Optimal technique	IAT; enzyme IAT
Complement binding	Yes; some hemolytic

Clinical significance of alloanti-Vel

Transfusion reaction	No to severe/hemolytic
HDN	Positive DAT to severe[1]

Autoanti-Vel

Yes.

Comments

Three of 14 anti-Vel did not react with 4 Ge:−2,−3,4 samples[3].
A disproportional number of Vel− samples have the P_2 phenotype.
Six of eight Vel− RBC samples were weakly reactive and one was non-reactive with anti-ABTI[4].

References

1 Le Masne, A. et al. (1992) Arch. Fr. Pediatr. 49, 899–901.
2 Issitt, P.D. (1985) Applied Blood Group Serology, 3rd Edition, Montgomery Scientific Publications, Miami, FL.
3 Issitt, P. et al. (1994). Transfusion 34 (Suppl.), 60S (abstract).
4 Schechter, Y. et al. (1996) Transfusion 36 (Suppl.), 25S (abstract).

Lan ANTIGEN

Terminology

ISBT symbol (Number)	Lan (901.002)
Other names	Langereis; Gn^a; Gonsowski; So; 900.003
History	Reported in 1961; named after the first antigen-negative proband to make anti-Lan

Occurrence

All populations: >99%.
   The Lan– phenotype occurs in about 1 in 20000 people; found in Blacks,[1,2] Caucasians and Japanese.

Expression

Cord RBCs	Expressed
Altered	A weak form of Lan has been reported[3]

Effect of enzymes/chemicals on Lan antigen on intact RBCs

Ficin/papain	Resistant
Trypsin	Resistant
α-Chymotrypsin	Resistant
Pronase	Resistant
Sialidase	Resistant
DTT 200 mM	Resistant
Acid	Resistant

In vitro characteristics of alloanti-Lan

Immunoglobulin class	IgG
Optimal technique	IAT
Complement binding	Some

Clinical significance of alloanti-Lan

Transfusion reaction	No to severe/hemolytic
HDN	No to mild

Autoanti-Lan

One example in a patient with depressed Lan antigens.

References
[1] Sturgeon, J.K. et al. (2000). Transfusion 40 (Suppl.), 115S (abstract).
[2] Ferraro, M.L. et al. (2000). Transfusion 40 (Suppl.), 121S (abstract).
[3] Storry, J.R. and Øyen, R. (1999) Transfusion 39, 109–110.

At^a ANTIGEN

Terminology

ISBT symbol (Number)	At^a (901.003)
Other names	August; Augustine; El; Elridge; 900.006
History	Reported in 1967 after the first antigen-negative proband to make anti-At^a

Occurrence

All populations: >99%. All people with At(a−) RBCs are Black.

Expression

Cord RBCs	Expressed

Effect of enzymes/chemicals on At^a antigen on intact RBCs

Ficin/papain	Resistant
Trypsin	Resistant
α-Chymotrypsin	Resistant
Pronase	Resistant
Sialidase	Resistant
DTT 200 mM	Resistant
Acid	Resistant

In vitro characteristics of alloanti-At^a

Immunoglobulin class	IgG
Optimal technique	IAT
Complement binding	No

Clinical significance of alloanti-At[a]

Transfusion reaction	No to severe[1-3]
HDN	Most At(a+) babies born to At(a−) mothers were not affected; only one mild case

References
1 Sweeney, J.D. et al. (1995) Transfusion 35, 63–67.
2 Ramsey, G. et al. (1995) Vox Sang. 69, 135–137.
3 Cash, K.L. et al. (1999) Transfusion 39, 834–837.

Jr[a] ANTIGEN

Terminology

ISBT symbol (Number)	Jr[a] (901.005)
Other names	Junior; 900.012
History	The first five examples of anti-Jr[a] were reported in 1970. Named for the first maker of anti-Jr[a]

Occurrence

All populations: >99%.

Approximately half of the known Jr(a−) persons are Japanese. The Jr(a−) phenotype has also been found in persons of northern European extraction, Bedouin Arabs and in one Mexican.

Expression

Cord RBCs	Expressed

Effect of enzymes/chemicals on Jra antigen on intact RBCs

Ficin/papain	Resistant (↑)
Trypsin	Resistant
α-Chymotrypsin	Resistant
Pronase	Resistant
Sialidase	Resistant
DTT 200 mM	Resistant
Acid	Resistant

In vitro characteristics of alloanti-Jr^a

Immunoglobulin class	IgG more common than IgM
Optimal technique	IAT
Complement binding	Some

Clinical significance of alloanti-Jr^a

Transfusion reaction	^{51}Cr cell survival studies indicated reduced RBC survival;[1] patient with anti-Jr^a developed rigors after transfusion of 150 mL of crossmatch incompatible blood.
HDN	Positive DAT, but no clinical HDN

Reference
[1] Kendall, A.G. (1976) Transfusion 16, 646–647.

Emm ANTIGEN

Terminology

ISBT symbol (Number)	Emm (901.008)
Other names	Emma; 900.028
History	Reported in 1987 and named after the first antigen-negative proband to make anti-Emm

Occurrence

All populations: >99%.

Six probands with the Emm− phenotype found in a French Madagascan, a white American, a Pakistani, a French Canadian and two New Yorkers.

Expression

Cord RBCs	Expressed

Effect of enzymes/chemicals on Emm antigen on intact RBCs

Ficin/papain	Resistant
Trypsin	Resistant
α-Chymotrypsin	Resistant
Pronase	Resistant
Sialidase	Resistant

DTT 200 mM   Resistant
Acid   Resistant

In vitro characteristics of alloanti-Emm

Immunoglobulin class   IgG more common than IgM (four of five)
Optimal technique   IAT; 4°C (original anti-Emm)
Complement binding   Some (2 of 5)

Clinical significance of alloanti-Emm

No data are available. Six of the seven examples of anti-Emm were in non-transfused males[1].

Comments

Emm is carried on a GPI-linked protein in the RBC membrane[2].

References
[1] Daniels, G.L. et al. (1987) Transfusion 27, 319–321.
[2] Telen, M.J. et al. (1990) Blood 75, 1404–1407.

AnWj ANTIGEN

Terminology

ISBT symbol (Number)   AnWj (901.009)
Other names   Anton; Wj; 005.015; Lu15
History   Reported in 1982 as an alloantibody to an antigen called Anton, and in 1983 as an autoantibody to an antigen called Wj. In 1985, it was shown that both antibodies detected the same antigen and the name AnWj was applied

Occurrence

All populations: >99%.
    Genetic form of AnWj-negative in two Israeli women and one Arab-Israeli family.

Expression

Cord RBCs   Not expressed
Altered   Weak on dominant Lu(a−b−) RBCs
             Expression varies from person to person

Molecular basis associated with AnWj antigen

Carried on a CD44 proteoglycan[1,2]. The AnWj epitope is likely to reside in the glycosylated region encoded by exons 5 and 15. If this is true, all information regarding CD44 (Indian blood group system, [IN]) would apply.

Effect of enzymes/chemicals on AnWj antigen on intact RBCs

Ficin/papain	Resistant
Trypsin	Resistant
α-Chymotrypsin	Resistant
Pronase	Resistant
Sialidase	Resistant
DTT 200 mM	Variable
Acid	Resistant

In vitro characteristics of alloanti-AnWj

Immunoglobulin class	IgG
Optimal technique	IAT
Complement binding	Rare

Clinical significance of alloanti-AnWj

Transfusion reaction	Severe in one case[3]
HDN	No

Autoanti-AnWj

Yes, may appear to be an alloantibody because of transient suppression of AnWj antigen.

Comments

Only two examples of alloanti-AnWj (both in Israeli women) have been described. The AnWj− phenotype is usually the result of transient (often long-term) suppression of AnWj.

AnWj antigen is the receptor for *Haemophilus influenzae*[4].

References

1 Telen, M.J. et al. (1993). Clin. Res. 41, 161A (abstract).
2 Udani, M. et al. (1995). Blood 86 (Suppl. 1), 472a (abstract).
3 de Man, A.J. et al. (1992) Vox Sang. 63, 238.
4 van Alphen, L. et al. (1986) FEMS Microbiol. Lett. 37, 69–71.

Sdᵃ ANTIGEN

Terminology

ISBT symbol (Number)	Sdᵃ (901.012)
Other names	Sid; Tamm-Horsfall glycoprotein; uromodulin
History	Reported in 1967 after many years of investigation. Named for Sidney Smith, head of the maintenance department at the Lister Institute in London. For many years, his RBCs were frequently used as a Sd(a+) control

Occurrence

All populations: 91% of RBC samples express Sdᵃ; however, 96% of urine samples have Sdᵃ substance. Four percent of people are truly Sd(a−).

Expression

Soluble form	Urine (Tamm-Horsfall glycoprotein)
Newborns	Not expressed on RBCs. Expressed in saliva, urine and meconium
Adult RBCs	Strength of expression varies greatly; the strongest expression is on Cad phenotype RBCs
Altered	Marked reduction of Sdᵃ expression occurs in pregnancy
Other tissues	Stomach, colon, kidney, lymph nodes

Molecular basis associated with antigen[1]

Sdᵃ-active Pentasaccharide from GPA

$$\text{GalNAc} \xrightarrow{\beta1-4} \text{Gal} \xrightarrow{\beta1-3} \text{GalNAc} \longrightarrow \text{Ser/Thr}$$
$$\qquad\qquad |\alpha2-3 \qquad\qquad |\alpha2-6$$
$$\qquad\qquad \text{NeuAc} \qquad\quad \text{NeuAc}$$

Sdᵃ-active ganglioside

$$\text{GalNAc} \xrightarrow{\beta1-4} \text{Gal} \xrightarrow{\beta1-4} \text{GlcNAc} \xrightarrow{\beta1-3} \text{Gal} \xrightarrow{\beta1-4} \text{Glc-Cer}$$
$$\qquad\qquad |\alpha2-3$$
$$\qquad\qquad \text{NeuAc}$$

Sdᵃ-active Tamm-Horsfall glycoprotein

$$\text{GalNAc} \xrightarrow{\beta1-4} \text{Gal} \xrightarrow{\beta1-4} \text{GlcNAc} \xrightarrow{\beta1-3} \text{Gal}$$
$$\qquad\qquad |\alpha2-3$$
$$\qquad\qquad \text{NeuAc}$$

Effect of enzymes/chemicals on Sd^a antigen on intact RBCs

Ficin/papain	Resistant ($\uparrow$)
Trypsin	Resistant ($\uparrow$)
α-Chymotrypsin	Resistant ($\uparrow$)
Pronase	Resistant ($\uparrow$)
Sialidase	Usually resistant
DTT 200 mM	Resistant
Acid	Resistant

In vitro characteristics of alloanti-Sd^a

Immunoglobulin class	IgM more common than IgG
Optimal technique	RT; IAT
Neutralization	Urine (guinea pig and human)
Complement binding	Yes, some

Clinical significance of alloanti-Sd^a

Transfusion reaction	Two cases reported associated with transfusion of Sd(a++) RBCs
HDN	No

Comments

Agglutinates are typically small and refractile in a sea of free RBCs. Anti-Rx (formerly anti-Sd^x) can be confused with anti-Sd^a because of similar type of agglutination[2].

Tamm-Horsefall protein in urine binds specifically to type 1[3] fimbriated *E. Coli*, thereby preventing adherence of pathogenic *E. Coli* to urothelial receptors.

Hemagglutination inhibition tests with urine are the most reliable way of determining Sd^a status.

Urine inhibition tests (particularly using guinea pig urine) are useful for the identification of anti-Sd^a.

Encoded by a gene at 16p12.3–p13.1[4].

References

1 Watkins, W.M. (1995) In: Molecular Basis of Human Blood Group Antigens (Cartron, J.-P. and Rouger, P. eds), Plenum Press, New York, pp. 351–375.

2 Issitt, P.D. (1991) In: Blood groups: P, I, Sd^a and Pr (Moulds J.M. and Woods L.L. eds), American Association of Blood Banks, Arlington, VA, pp. 53–71.

3 Pak, J. et al. (2001) J. Biol. Chem. 276, 9924–9930.

4 Pook, M.A. et al. (1993) Ann. Hum. Genet. 57, 285–290.

Duclos ANTIGEN

Terminology

ISBT symbol (Number)	Duclos (901.013)
Other names	RH38
History	Reported in 1978, named after the first and only producer of the alloantibody and thought to be an Rh antigen

Occurrence

All populations	100%

Expression

Cord RBCs	Presumed expressed

Molecular basis associated with Duclos antigen

Antigen is lacking from Rh_{null} U−, Rh_{mod} U− RBCs and those of the antibody maker. Monoclonal antibody MB-2D10 has similar specificity. MB-2D10 epitope is expressed by Rh glycoprotein (RhAG)[1]. Interaction between GPB and MB-2D10 carrying glycoprotein may be required for the expression of Duclos[1].

Effect of enzymes/chemicals on Duclos antigen on intact RBCs

Ficin/papain	Resistant (↑↑)
Trypsin	Resistant (↑↑)
α-Chymotrypsin	Weakened
Pronase	Sensitive
Sialidase	Resistant (↑)
DTT 200 mM	Presumed resistant
Acid	Presumed resistant

In vitro characteristics of alloanti-Duclos

Immunoglobulin class	IgG
Optimal technique	IAT

Clinical significance of alloanti-Duclos

No data are available.

Reference
[1] Mallinson, G. et al. (1990) Transfusion 30, 222–225.

PEL ANTIGEN

Terminology

ISBT symbol (Number)	PEL (901.014)
Other names	Pelletier
History	Identified in 1980 and reported in 1996 when the antigen was named after the first antigen-negative proband who made anti-PEL

Occurrence

All populations: >99%. PEL− phenotype found in two French Canadian families.

Expression

Cord RBCs	Expressed
Altered	Weak expression (shown by absorption studies) in two French Canadian families[1].

Effect of enzymes/chemicals on PEL antigen on intact RBCs

Ficin/papain	Resistant
Trypsin	Resistant
α-Chymotrypsin	Resistant
Pronase	Resistant
Sialidase	Resistant
DTT 200 mM	Resistant
Acid	Not known

In vitro characteristics of alloanti-PEL

Immunoglobulin class | Presumed IgG
Optimal technique | IAT
Complement binding | Not known

Clinical significance of alloanti-PEL

Transfusion reaction | Reduced survival of ^{51}Cr-labelled RBCs
HDN | No

Comments

An antibody with similar specificity to anti-PEL was made by the probands from the two French Canadian families with suppressed PEL expression. Antibody provisionally named anti-MTP[1].

Reference

[1] Daniels, G.L. et al. (1996) Vox Sang. 70, 31–33.

ABTI ANTIGEN

Terminology

ISBT symbol (Number) | ABTI (901.015)
History | Anti-ABTI reported in 1996 in three multiparous women, members of an inbred Israeli Arab family. Named after the family[1]

Occurrence

All populations: >99%. ABTI-negative phenotype found in one Israeli Arab family[1] and one Bavarian lady[2].

Expression

Cord RBCs | Presumed expressed
Altered | Vel-negative RBCs are ABTI+^{W} (1 was ABTI−)

515

Effect of enzymes/chemicals on ABTI antigen on intact RBCs

Ficin/papain	Resistant
Trypsin	Resistant
α-Chymotrypsin	Resistant
Pronase	Presumed resistant
Sialidase	Resistant
DTT 200 mM	Resistant
Acid	Presumed resistant

In vitro characteristics of alloanti-ABTI

Immunoglobulin class	IgG (IgG1 plus IgG3)
Optimal technique	IAT

Clinical significance of alloanti-ABTI

Transfusion reaction	No data
HDN	No

Comments

ABTI− RBCs have a weak expression of Vel.

Reference
[1] Schechter, Y. et al. (abstract). Transfusion 1996, 36 (Suppl.), 25S.
[2] Poole, J. (2003) Personal Communication.

MAM ANTIGEN

Terminology

ISBT symbol (Number)	MAM (901.016)
History	Reported in 1993; assigned to the 901 Series in 1999; name is derived from the initials of the first antigen-negative proband to make anti-MAM[1]

Occurrence

All populations: > 99%. MAM− probands: Irish and Cherokee descent; Arab; Jordanian.

Expression

Cord RBCs	Expressed
Other blood cells	Lymphocytes, granulocytes, monocytes, and probably on platelets

Molecular basis associated with MAM antigen[2]

M_r on SDS-PAGE of approximately 30000–40000. Possibly on an N-linked carbohydrate-based antigen[3].

Effect of enzymes/chemicals on MAM antigen on intact RBCs

Ficin/papain	Resistant
Trypsin	Resistant
α-Chymotrypsin	Resistant
Pronase	Resistant
Sialidase	Resistant
DTT 200 mM	Resistant
Acid	Presumed resistant

In vitro characteristics of alloanti-MAM

Immunoglobulin class	IgG
Optimal technique	IAT

Clinical significance of alloanti-MAM

Transfusion reaction	Monocyte monolayer assay suggests anti-MAM is potentially clinically significant
HDN	No to severe

Comments

Anti-MAM may also cause neonatal thrombocytopenia, however one MAM+ baby born to a woman with anti-MAM had no thrombocytopenia nor anemia[4].

References

1 Anderson, G. et al. (1993). Transfusion 33 (Suppl.), 23S (abstract).
2 Montgomery, W.M. Jr. et al. (2000) Transfusion 40, 1132–1139.
3 Li, W. and Denomme, GA. (2002) Transfusion 42 (Suppl.), 10S (abstract).
4 Denomme, G.A. et al. (2000) Transfusion 40 (Suppl.), 28S (abstract).

OTHER USEFUL FACTS

Antigen-based facts

Usefulness of the effect of different enzymes on antigens in antibody identification[1,2]

The following table shows patterns of reactions; for more detail, see individual antigen sheets.

Possible antibody specificity is based on reactions against enzyme-treated RBCs (assuming no anti-enzyme is present or an eluate is used). This is particularly important when using pronase since the occurrence of "anti-pronase" is so high.

Ficin/ papain	Trypsin	α-Chymotrypsin	Pronase	Possible specificity
Neg	Neg	Neg	Neg	Bp^a; Ch/Rg; Indian; JMH; Xg
Neg	Neg	Pos	Neg	M, N, En^{a}TS; Ge2, Ge4
Neg	Pos	Neg	Neg	'N'; Fy^a, Fy^b, Fy6
Variable	Pos	Neg	Neg	S, s; Yt^a
Neg	Pos	Pos	Neg	En^{a}FS
Pos	Neg	Neg	Neg	Lutheran, MER2
Pos – Papain Weak or neg – Ficin	Neg	Neg	Pos	Knops
Pos	Neg	Pos/Weak	Neg	Ge3; Dombrock
Pos	Pos	Neg	Neg	Some DI antigens, Cromer
Pos	Pos	Pos/Weak	Neg	LW
Pos	Pos	Pos	Weak/ Neg	Scianna
Pos	Pos	Pos	Pos	A, B; H; P1; Lewis; Rh; Kidd; Kell*; En^{a}FR, U; Fy3; Di^a, Di^b, Wr^a, Wr^b, some DI antigens; Colton; Kx; Ok^a; RAPH; I,i; P, P^k, LKE; AnWj; At^a; Cs^a; Emm; Er; Jr^a; Lan; Vel; Sd^a; PEL; MAM; ABTI

* Kell blood group system antigens are sensitive to treatment with a mixture of trypsin and α-chymotrypsin.

Substrate specificity of selected enzymes for peptide and CHO bonds

Classification	Enzyme	Substrate specificity (in order of preference)
Thiol endoprotease (has an essential cysteine in the active site and may require a sulfhydryl compound to activate it)	Bromelin (Pineapple)	Hydrolyses C-terminal peptide bond of Lys, Ala, Tyr, Gly
	Ficin (Fig tree latex)	Hydrolyses C-terminal peptide bond of Lys, Ala, Tyr, Gly, Asp, Leu, Val
	Papain (Papaya)	Hydrolyses C-terminal peptide bond of Arg, Lys and bond next but one to Phe
Metallo endoprotease (requires a specific metal ion in the active site)	Pronase (*Streptomyces griseus*)	Hydrolyses C-terminal peptide bond of any hydrophobic amino acid
Serine endoprotease (serine and histidine residues are essential at the enzyme site for enzymatic activity)	α-chymotrypsin (Bovine pancreas)	Hydrolyses C-terminal peptide bond of Phe, Trp, Tyr, Leu
	Proteinase K (*Tritirachium album*)	Hydrolyses C-terminal peptide bond of aromatic or hydrophobic amino acids
	Trypsin (Bovine or Porcine pancreas)	Hydrolyses C-terminal peptide bond of Arg, Lys
	V8 protease (*Staphylococcus aureus* strain V8)	Hydrolyses C-terminal peptide bond of Glu, Asp
Carboxyl endoprotease (has an essential COOH in the active site)	Pepsin (Porcine stomach mucosa)	Hydrolyses C-terminal peptide bond of Phe, Leu, Trp, Tyr, Asp, Glu
Exoglycosidase	Sialidase/ Neuraminidase (*Vibrio cholerae*)	Hydrolyses glycosidic bonds between NeuAc in any linkage to any sugar
Endoglycosidase	Endo β (*Pseudomonas*)	Hydrolyses Gal in polylactosamine units
	Endo F (*Flavo-bacterium meningo-septricum*)	Hydrolyses GlcNAc from GlcNAc when attached to Asn
	Pepidyl-N-glycanase	Hydrolyses GlcNAc from Asp

Endo = internal substrate bonds; Exo = terminal substrate bonds.
Note: Bacterial deacetylases may modify the side-chains of sugars, for example, the acquired B phenomenon results from deactylation of *N*-acetylgalactosamine to galactosamine. Organisms such as *E. coli, Clostridium tertium* and *Proteus mirabilis* have been implicated in this phenomenon.

Usefulness of the effect of enzymes and DTT on antigens in antibody identification

Ficin/papain	DTT (200 mM)	Possible specificity
Negative	Positive	M,N,S,s*; Ge2,Ge4; Xg^a; Fy^a,Fy^b; Ch/Rg
Negative	Negative	Indian; JMH
Positive	Weak	Cromer; Knops (weak or negative in ficin); Lutheran; Dombrock; AnWj; MER2
Variable	Negative	Yt^a
Positive	Negative	Kell; LW; Scianna
Positive	Positive	A,B; H; P1; Rh; Lewis; Kidd; Fy3; Diego; Co; Ge3; Ok^a; I,i; P, LKE; At^a; Cs^a; Emm; Er^a; Jr^a; Lan; Vel; Sd^a; PEL
Positive	Enhanced	Kx

* s variable with ficin/papain.

Reactivity of antigen-positive RBCs after treatment with ficin/papain or DTT

Possible antibody specificity is based on reactions against enzyme-treated RBCs (assuming no anti-enzyme is present or an eluate is used). This is particularly important when using pronase since the occurrence of "anti-pronase" is so high.

Ficin/ Papain	Trypsin	α-chymotrypsin	DTT (200 mM)	Possible specificity
Neg	Neg	Neg	Pos	Bp^a; Ch/Rg; Xg
Neg	Neg	Neg	Neg	Indian, JMH
Neg	Neg	Pos	Pos	M,N, En^{a}TS; Ge2,Ge4
Neg/Variable	Pos	Neg	Pos	'N' S,s; Fy^a,Fy^b,Fy6
Variable	Pos	Neg	Neg	Yt^a
Neg	Pos	Pos	Pos	En^{a}FS
Pos	Neg	Neg	Weak	Lutheran, MER2
Pos – Papain Weak or neg – Ficin	Neg	Neg	Weak	Knops
Pos	Pos	Pos	Neg	Kell**; Scianna
Pos	Neg	Pos/Weak	Neg	Ge3; Dombrock
Pos	Pos	Neg	Neg	Some DI antigens, Cromer
Pos	Pos	Pos/Weak	Neg	LW
Pos	Pos	Pos	Enhanced	Kx
Pos	Pos	Pos	Pos	A,B; H; P1; Lewis; Rh; Kidd; En^{a}FR, U; Fy3; Di^a,Di^b,Wr^a,Wr^b, some DI antigens; Colton; Ok^a; RAPH, I,i; P,P^k,LKE; AnWj; At^a; Cs^a; Emm; Er; Jr^a; Lan; Vel; Sd^a; PEL, MAM, ABT

* s variable with ficin/papain
** Kell blood group system antigens are sensitive to treatment with a mixture of trypsin and α-chymotrypsin.

Effect of acid on antigen expression

EDTA/acid/glycine-treated RBCs do not express antigens in the Kell blood group system, the Er collection or Bg antigens, and antigens of the Jk blood group system may be weakened.

Effect of chloroquine diphosphate on antigen expression

A modified technique of treating RBCs with chloroquine for 30 min at 37°C weakens: Mt^a, Fy^b, Lu^b, Yt^a and antigens of the Rh, Dombrock, Knops, and JMH blood group systems.

Antigens in soluble form

Human source (from antigen-positive people)

A, B and H (in secretors), Le^a and Le^b	Saliva, serum/plasma
I and i	Human milk, serum/plasma (low levels)
Sd^a	Urine (Tamm-Horsfall glycoprotein)
Ch (C4B) and Rg (C4A)	Serum/plasma
Cromer antigens (DAF)	Urine, serum/plasma (low levels)
In^b (CD44)	Serum/plasma (low levels)
Knops antigens (CR1)	Serum/plasma (low levels)

Non-human source

P1	Pigeon egg white and hydatid cyst fluid
Sd^a	Guinea pig urine

Many antigens can be synthesized by recombinant techniques.

Cord RBCs are

Negative for	Le^a, sometimes Le^b; Sd^a; Ch, Rg; AnWj
Weak for	A,B; H; I; P1; Le^b; Lu^a, Lu^b; Yt^a; sometimes Xg^a; Vel; Bg; Knops and Dombrock system antigens; Fy3 as detected by anti-Fy3 made by blacks
Strong for	LW system antigens; i antigen

Antigens with variable expression on different RBCs in the same sample and on RBCs from different donors (presumed to be due to different antigen copy number)

Carbohydrate antigens:	A,B; H; I; P1; Sd^a
Plasma adsorbed antigens:	Ch, Rg
Protein antigens:	Lu^a, Lu^b; Xg^a; Kn^a,McC^a,Sl^a,Yk^a; MER2; JMH; Vel; Lan; Jr^a; AnWj

Mixed field agglutination may be observed in

Transfused patients
Fetal maternal hemorrhage

Transplant (bone marrow; stem cell) recipient patients

Chimera (Genetic)

Genetic variants of antigen, for example, A_3, A_{mos}, B_{mos}

Chromosomal abnormalities resulting in two populations of RBCs e.g., ABO and Rh in leukemia

Low density of antigen sites, for example, Xg^a, Sd^a, Lu^a

Polyagglutination, for example, Tn

X-inactivation, Kx (in female carriers)

Blood group antigens absent (altered) on selected RBC phenotypes

(Listed in ISBT system order)

Phenotype	Absent or altered, usually reduced (in parentheses) antigens
O	ABO system
O_h (Bombay)	ABO system and H; rarely Lewis
En(a−)Fin	M, N, GPA-associated and Wr^a, Wr^b
U-	S, s, U, He, GPB-associated
M^kM^k	MNS system and Wr^a, Wr^b (Some antigens, not in the MNS system, may appear to be enhanced due to reduced sialic acid)
p [Tj(a−)]	P, P1, P^k
Rh_{null}	Rh and LW systems, Fy5, Duclos and other Rh-related proteins (S,s and U may be weak)
Rh_{mod}	(Weak Rh and LW systems; S,s,U; Fy5; Duclos and other Rh-related proteins)
Recessive Lu(a−b−)	Lutheran system
Dominant Lu(a−b−) (In(Lu))	(Weak Lutheran, Knops and Indian systems, P1, AnWj, MER2)
Sex-Linked Lu(a−b−)	(Weak Lutheran system, I)
K_o	Kell system (Kx increased)
K_{mod}	(Weak Kell system; Kx increased)
Kp(a+b−)	(Weak Kell system; Kx slightly increased)
Le(a−b−)	Le^a, Le^b
Fy(a−b−)	Duffy system
Fy^X	Fy^b (Weak Fy^b or often silenced)
Recessive Jk(a−b−)	Kidd system
Dominant Jk(a−b−) (In(Jk))	(Weak Jk antigens)
Sc:−1,−2	Scianna system
Gy(a−)	Dombrock system
Hy-	Hy, Jo^a (Weak Gy^a, Do^b)
Co(a−b−)	Colton system
LW(a−b−)	Landsteiner–Wiener system
Ch-Rg-	Ch/Rg system

Phenotype	Absent or altered, usually reduced (in parentheses) antigens
McLeod	Kx (Weak Kell system)
Leach (Ge:−2,−3,−4)	Gerbich system (Weak Kell system)
Gerbich (Ge:−2,−3,4)	Ge2, Ge3 (Weak Kell system)*
Yus (Ge:-2,3,4)	Ge2
INAB	Cromer system
Dr(a−)	Dr^a (Dramatically weak Cromer system)
Helgeson	Knops system, Cs^a
Gil−	GIL system

* Most but not all.

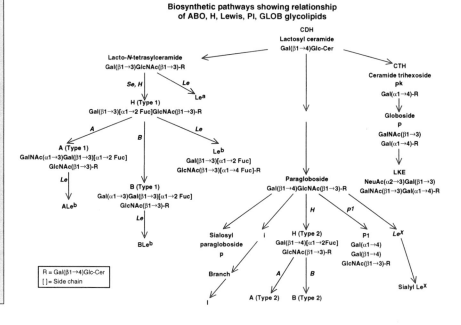

Biosynthetic pathways showing relationship of ABO, H, Lewis, PI, GLOB glycolipids

High prevalence antigens absent (and selected phenotypes)

Antigen or phenotype	Population
O$_h$ (Bombay)	Indian > Japanese > any
Para-Bombay	Reunion Island > Indians > any
U	Blacks
En^a	Finns > Canadian > English > Japanese
M^{k}M^k	Swiss > Japanese
hr^S	Blacks
hr^B	Blacks

Antigen or phenotype	Population
Lu20	Israeli (1)
Lu21	Israeli (1)
k	Caucasians $\gg$ Blacks > any
Kp^b	Caucasians > Japanese
Js^b	Blacks
K12	Caucasians (4)
K14	French-Cajun (3)
K22	Israelis (2)
K_0 (K_{null})	All
Fy(a$-$b$-$)	Blacks$\gg$Arabs/Jews>Mediterranean$\gg$Caucasians
Lu(a$-$b$-$)	All
Jk(a$-$b$-$)	Polynesians $\gg$ Finns > Japanese
Di^b	S. American > Native Americans > Japanese
LW^a	Baltic Sea region
LW(a$-$b$-$)	(Transient) $\gg$ Canadian
Yt^a	Arabs > Jews
Gy^a	Eastern European (Romany) > Japanese
Hy	Blacks
Jo^a	Blacks
Ge:2,$-$3 (Yus phenotype)	Mexican > Israelis > Mediterrean > any
Ge:$-$2,$-$3 (Gerbich phenotype)	Papua New Guinea > Melanesians > Caucasians > any
Ge:$-$2,$-$3,$-$4 (Leach phenotype)	All
Cr^a	Blacks
Tc(a$-$b$+$c$-$)	Blacks
Tc(a$-$b$-$c$+$)	Caucasians
Dr^a	Jews from the Bukaran area of Uzbekistan > Japanese
WES^a	Finns > Blacks > any
IFC (Cr_{null}, Inab)	Japanese > any
UMC	Japanese
Es^a	Mexican (1), Black (1), S. American (1)
GUTI	Chilean
Kn^a	Caucasians > Blacks > any
McC^a	Blacks > Caucasians > any
Sl^a	Blacks $\gg$ Caucasians > any
Yk^a	Caucasians > Blacks > any
In^b	Indians
Ok^a	Japanese
MER2	Indian Jews
At^a	Blacks
Jr^a	Japanese > Mexicans > any
Lan	Caucasian
AnWj	Israeli Arabs > (Transient in all)
PEL	French-Canadians (2)

Numbers in parentheses refer to probands.

Low prevalence antigens present in certain ethnic populations

Antigen or phenotype	Population
He	Natals > Blacks
Mi^a	Chinese > SE Asian > any
M^c	Europeans
Vw	SE Switzerland > Caucasians
Mur	Thais > Taiwanese > Chinese > any
M^g	Swiss > Sicilian > any
Vr	Dutch
Mt^a	Thais > Swiss > Caucasians > Blacks
St^a	Japanese > Asian > Caucasians
Cl^a	Scottish (1), Irish (1)
Ny^a	Norwegian > Swiss > any
Hil	Chinese > any
Mit	Western Europeans
Dantu	Blacks
Hop	Thais > any
Or	Japanese (2), Australian (1), Black (1), Jamaican (1)
DANE	Danes
MINY	Chinese > any
MUT	Chinese
SAT	Japanese
Os^a	Japanese
HAG	Isreali (2)
MARS	Choctaw tribe of Native Americans
C^W	Latvians > Finns > Caucasians
C^X	Finns > Caucasians
V	Blacks > Caucasians
E^W	Germans > any
VS	Bantu > Blacks
Go^a	Blacks
Rh32	Blacks > Caucasians > Japanese
Rh33	Germans > Caucasians
Rh35	Danes
Rh36	German, Poles
Evans	Celts
Rh42	Blacks
JAL	English, French-speaking Swiss, Brazilians, Blacks
STEM	Blacks
FPTT	All
DAK	Blacks > Caucasians
Lu14	English > Danes > any
K	Arabs > Iranian Jews > Caucasians
Kp^a	Caucasians

Antigen or phenotype	Population
Js^a	Blacks
Ul^a	Finns > Japanese
Kp^c	Japanese
K24	French-Cajun (3)
Di^a	S. American Indians > Japanese > N. American Indians > Chinese, Poles
Wd^a	Hutterites
WARR	Native Americans
Wu	Scandinavian > Dutch > Black
Bp^a	English (1), Italian (1)
Mo^a	Belgian (2), Norwegian (1)
Hg^a	Welsh > Australians
NFLD	French Canadian (2), Japanese (2)
Jn^a	Polish (1), Slovakian (1)
KREP	Polish (1)
Fr^a	Mennonites
Sc	Mennonites > N. Europeans
Rd	Danes > Canadians > Jews > Blacks > any
Yt^b	Arabs > Jews > Europeans
LW^b	Estonians > Finns > Baltic > Europeans
Wb	Welsh > Australians
Ls^a	Blacks > Finns > any
An^a	Finns
Tc^b	Blacks
Tc^c	Caucasians
WES^b	Finns > Blacks
Kn^b	Caucasians > Blacks
In^a	Arabs > Iranians > Indians > any

Numbers in parentheses refer to probands.

References
[1] Judd, W.J. (1994) Methods in Immunohematology, 2nd Edition, Montgomery Scientific Publications, Durham, NC.
[2] American Red Cross National Reference Laboratory Methods Manual Committee. (1993) Immunohematology Methods. American Red Cross National Reference Laboratory, Rockville, MD.

Warm autoantibodies have been described with activity against the following blood group antigens

AnWj	Kp^b, Js^b, K13, Kell protein (Ku)
Di^b, Wr^b	LW^a, LW^{ab}
En^a, U, N	Rh, in particular e, RhE protein and RhD protein
Fy^b	Sc1
Ge2, Ge3	Vel
Jk^a, Jk^b, Jk3	Xg^a
JMH	

In some autoimmune cases, the target antigen may be weakened to the extent that the patient's RBCs are negative in the DAT. The following antigens have been implicated (listed alphabetically)[1]:

AnWj	Ge3	JMH	Rh	U
Co3	Jk^a	Kp^b	Sc1	Vel
En^a	Jk^b	LW	Sc3	

Antibodies to the following drugs have been described[1–5]

Mechanisms implicated in drug induced positive direct antiglobulin tests
Autoantibody production, drug adsorption, membrane alteration and the so-called immune complex mechanisms have been implicated in causing a positive direct antiglobulin test and in some cases of *in vivo* hemolysis (see * in following table). Because some drugs react by more than one mechanism it is difficult to categorize them.

Generic chemical name	*Associated trade names*
Acetaminophen	Numerous drugs, e.g., Anacin, Codeine, Contac Sinus Pain, Dristan, Panadol, Sinutab, Sudafed, Tylenol
Aceclofenac	
Acetohexamide	Dimelor
Aldrin	Aldrin (chlorinated hydrocarbon)
Amphotericin B	Fungizone
Amoxicillin	Amoxil, Apo-Amoxi, Novamoxin, Nu-Amoxi, Clavulin
Ampicillin	Ampicin, Apo-Ampi, Novo-Ampicillin, Nu-Ampi, Penbritin
Antazoline	
Apronal	
Aspirin	Numerous drugs
Astemazole	Hismanal
Azapropazone	Apazone
Brompheniramine	Dimetapp

Generic chemical name	Associated trade names
Butizide	
Carbenicillin	Geocillin
Carbimazole	
Carbromal	Carbrital
Catergen	
Cefamandole naftate	Mandole
Cefazolin	Ancef, Kefzol
Cefotaxime	Claforan
Cefotetan*	Cefotan
Cefoxitin	Mefoxin
Ceftazidime	Fortaz
Ceftizoxime	
Ceftriaxone	Rocephin
Cephalexin	Apo-Cephalex, Ceporex, Keflex, Novo-Lexin, Nu-Cephalex
Cephaloridine	Laradine
Cephalothin*	Keflin
Chaparral	None, but used in a tea
Chlorinated hydrocarbons	Chlordane, insecticides
Chlorpheniramine	Numerous cold remedies
Chlorpropamide	Apo-Chlorpropamide, Diabinese, Novo-Propamide
Chlorpromazine	Chlorpromanyl, Largactil, Novo-Chlorpromazine
Cianidanol	
Cisplatin	cis-platinum, Platinol
Clavulanate potassium	(e.g., in timentin)
Cloxacillin	Apo-Cloxi, Novo-Cloxin, Nu-Cloxi, Orbenin, Tegopen
Cotrimoxazole	Sulfamethoxazole/Trimethoprim-Apo-Sulfatrim, Bactrim Roche, Novo-Trimel, Nu-Cotrimix, Roubac, Septra
Cyanidanol	
Cyclofenil	
Cyproheptadine	Periactin
DDT	DDT (chlorinated hydrocarbon)
Dexbrompheniramine	Drixoral
Diclofenac, sodium	Voltaren
Diethylstilbestrol	
Diglycoaldehyde	
Diphenhydramine	Allerdryl, Benadryl, Insomnal, PMS-Diphenhydramine, Benylin Decongestant Syrup
Dipyrone	Dirone, Novolone, Novolate
Endosulfan	Thiodan (chlorinated hydrocarbon)

Generic chemical name	Associated trade names
Erythromycin	Apo-Erythro, E-Mycin, Erybid, Eryc, Erythromid, Novo-Rythro, PCE, Ilosone, EES-200, Eryped, Erythrocin, Ilotycin, Sans-Acne, Pediazole
Etodolac	
Fenoprofen	Numerous drugs contain Fenoprofen
Fludarabine	Fludara
Fluorescein	
Fluorouracil (5-FU)	Adrucil, Efudex
Furosemide	Apo-Furosemide, Lasix, Novo-Semide, Uritol
Gamma BHC	Lindane (chlorinated hydrocarbon)
Glafenine	
Glyburide	DiaBeta, Euglucon
Hydralazine	Apo-Hydralazine, Apresoline, Novo-Hyzin
Hydrochlorothiazide	Numerous drugs, e.g., Aldoril, Dyazide, Moduret, Timolide
Hydroxyzine	Apo-Hydroxyzine, Atarax, Histantil, Multipax, Novo-Hydroxyzin, Phenergan, PMS Hydroxyzine, PMS Promethazine
9-Hydroxymethyl-ellipticinium Ibuprofen	Actiprofen, Advil, Apo-Ibuprofen, Medipren, Motrin, Novo-profen
Ibuprofen	
Indene derivatives	(e.g., Sulindac)
Insulin	Beef, Pork and Human insulin products
Isoniazid	Isotamine, PMS Isoniazid
Latamoxef	
Levodopa	Laradopa, Prolopa, Sinemet
Mefenamic acid	Ponstan
Melphalan	Alkeran
Methadone	Dolophine, Physeptone
Methicillin	
Methotrexate	Methotrexate
Methoxychlor	Methoxychlor (chlorinated hydrocarbon)
Methyldopa	Aldomet, Apo Methyldopa, Dopamet, Novomedopa, Aldoril, Aldovil, Apo-Methazide, Novodoparil, Nu-Medopa, PMS dopazide, Supres
Methysergide	Sansert
Nafcillin, sodium	Unifen
Nomifensine	No longer used. Banned internationally
p-Aminosalicylic Acid	
Penicillin G*	Megacillin, Crystapen
Penicillin G Benzanthine	Bicillin

Generic chemical name	Associated trade names
Penicillin V	Phenoxymethyl Penicillin-Apo Pen VK, Ledercillin VK, Nadopen-V, Novo-Pen-VK, Nu-Pen-VK, Pen Vee, PVF benzan-thine, PVF K, VC-K 500, V-Cillin K
Perthane	Perthane (chlorinated hydrocarbon)
Phenacetin	Banned in USA. Causes kidney problems
Phenyltoloxamine	Sinutab
Piperacillin	Piperacil
Podophyllotoxin	
Probenecid	Benemid, Ampicin-PRB
Procainamide	Procan-SR, Pronestyl, Pronestyl SR
Promethazine	Phenergan Expectorant
Pyramidon	
Quinidine	Biquin Durules, Quinaglute Duratabs, Quinate, Natisedine, Quinobarb, Cardioquin, Quinidex Extentabs
Quinine	Novo-Quinine, Quinobarb
Ranitidine hydrochloride	Zantac
Rifampicin	Rifadin, Rimactane
Sodium pentothal	
Sodium thiopental	Pentothal Sodium
Stibophen	
(Diethylstilbestrol)	Honvol, Stilboestrol
Streptomycin (sulfate)	Streptomycin
Sulbactam sodium	(e.g., in Unasyn)
Sulfadiazine	Co-trimazine
Sulfasalazine	PMS Sulfasalazine, Salazopyrin, SAS-Enema, SAS-500
Sulfisoxazole	Novo-Soxazole, Azo Gantri sin, Pediazole
Sulfonamide	
Sulindac (Indene)	Clinoril
Suprofen	
Tazobactam	
Teicoplanin	
Teniposide	Vumon
Terfenadine	Contac Allergy Formula, Seldane
Tetracycline	Achromycin, Apo-Tetra, Novo-Tetra, Nu-Tetra, Tetracyn
Thiopental	Pentothal Sodium
Ticarciuine	Ticar, Timentine
Tolbutamide	Apo-Tolbutamide, Mobenol, Novo-Butamide, Orinase
Tolmetin	Tolectin

Generic chemical name	Associated trade names
Triamterene	Dyrenium, Apo-Triazide, Dyazide, Neo-Diurex, Novo-Triamzide, Nu-Triazide
Trimellitic anhydride	
Zomepirac	

References
1 Arndt, P.A. et al. (1999) Transfusion 39, 1239–1246.
2 Garratty, G. (1994) Immunohematology 10, 41–50.
3 Campbell, S. et al. (1992) CSTM Bull. 4, 40–44.
4 Myint, H. et al. (1995) Br. J. Haematol. 91, 341–344.
5 Petz, L.D. and Garratty G. (2003) Acquired Immune Hemolytic Anemias, 2nd Edition, Churchill Livingstone, New York.

Clinically useful information

Clinical significance of some alloantibodies to blood group antigens[1]

Usually clinically significant	Sometimes clinically significant	Clinically insignificant if not reactive at 37°C	Generally clinically insignificant
A and B	At^a	A1	Chido/Rodgers
Diego	Colton	AnWj	Cost
Duffy	Cromer	H	JMH
H in O$_h$	Dombrock	Le^a	HLA/Bg
Kell	Gerbich	Lutheran	Knops
Kidd	Indian	M,N	Le^b
P, PP1P^k	Jr^a	P1	Xg^a
Rh	Lan	Sd^a	
S,s,U	Landsteiner–Wiener		
Vel	Scianna		
	Yt		

Characteristics of some blood group alloantibodies

Antibody specificity	IgM (direct)	IgG (indirect)	Clinical transfusion reaction	HDN
ABO	Most	Some	Immediate Mild to severe	Common Mild to moderate
Rh	Some	Most	Immediate/delayed Mild to severe	Common Mild to severe
Kell	Some	Most	Immediate/delayed Mild to severe	Sometimes Mild to severe
Kidd	Few	Most	Immediate/delayed Mild to severe	Rare; mild
Duffy	Rare	Most	Immediate/delayed Mild to severe	Rare; mild
M	Some	Most	Delayed (rare)	Rare; mild
N	Most	Rare	None	None
S	Some	Most	Delayed/mild	Rare; mild to severe
s	Rare	Most	Delayed/mild	Rare; mild to severe
U	Rare	Most	Immediate/delayed Mild to severe	Rare; severe
P1	Most	Rare	None (rare)	None
Lutheran	Some	Most	Delayed	Rare; mild
Le^a	Most	Few	Immediate (rare)	None
Le^b	Most	Few	None	None
Diego	Some	Most	Delayed; None to severe	Mild to severe

Antibody specificity	IgM (direct)	IgG (indirect)	Clinical transfusion reaction	HDN
Colton	Rare	Most	Delayed; mild	Rare; Mild to severe
Dombrock	Rare	Most	Immediate/delayed Mild to severe	Rare; mild
LW	Rare	Most	Delayed; none to mild	Rare; mild
Yt^a	Rare	Most	Delayed (rare); mild	None
I	Most	Rare	None	None
Ch/Rg	Rare	Most	Anaphylactic (3)	None
JMH	Rare	Most	Delayed (rare)	None
Knops	Rare	Most	None	None
Xg^a	Rare	Most	None	None

Antigen-negative incidence for common polymorphic antigens

System	Antigen	Incidence Caucasian	Black
Rh	D	0.15	0.08
	C	0.32	0.73
	E	0.71	0.78
	c	0.20	0.04
	e	0.02	0.02
	f	0.35	0.08
	C^W	0.98	0.99
	V	>0.99	0.70
	VS	>0.99	0.73
	CE	>0.99	0.99
	cE	0.72	0.78
	Ce	0.32	0.73
MNS	M	0.22	0.26
	N	0.30	0.25
	S	0.48	0.69
	s	0.11	0.06
	M−S−	0.15	0.19
	M−s−	0.01	0.02
	N−S−	0.10	0.16
	N−s−	0.06	0.02
P	P1	0.21	0.06
Lewis	Le^a	0.78	0.77
	Le^b	0.28	0.45

		Incidence	
System	Antigen	Caucasian	Black
Lutheran	Lu^a	0.92	0.95
	Lu^b	<0.01	<0.01
Kell	K	0.91	0.98
	k	0.002	<0.001
	Kp^a	0.98	>0.99
	Kp^b	<0.01	<0.01
	Js^a	>0.99	0.80
	Js^b	<0.001	0.01
Duffy	Fy^a	0.34	0.90
	Fy^b	0.17	0.77
Kidd	Jk^a	0.23	0.08
	Jk^b	0.26	0.51
Dombrock	Do^a	0.33	0.45
	Do^b	0.18	0.11
Colton	Co^a	<0.001	<0.001
	Co^b	0.90	0.90

To calculate the incidence of compatible donors, multiply the incidence of antigen-negative donors for each antibody, e.g., the incidence of K−, S−, Jk(a−) donors in the general donor pool is (0.91) × (0.48) × (0.23) = 0.10 in 100 or 1 in 10.

Potentially useful information for problem-solving in immunohematology

Available information	Considerations
Patient demographics	Diagnosis, age, sex, ethnicity, transfusion and/or pregnancy history, drugs, IV fluids (Ringer's lactate, IV-IgG, ALG, ATG), infections, malignancies, hemoglobinopathies, stem cell transplantation
Initial serological results	ABO, Rh, DAT, phenotype, antibody detection results, autologous control, crossmatch results
Hematology/ chemistry values	Hemoglobin, hematocrit, bilirubin, LDH, reticulocyte count, haptoglobin, hemoglobinuria, albumin:globulin ratio, RBC morphology
Sample characteristics	Site and technique of collection, age of sample, anticoagulant, hemolysis, lipemic, color of serum/ plasma, agglutinates/aggregates in the sample
Other	Check records in current and previous institutions for previously identified antibodies

Available information	Considerations
Antibody identification	Auto control, phase of reactivity, potentiator (saline, albumin, LISS, PEG), reaction strength, effect of chemicals on antigen (proteases, thiol reagents), pattern of reactivity (single antibody or mixture of antibodies), characteristics of reactivity (mixed field, rouleaux), hemolysis, preservatives/antibiotics in reagents

Alloantibodies that may have *in vitro* hemolytic properties

Anti-A, -B, -H (in O_h people), -A,B, -I, -i, -Le^a, -Le^b, -PP1P^k, -P, -Jk^a, -Jk^b, -Jk3, -Ge3, -Vel and rare examples of anti-Sc1, -Lan, -Jr^a, -Co3, -Emm and -Milne

Conditions associated with suppression (sometimes total) or with alteration of antigen expression

Condition	Antigens affected
Pregnancy	A, B, H, I, Lewis, LW, P1, JMH, Sd^a, some Jk^a, Gy^a and AnWj
Carcinoma	A, B, H, I, P1, Knops
Leukemia	A, B, H, I, Rh, Yt^a, Colton
Infection	A, B, H, I, A with appearance of Tn, A with appearance of acquired B, K
Hodgkins disease	A, B, H, LW, Colton
LADII (CDG-II)	A, B, H, Lewis
Infectious mononucleosis	Lewis
Alcoholic cirrhosis	Lewis
Thalassemia	I
PNH	Cromer, Yt, Dombrock, MER2, JMH, Emm
Splenic infarctions	Cromer
AIHA	En^a,U; Rh; Kell; Jk^a; LW; Gerbich; Scianna; Vel; Diego; AnWj
SLE	Chido/Rodgers; Knops; Yt^a
AIDS	Knops
Hematopoietic stress	A, B, H, I (concomitant increased expression of i)
Diseases with increased clearance of immune complexes:	Knops
Old age	JMH, A, B, H
SE Asian ovalocytes	En^a, S, s, U, Di^b, Wr^b, D, C, e, Kp^b, Jk^a, Jk^b, Xg^a, LW, Sc1[2]

Causes of apparent *in vivo* hemolysis

Immune
ABO incompatibility
Clinically significant alloantibody
Anamnestic alloantibody response
Autoimmune hemolytic anemia
Cold agglutinin disease
HDN
Drug-induced hemolytic anemia
Polyagglutination (sepsis T-active plasma)
Paroxysmal cold hemoglobinuria
TTP/HUS*-microangiopathic process

Non-immune
1. **Mechanical**
   Poor sample collection
   Small-bore needle used for infusion
   Excessive pressure during infusion
   Malfunctioning blood warmer
   Donor blood exposed to excessive heat or cold
   Urinary catheter
   Crush trauma
   Prosthetic heart valves
   Aortic stenosis
   March hemoglobinurea
2. **Microbial**
   Sepsis
   Malaria
   Contamination of donor blood
3. **Chemical**
   Inappropriate solutions infused
   Drugs infused
   Serum phosphorus < 0.2 mg/dl
   Water irrigation of bladder
   Azulfidine
   Dimethyl sulfoxide
   Venom (snake, bee, Brown Recluse spider)[3]
   Certain herbal preparations, teas, enemas
4. **Inherent RBC Abnormalities**
   Paroxysmal nocturnal hemoglobinuria
   Sickle cell anemia
   Spherocytosis
   Hemoglobin H
   G6PD deficiency

* Thrombotic thrombocytopenic purpura/hemolytic uremic syndrome

Normal ranges of hematology values in adults[4,5]

		SI units	Conventional units
Albumin		39–46 g/l	3.9–4.6 g/dl
Bilirubin total	Adult	1.7–20.5 μmol/l (or 2–21)	0.1–1.2 mg/dl
	Newborns	17.1–205.0 μmol/l	1–12 mg/dl
Globulin		23–36 g/l	2.3–3.5 g/100 dl
Haptoglobin		0.6–2.7 g/l	60–270 mg/dl
Hematocrit	Females	0.38–0.47	38–47%
	Males	0.40–0.54	40–54%
Hcmoglobin	Females	120–160 g/l	12.0–16.0 g/dl
	Males	135–180 g/l	13.5–18.0 g/dl
	Newborn	160–280 g/l	16.0–28.0 g/dl
Immunoglobulin	IgG	8.0–18.0 g/l	800–1801 mg/dl
	IgA	1.1–5.6 g/l	113–563 mg/dl
	IgM	0.54–2.2 g/l	54–222 mg/dl
	IgD	5.0–30 mg/l	0.5–3.0 mg/dl
	IgE	0.1–0.4 mg/l	0.01–0.04 mg/dl
Platelet count		150–450×10^9/l	150–450×10^3/mm^3
Red blood cells	Females	4.2–5.4×10^{12}/l	4.2–5.4×10^6/mm^3
	Males	4.6–6.2×10^{12}/l	4.6–6.2×10^6/mm^3
Reticulocytes		25–75×10^9/l	25–75×10^3/mm^3
Total protein		65–77 g/l	6.5–7.7 g/dl
White blood cells		4.5–11.0×10^9/l	4.5–11.0×10^3/mm^3

References

[1] Petz, L.D. et al. eds. (1996) Clinical Practice of Transfusion Medicine, 3rd Edition, Churchill Livingstone, New York.

[2] Booth, P.B. et al. (1977) Vox Sang. 32, 99–110.

[3] Williams, S.T. et al. (1995) Am. J. Clin. Pathol. 104, 463–467.

[4] Vengelen-Tyler, V. (1996) Technical Manual, 12th Edition, American Association of Blood Banks, Bethesda, MD.

[5] Henry, J.B. et al. eds. (2001) Clinical Diagnosis and Management by Laboratory Methods, 20th Edition, WB Saunders, Philadelphia, PA.

Blood group system and protein-based facts

Carbohydrate-based blood group systems

System name	Gene product
ABO	Glycosyltransferases
P1	Galactosyltransferase
Lewis	Fucosyltransferase
Hh	Fucosyltransferase
I	Acetylglucosaminyltransferase
P	Acetlygalactosaminyltransferase

Blood group systems located on single-pass membrane protein

System name	Gene product	Number of amino acids	N-terminus	Function in RBCs
MNS	Glycophorin A	131	Exofacial	Carrier of sialic acid;
	Glycophorin B	72	Exofacial	Complement regulation
Gerbich	Glycophorin C	128	Exofacial	Carrier of sialic acid;
	Glycophorin D	107	Exofacial	Interacts with band 4.1 and p55 in RBC membrane
Kell	Kell glycoprotein	732	Cytoplasmic	Cleaves big endothelin
Lutheran	Lutheran glycoprotein	597	Exofacial	Binds laminin
Xg	Xg^a glycoprotein	180	Exofacial	Possible adhesion molecule
	CD99	163	Exofacial	Adhesion molecule
LW	LW glycoprotein	241	Exofacial	Ligand for integrins
Indian	CD44	341	Exofacial	Possible adhesion
Knops	CD35 (CR1)	1998	Exofacial	Complement regulation
Ok^a	CD147	248	Exofacial	Possible cell–cell adhesion
Scianna	ERMAP	475	Exofacial	Possible adhesion

Blood group systems located on multi-pass membrane proteins

System name	Gene product	Number of amino acids	Predicted number spans	Function
Rh	D protein	417	12	Possibly ammonium transport
	CcEe protein	417	12	
Duffy	Fy glycoprotein	338	7*	Malaria/cytokine receptor
Diego	Band 3 (AE1)	911	14	Anion transport
Colton	CHIP-1 (AQP1)	269	6	Water transport
Kidd	Urea transporter	391	10	Urea transport
Kx	Kx glycoprotein	444	10	Possible neurotransmitter
GIL	AQP3	342	6	Glycerol/water/urea transport

* N-terminus oriented to exofacial surface, C-terminus to cytoplamic surface. All others are predicted to be oriented with both their N- and C-termini to the cytoplasmic aspect of the RBC membrane

Blood group systems carried on glycosylphosphatidylinositol-linked proteins

System name	Gene product	Number of amino acids	Function
Yt	Acetylcholinesterase	557	Enzymatic
Cromer	CD55 (DAF)	347	C' regulation
Dombrock	Do glycoprotein	314	Possibly enzymatic
JMH	CD108	656	Possible adhesion

Blood group systems located on proteins adsorbed from the plasma

System name	Component	Antigen location	Function
Chido/Rodgers	C' component 4 (C4)	C4d fragment	C' regulation

Other RBC membrane proteins

Protein component	Comments
1. Chromosomal location known	
Glucose transporter (band 4.5; GLUT1)	Chromosome 1; M_r 45 000–65 000; 492 amino acids; 200 000–700 000 copies/RBC; 12 membrane passes; Carries A, B, H and I antigens; broad tissue distribution
Band 7.2 protein (Stomatin [7.2b]) cation transporter	Chromosome 9q33–q34.2 (7.2b); M_r 31 000; 288 amino acids (7.2b); 10 000 copies/RBC; single pass type I protein
Rh associated glycoprotein (RhAG; RH50; 2D10)	Chromosome 6p21.1–p11; M_r 45 000–100 000; 409 amino acids; 100 000–200 000 copies/RBC; 12 membrane passes; carries A, B, H, I and 2D10 antigens; broad tissue distribution
CD47 [BRIC 125, integrin-associated protein (IAP), AgOAB, ID8]	Chromosome 3q13; M_r 47 000–52 000; 323 amino acids; 10 000–50 000 copies/RBC, 5 membrane passes, heavily glycosylated. Extracellular N-terminal domain, member of immunoglobulin superfamily. May be involved in intracellular calcium concentration.
LFA3 (CD58)	Chromosome 1p13; M_r 55 000–70 000; 222 amino acids; 3000–8000 copies/RBC; cell adhesion molecule, heavily glycosylated, most GPI-linked but some transmembrane with cytoplasmic domain, ligand for T-lymphocyte surface molecule CD2 which is important in T-cell activation; function in RBC unknown; broad tissue distribution
CD59 (MIRL, H19, P18, Fib 75, HRF_{20}, MACIF, protectin)	Chromosome 11p13–p14, M_r 18 000–20 000, 128 amino acids; 20 000–40 000 copies/RBC; Complement-regulatory protein (may inhibit insertion of C9 into membrane), GPI-linked, member of Ly-6 superfamily, heavily glycosylated
2. Proteins that are more abundant on reticulocytes than on erythrocytes	
Transferrin receptor (CD71)	Chromosome 3q26.2-qter; M_r 180 000–190 000 (90 000–95 000 reduced); 760 amino acids; type II membrane protein that exists as a homodimer
4F2 (2F3; CD98)	M_r 125 000 (85 000 and 40 000 chains); type II membrane protein
α4β1 [Intergrin; very late antigen-4 (VLA-4)]	α4 is CD49d; β1 is CD29. Increased expression on stress reticulocytes and those containing HBSS; important in hematopoiesis

Protein component	Comments
Flotillin-1 (reggie-2)[1]	
Flotillin-2 (reggie-1)[2]	
MHC class I (HLA 1; HLA 2; HLA-A,B,C)	The light chain of this complex is β2 macroglobulin
Erythroblast macrophage protein (Emp)	M_r 43 000. 395 amino acids[3]

3. Others

Ca-Activated K channel (Gardos channel)	1200 amino acids; 200 copies/RBC
Nucleoside transporter	M_r 55 000; 10 000 copies/RBC
C8-binding protein (homologous restriction factor (HRF$_{60}$), MIP)	M_r 55 000; inhibits formation of membrane attack complex
Na/K-ATPase	Oubain inhibitable voltage-gated K+ channels
Ca-ATPase, Ca/Mg-ATPase Mg-ATPase	M_r 140 000
CD75	Carbohydrate antigen involved in differentiation
K-Cl cotransporter (KCC1)	
Na-H$^+$ exchanger (NHE1)	

References

[1] Morrow, I.C. et al. (2002) J. Biol. Chem. 277, 48834–48841.
[2] Solomon, S. et al. (2002) Immunobiology 205, 108–119.
[3] Hanspal, M. et al. (1998) Blood 92, 2940–2950.

Lectins and polyagglutination information

Some lectins and their specificities[1]

Lectin	Common name	Carbohydrate-binding specificity
Arachis hypogaea	Peanut	D-Galβ(1–3)GalNAc > α-D-Gal
Dolichos biflorus	Horsegram	α-D-GalNAc >> α-D-Gal
Glycine max	Soybean	α-D-GalNAc > β-D-GalNAc > α-D-Gal
Griffonia simplicifolia I*	GS1	α-D-Gal > α-D-GalNAc
Griffonia simplicifolia II*	GS2	α-D-GlcNAc = β-D-GlcNAc
Helix pomatia	Edible snail	α-D-GalNAc > α-D-GlcNAc > α-D-Gal
Leonurus cardiaca	Motherwort	α/β-D-GalNAc
Phaseolus lunatus	Lima bean	α-D-GalNAc
Salvia horminum	Clary**	α-D-GalNAc > β-D-GalNAc
Salvia sclarea	Clary**	α-D-GalNAc
Ulex europaeus I	Gorse/furze	α-L-Fuc
Vicia cretica		D-Gal

* Previously known as *Bandeiraea simplicifolia* (BS) lectins.
** Both *Salvia horminum* and *Salvia sclarea* are commonly known as clary, but they are botanically different.

Polyagglutination types and the expected reactions with group O* RBCs and lectins[1,2]

Lectin	Acquired B	T	Tk	Th	Tx	Tn	Tr	Cad	Nor	VA	HEMPAS	HbM**
Arachis hypogaea	0	+	+	+	+	0	↓	0	0	0	0	↓↓
Dolichos biflorus	↓	0	0	0	0	+	0	+	0	0	0	0
Glycine max	0	+	0	0	0	+	+	0	0	0	0	+
GSI	+	0	0	0	0	+	+	0	0	0	0	0
GSII	0	0	+	0	0	0	+	0	0	0	0	+
Helix pomatia	+	+	0	ND	ND	+	↓	+	0	+	+	+
Leonurus cardiaca	0	0	0	0	0	0	ND	+	0	0	0	0
Phaseolus lunatus	+	0	0	0	0	0	0	0	0	0	0	ND
Salvia horminum	0	0	0	0	0	+	↓	+	0	0	0	↓↓
Salvia sclarea	0	0	0	0	0	+	↓	0	0	0	0	0
Ulex europaeus	+	↑↑	↓	+	+	↑	+	↓	+	↓	↓	↑↑
Vicia cretica	0	+	0	+	0	0	ND	0	0	0	0	↓↓

↓ (↓↓) Weaker than normal RBCs.
↑ (↑↑) Stronger than normal RBCs.
+ Agglutination.
0 No agglutination.
ND Not done.
* Except acquired B.
** HbM-Hyde Park.

545

Polyagglutination

T Neuraminidase made by some organisms (*Vibrio cholerae, Clostridium perfringens*, pneumococci, influenza virus) cleaves sialic acid (NeuAc) from RBCs. Galβ1–3GalNAc-Ser/Thr (disialylated alkali-labile tetrasaccharide)

Tk β-Galactosidases produced by some organisms (*Bacteroides fragilis, Aspergillus niger, Serratia marcescens, Candida albicans*) cleaves Gal from GlcNAc in complex carbohydrate structures including A, B, H, P1, I active carbohydrate chains.

A diagrammatic representation of Cad, Tn, T, Tk, and acquired B is shown in the figure.

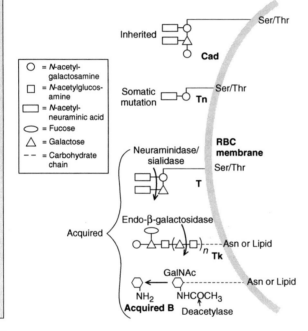

Lectins, inhibitory sugar and RBC antigen[3]

Lectin	Common name	Inhibitory sugar	RBC antigen
Amarillis lectin			St[a]
Anguilla anguilla	Freshwater eel	L-Fuc	H
Arachis hypogaea	Peanut	D-Galβ(1–3)GalNAc	T, Tk, Th, Tx
Canavalia ensiformis (Con A)	Jack bean	D-Man or D-Glc	
Dolichos biflorus	Horsegram	Terminal α-D-GalNAc	A, Tn, Cad
Euonymus europaeus	Spindle tree	Not inhibited by simple sugars	B, H, A₂

Lectin	Common name	Inhibitory sugar	RBC antigen
Fomes fomentarius		D-Gal	B, P
Glycine max	Soybean	Terminal GalNAc	$A_1 > A_2 >> B$, T,Tn
Griffonia simplicifolia I		D-Gal	B, Tn
Griffonia simplicifolia II		D-GlcNAc	Tk
Helix aspersa	Garden snail	GalNAc >> GlcNAc	A
Helix pomatia	Edible snail	GalNAc	A
Iberis amara	Candytuft	Not inhibited by simple sugars	M
Leonurus cardiaca	Motherwort	GalNAc	Cad
Lotus tetragonolobus	Asparagus pea	L-Fuc	H
Lycopersicon esculentum	Tomato	GlcNAcβ(1–4) GlcNAc oligomers	O > A and B
Phaseolus lunatus	Lima bean	GalNAc	A
Phaseolus vulgaris (PHA)	Black kidney bean	GalNAc	
Phytolacca americana	Pokeweed	β(1–4)GlcNAc oligomers, *N*-acetyllactosamine	
Salvia horminum	Clary	GalNAc	Tn, Cad
Salvia sclarea	Clary	GalNAc	Tn
Sophora japonica	Japanese pagoda tree	GalNAc > Gal	B >> A
Triticum vulgaris (WHA)	Wheat germ agglutinin	sialic acid, GlcNAc	
Ulex europaeus	gorse/furze	L-Fuc	H, Th, Tx, Nor
Vicia cretica		D-Gal	T, Th
Vicia graminea		O-linked Galβ(1–3)-GalNAc	N

References

[1] Judd, W.J. (1992) Immunohematology 8, 58–69.
[2] Mollison, P.L. et al. (1997) Blood Transfusion in Clinical Medicine, 10th Edition, Blackwell Science, Oxford.
[3] EY Laboratories I. (1992) Lectins and Lectin Conjugates. EY Laboratories, Inc., San Mateo, CA.

Gene-based facts

Mechanisms of silencing genes encoding blood groups

Gene deletion
Exon deletion with frameshift
Nucleotide(s) deletion with frameshift
Nucleotide(s) insertion with frameshift
Nonsense mutation
Nucleotide substitution in donor splice site
Nucleotide substitution in acceptor splice site
Nucleotide substitution in GATA box

Causes of pseudo-discrepancies between genotype and phenotype

Event	Mechanism	Blood group phenotype (examples)
Transcription	Nt change in GATA box	Fy(b−)
Alternative splicing	Nt change in splice site: Partial/complete skipping of exon	S−s−; Gy(a−)
	Deletion of nts	Dr(a−)
Premature stop codon	Deletion of nt(s) → frameshift	Fy(a−b−); D−; Rh$_{null}$;
	Insertion of nt(s) → frameshift	Ge: −2,−3,−4; Gy(a−)
	Nt change	K$_0$; McLeod; D−; Co(a−b−); Fy(a−b−); r′; Gy(a−)
Amino acid change	Missense mutation	D−; Rh$_{null}$; K$_0$; McLeod
Reduced amount of protein	Missense mutation	Fy^X; Co(a−b−)
Hybrid genes	Crossover	GP.Vw; GP.Hil; GP.TSEN
	Gene conversion	GP.Mur; GP.Hop; D− −; R$_0$^{Har}
Interacting protein	Absence of RhAG	Rh$_{null}$
	Absence of Kx	Kell antigens are weak
	Absence of aas 59 to 76 of GPA	Wr(b−)
	Absence of protein 4.1	Ge antigens are weak
Modifying gene	In(Lu)	Lu(a−b−)
	In(Jk)	Jk(a−b−)

Uses of DNA in the clinical laboratory

To type patients who have been recently transfused
To identify a fetus at risk for hemolytic disease of the newborn
To type patients whose RBCs are coated with immunoglobulin
To determine which phenotypically antigen-negative patients can receive antigen-positive RBCs

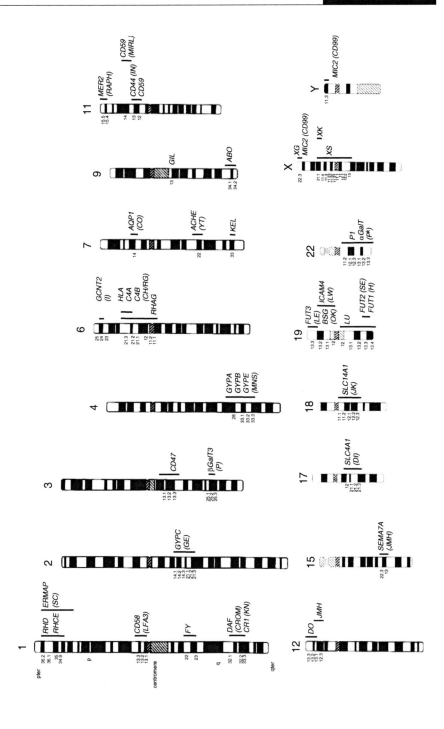

To type patients who have an antigen that is expressed weakly on RBCs

To determine *RHD* zygosity

To mass screen for antigen-negative donors

To resolve blood group A, B, and D discrepancies

To determine the origin of engrafted leukocytes in a stem cell recipient

To determine the origin of lymphocytes in a patient with graft-versus-host disease

For tissue typing

For paternity and immigration testing

For forensic testing

A chromosome is depicted on a page from top to bottom pter-p-centromere-q-qter. "ter" represents terminus. The p arm is the shorter arm.

Useful definitions

Absorbed	From; away
Adsorbed	Onto
	Thus, an antibody is *absorbed* from serum but *adsorbed* onto RBCs. Another definition is that *absorbed* is a non-specific term (as in 'absorbed' by a sponge) while *adsorbed* is a specific reaction
Allele	Alternative form(s) of a *gene* at a given locus (antigens cannot be allelic)
Antithetical	Refers to *antigens* produced by alleles (alleles cannot be antithetical)
Haplotype	A set of alleles of a group of closely linked genes, which are usually inherited together. People have haplotypes, RBCs do not.
Frequency	Used to describe prevalence on the genetic level, i.e. occurance of an allele (gene) in a population.
Incidence	Used to describe prevalence of a condition (phenotype) that changes over time (e.g. a disease)
Prevalence	Used to describe the prevalence of a permanent/inherited characteristic on the phenotypic level (e.g. a blood group)
Allogeneic	Homologous (iso) (same species – genetically different)
Autogeneic	Autologous (own genes)
Syngeneic	Between same strain of animal or identical twin (same genes)
Xenogeneic	Different species
Propositus	Singular male or index case (singular) regardless of sex
Propositi	Plural male or index cases (plural) regardless of sex
Proposita	Singular female
Propositae	Plural female
Proband	Index case regardless of sex
Probands	Plural for index cases regardless of sex
Transition	Change of purine (A, G) to purine, or pyrimidine (T, C) to pyrimidine
Transversion	Change between purine and pyrimidine
Missense mutation	Nucleotide change leading to a change of amino acid
Nonsense mutation	Nucleotide change leading to a stop codon
Silent mutation	Nucleotide change that, due to redundancy in the genetic code, does not change the amino acid
Northern blot	Analysis of RNA
Southern blot	Analysis of DNA
Western blot	Analysis of proteins

Index

A

A antigen, 23–5
A1 antigen, 26–7
ABO blood group system, 19–28
  amino acid:
    changes associated with hybrid
      transferases, 22
    sequence, 20
  antigens, number of, 19
  carrier molecule description, 20
  database accession numbers, 20
  disease association, 20
  expression, 19
  gene, 19
  molecular basis:
    hybrid transferase genes, 22
    O phenotype, 22
    variant A transferases, 21
    variant B transferases, 22
  phenotypes, 21
  terminology, 19
ABTI antigen, 515–16
Acid:
  effect on antigen expression, 524
  treatment, effect on antigens, 13
Agglutination, mixed field, 524
ALe^b antigen, 274–5
Alloantibodies:
  clinical significance of, 13
    blood group antigens, 535
  in vitro:
    characteristics, 13
    hemolytic properties, 538
  see also individual blood group antigens
Amino acids:
  codes, three-letter and single-letter, 9
  sequence, 9
2-Aminoethylisothiouronium bromide
  (AET), 13
An^a antigen, 415–16
Antibodies, 4
Antigen/s, 4
  absent (altered) on selected RBC
    phenotypes, 524–5
  antithetic, 12
  assigned to each system, 7
  autoantibodies with activity against, 530
  carrier proteins, model of, 4
  clinical significance of
    alloantibodies, 535
  for Caucasians, 12
  copy number, 524
  effect of Dithiothreitol (DTT) in antibody
    identification, 523

effect of enzymes in antibody
  identification, 523
ethnic differences, 12
expression conditions associated with
  suppression or alteration, 538
molecular basis, of 10, 12
occurrence, 11
in soluble form, 524
terminology, 11
with variable expression on RBCs, 524
see also individual antigens
Antigen-based facts, 521
Antigen-negative incidence for common
  polymorphic antigens, 535
Anton antigen see AnWj antigen
AnWj antigen, 509–10
At^a antigen, 506–7
Au^a antigen, 218–19
Au^b antigen, 220–1
Autoantibodies, 14, 530–4
  with activity against antigens, 530
  see also individual antigens
Avis antigen see ENAV antigen

B

B antigen, 25–6
BARC antigen, 187–8
Bastiaan antigen see HR^B antigen
Be^a antigen, 170
Bigelow antigen see LW^{ab} antigen
BLe^b antigen, 276–7
Blood group alloantibodies, characteristics
  of, 535–6
Blood Group Antigen Gene Mutation
  Database, 9
Blood group antigens see antigens
Blood group collections, 14, 482
  see also specific blood group
    collections
Blood group systems, 7
  antigens assigned to, 7
  carbohydrate-based, 541
  carried on glycosylphosphatidylinositol-
    linked proteins, 542
  with gene name and location, 3
  located on multi-pass membrane
    proteins, 542
  located on protein adsorbed from
    plasma, 542
  located on single-pass membrane
    protein, 541
  and protein-based facts, 541–4
BOW antigen, 322–3
Bp^a antigen, 315–16

Bu^a antigen *see* Sc2 antigen
Bullee antigen *see* Sc2 antigen

C

C antigen, 131–33
c antigen, 135–6
Carbohydrate antigens, 10
Carbohydrate-based blood group
  systems, 541
Carrier:
  molecule, 10
  proteins:
    function of, 11
    model of, 4
Cartwright antigen *see* YT^a antigen
Cartwright blood group system *see* YT blood
  group system
CD numbers, 7
CD99 antigen, 343
Ce antigen, 140–1
CE antigen, 154–5
cE antigen, 158–9
Ce-like antigen *see* Rh41 antigen
Centauro antigen *see* K23 antigen
C^G antigen, 154
Ch1 antigen, 384
Ch2 antigen, 385–6
Ch3 antigen, 386
Ch4 antigen, 386
Ch5 antigen, 387
Ch6 antigen, 387–8
Chemicals, effect on intact RBCs, 12
Chido/Rodgers blood group system, 381–90
  antigens, number of, 381
  carrier molecule, 381
  database accession numbers, 381
  disease association, 382
  expression, 381
  function, 382
  gene, 381
  molecular basis of antigens, 382
  phenotypes, 383
  terminology, 381
Cl^a antigen, 63–4
C-like antigen *see* RH39 antigen
Chloroquine diphosphate, effect on antigen
  expression, 524
Co3 antigen, 369–71
Co^a antigen, 367–8
Co^b antigen, 368
Colton blood group system, 364–71
  amino acid sequence, 365
  antigens, number of, 364

carrier molecule, 365
database accession numbers, 364
disease association, 366
expression, 364
function, 365
gene, 364
molecular basis:
  of antigens, 365
  of Co(a − b−) phenotype, 366
  of Co(a − b + ^{w}), 366
phenotypes, 366
terminology, 364
Complement binding, 13
Cord RBCs, 524
Cost blood group collection, 14, 483–5
  antigens, number of, 483
  phenotypes, 483
  terminology, 483
Cr^a antigen, 422–3
Crawford antigen, 176–7
Cromer blood group system, 419–38
  amino acid sequence, 420
  antigens, number of, 419
  carrier molecule, 420
  database accession numbers, 419
  disease association, 421
  expression, 419
  function, 421
  gene, 419
  molecular basis:
    of antigens, 421
    of phenotypes, 421–2
  phenotypes, 421
  terminology, 419
Cs^a antigen, 483–4
Cs^b antigen, 485
C^W antigen, 141–3
C^X antigen, 143–5

D

D antigen, 121–31
  DEL (D$_{el}$) phenotype, 129
  D-negative phenotypes, molecular basis
    of, 130
  molecular basis:
    of partial D phenotypes, 124–7
    of weak D, 128
  molecular and phenotypic
    information, 122–4
  partial D phenotypes, 127
  weak D phenotype, 129
DAK antigen, 190–1

DANE antigen, 85–7
Dantu antigen, 75–6
Database accession numbers, 9
Dav antigen, 180–1
Deal antigen *see* Rh26 (c-like) antigen
Definitions of terms, 551
Dh^a antigen, 417–18
Di^a antigen, 302–3
Di^b antigen, 303–5
Diego blood group system, 298–331
    amino acid sequence, 299
    antigens, number of, 298
    carrier molecule, 299
    database accession numbers, 298
    disease association, 301
    expression, 298
    function, 300
    gene, 298
    molecular basis of antigens, 300
    phenotypes, 301
    terminology, 298
Disease association, 11
Dithiothreitol (DTT), 13
    usefulness of effect on antigens and
        antibody identification, 523
DNA, uses in clinical laboratory, 548
Do^a antigen, 356–7
Do^b antigen, 357–9
Dombrock blood group
        system, 353–63
    amino acid sequence, 353
    antigens, number of, 353
    carrier molecule, 354
    database accession numbers, 353
    disease association, 355
    expression, 353
    function, 354
    gene, 353
    molecular basis:
        of antigens, 354
        of Do$_{null}$ [Gy(a−)] phenotype, 355
        of Hy − and Jo(a−) phenotypes, 355
    phenotypes, 355
    terminology, 353
Dr^a antigen, 429–30
Dreyer antigen *see* s^D antigen
Drug induced positive direct antiglobulin
        tests, 530
Drugs, antibodies to, 530–4
Duclos antigen, 513–14
Duffy blood group system, 278–89
    amino acid sequence of Fy^b, 279
    antigens, number of, 278
    carrier molecule, 279

database accession numbers, 278
disease association, 280
expression, 278
gene, 278
molecular basis:
    of Fy(a−b−) phenotypes, 280
    of Fy^x [Fy(b + ^{w})] phenotype, 280
phenotypes, 280
terminology, 278
D^w antigen, 155–7

E

E antigen, 133–5
e antigen, 137–8
ELO antigen, 312–13
Emm antigen, 408–9
En^a antigen, 80–1
ENAV antigen, 101–2
ENEH antigen, 98–9
ENEP antigen, 97–8
ENKT antigen, 81–2
Enzymes:
    effect on antigens and antibody
        identification, 523
    effect on intact RBCs, 12–13
    substrate specificity for peptide and CHO
        bonds, 521
Er blood group collection, 14, 489–91
    antigens, number of, 489
    phenotypes, 489
    terminology, 489
Er^a antigen, 489–90
Er^b antigen, 490–1
ERIK antigen, 94–5
Erythrocyte blood group antigens,
        terminology, 4
Es^a antigen, 430–1
Ethnic populations, low prevalence
        antigens, 528–9
Evans antigen, 171–2
E^w antigen, 146–7

F

f antigen, 139–40
Far antigen, 71–2
Ficin, 13
FPTT antigen, 183–5
Fr^a antigen, 329–30
Fy antigens, molecular basis of, 279
    see also Duffy blood group system

Fy3 antigen, 284–6
Fy4 antigen, 286
Fy5 antigen, 286–8
Fy6 antigen, 288–9
Fyª antigen, 281–2
Fy^b antigen, 283–4

G

G antigen, 147–8
Ge2 antigen, 406–8
Ge3 antigen, 408–10
Ge4 antigen, 410–12
GenBank accession numbers, 9
Gene-based facts, 548–50
Genes, 7, 9
    encoding blood groups mechanisms
        of silencing, 548
Genotypes, 4
Gerbich blood group system, 403–18
    amino acid sequence, 404
    antigens, number of, 403
    carrier molecule, 404
    database accession numbers, 403
    disease association, 405
    expression, 403
    function, 405
    gene, 403
    glycophorin, 404
    molecular basis
        of antigens, 405
        of phenotypes, 405
    phenotypes, 405
    terminology, 403
GIL blood group system, 479–82
    amino acid sequence, 479
    antigens, 481
        number of, 479
    carrier molecule, 480
    database accession numbers, 479
    expression, 479
    function, 480
    gene, 479
    molecular basis of GIL$_{null}$ phenotype, 480
    phenotypes, 480
    terminology, 479
Gilfeather *see* M^g antigen
Globoside antigen *see* P antigen
Globoside blood group collection, 14, 492–8
    amino acid sequence, 493
    antigens, number of, 492
    carrier molecule, 493
    database accession numbers, 493
    disease association, 493

expression, 492
gene, 492
molecular basis of p phenotype due to
    mutations in α4Gal-T, 494
phenotypes, 494
terminology, 492
Globoside blood group system, 475–8
    amino acid sequence, 475
    antigens, number of, 475
    carrier molecule, 476
    database accession numbers, 475
    disease association, 476
    expression, 475
    function, 476
    gene, 475
    molecular basis of P^k phenotype due to
        mutations in *GLOB*
        (*β3GalNAc-T1*), 476
    phenotypes, 476
    terminology, 475
Globotriaosylceramide (Gb^{3}Cer) *see* P^k antigen
Glycosylation, 10
Glycosylphoshatidylinositol (GPI)-linked
    proteins, 10
Goª antigen, 161–2
Gonzales *see* Goª antigen
GPB^N *see* 'N' antigen
GUTI antigen, 437–8
Gyª antigen, 359–60

H

H antigen, 395–7
HAG antigen, 100–1
Har antigen *see* Rh33 antigen
He antigen, 43–6
Hematology values, normal ranges in
    adults, 540
Hemolysis, *in vivo*, 530
    causes of apparent, 539
Hemolytic disease of the newborn
    (HDN), 13
Hgª antigen, 318–19
Hh blood group system, 391–7
    amino acid sequence of isα-2-L-fucosyl-
        transferase, 392
    antigens, number of, 391
    Bombay (non-secretors), 393–5
    carrier molecule, 392
    database accession numbers, 391
    disease association, 392
    expression, 391
    function, 392
    gene, 391

Hh blood group system (*Continued*)
  molecular basis of H-deficient phenotypes
    due to mutations in *FUT1*, 393
  phenotypes, 392
  terminology, 391
High incidence antigens, 526–7
  901 series, 14
  902 series, 503–18
Hil antigen, 67–9
Hofanesian antigen *see* Lu14 antigen
Hop antigen, 76–8
Hr antigen, 150
hr' antigen *see* c antigen
hr" antigen *see* e antigen
Hr_0 antigen, 148–9
hr^B antigen, 162–4
Hr^B antigen, 167–8
hr^H antigen,159
hr^s antigen,151–2
hr^v antigen *see* V antigen
Hughes antigen *see* Hg^a antigen
7th Human Leucocyte Differentiation
    Antigens (HDLA) Workshop, 7
Hut antigen, 65–7
Hy antigen, 360–1

I

I antigen, 473–4
i antigen, 487–8
I blood group system, 471–4
  amino acid sequence for IgnTC
    β6GlcNAc-transferase, 472
  antigens, number of, 471
  carrier molecule, 472
  database accession numbers, 471
  disease association, 472
  expression, 471
  function, 472
  gene, 471
  molecular basis of I-negative phenotype, 473
  phenotypes associated with I antigen and
    reciprocal i antigen, 472
  terminology, 471
IFC antigen, 431–2
IgG, 13
IgM, 13
Ii blood group collection, 14, 486–8
  antigens, number of, 486
  carrier molecule, 486
  disease association, 486
  function, 486
  gene, 486
  phenotypes associated with i antigen and
    reciprocal I antigen, 486

terminology, 486
Immunoglobulin class of blood group
    antibody, 13
Imunohematology, problem-solving
    in, 537–8
IN^a antigen, 458–9
In^b antigen, 459–60
Indian blood group system, 455–60
  amino acid sequence, 456
  antigens, number of, 455
  carrier molecule, 456
  database accession numbers, 455
  disease association, 457
  expression, 455
  function, 457
  gene, 455
  molecular basis of antigens, 457
  phenotypes, 457
  terminology, 455
Indirect antiglobulin test (IAT), 13
International Society of Blood Transfusion
    (ISBT), 5, 7, 9
  Committee on Terminology for Red Cell
    Surface Antigens 3–4, 14
  number, 4, 7
  symbol, 4, 7, 9
International System for Gene Nomenclature
    (ISGN), 4

J

JAHK antigen, 188–9
JAL antigen, 182–3
Jarvis antigen *see* CE antigen
JK blood group system *see* Kidd blood
    group system
JK3 antigen, 296–7
Jk^a antigen, 293–4
Jk^b antigen, 294–6
JMH antigen, 469–70
JMH blood group system, 467–70
  amino acid sequence, 467
  antigens, number of, 467
  carrier molecule, 468
  database accession number, 467
  expression, 467
  function, 468
  gene, 467
  terminology, 467
Jn^a antigen, 325–6
Jo^a antigen, 362–3
Jr^a antigen, 507–8
Js^a antigen, 239–41
Js^b antigen, 241–2

K

K antigen, 231–3
k antigen, 233–4
K11 antigen, 244–5
K12 antigen, 245–6
K13 antigen, 246–7
K14 antigen, 248–9
K16 antigen, 249
K17 antigen, 249–50
K18 antigen, 251–2
K19 antigen, 252–3
K22 antigen, 256–7
K23 antigen, 258–9
K24 antigen, 259–60
K27 antigen *see* RAZ antigen
Karhula *see* Ul^a antigen
Kell blood group system, 225–64
  amino acid sequence, 226
  antigens, number of, 225
  carrier molecule, 226
  comparison of features of McLeod phenotype
    with normal and K$_0$ RBCs, 229
  database accession numbers, 225
  disease association, 227
  expression, 225
  function, 227
  gene, 225
  molecular basis:
    of antigens, 227
    of phenotypes, 230
  phenotypes, 228
  terminology, 225
Kell phenotypes, comparison of, 228
Kidd blood group system, 290–7
  amino acid sequence, 291
  antigens, number of, 290
  carrier molecule, 291
  database accession numbers, 290
  disease association, 291
  expression, 290
  function, 291
  gene, 290
  molecular basis:
    of antigens, 291
    of Jk(a − b−) phenotypes, 292
  phenotypes, 292
  terminology, 290
Km antigen, 253–5
KMW antigen *see* Sl3 antigen
Kn^a antigen, 443–4
Kn^b antigen, 445–6
Knops blood group system, 439-54
  amino acid sequence of CR1˙1, 440
  antigens, number of, 439

carrier molecule, 441
database accession numbers, 439
disease association, 441
expression, 439
function, 441
gene, 439
molecular basis:
  of antigens, 441
  of phenotypes, 442
phenotypes, 442
terminology, 439
Kp^a antigen, 234–6
Kp^b antigen, 236–7
Kp^c antigen, 255–6
KREP antigen, 326–7
Ku antigen, 238–9
Kx antigen, 401
Kx blood group system:
  amino acid sequence, 398
  antigens, number of, 398
  carrier molecular, 399
  database accession numbers, 398
  disease association, 399
  expression, 398
  function, 399
  gene, 398
  molecular bases of McLeod
    phenotype, 400–1
  phenotypes, 400
  terminology, 398

L

Lan antigen, 505–6
Landsteiner–Wiener blood group
  system, 372–80
  amino acid sequence, 373
  antigens, number of, 372
  carrier molecule, 373
  database accession numbers, 373
  disease association, 374
  expression, 372
  function, 374
  gene, 372
  molecular basis
    of antigens, 374
  of phenotypes, 375
  phenotypes, 374
  terminology, 372
Le^a antigen, 269–70
Le^{ab} antigen, 272–3
Le^b antigen, 270–1
Le^{bH} antigen, 273–4

Le^c antigen, 499
Lectins:
    inhibitory sugar and RBC antigen, 546–7
    and polyagglutination information, 545
    and their specificities, 545
Le^d antigen, 500
Levay antigen *see* Kp^c antigen
Lewis blood group system, 265–78
    amino acid sequence of $\alpha(1,3/1,4)$
        fucosyltransferase, 266
    antigens, number of, 265
    carrier molecule, 266
    database accession numbers, 265
    disease association, 266
    expression, 265
    function, 266
    gene, 265
    phenotypes, 267
    terminology, 265
    Le(a−b−) RBC phenotypes:
        due to FUT3 mutations, 267
        due to other mutations, 267
LKE antigen, 497–8
LOCR antigen, 191–2
Low incidence antigens, 115
    700 series, 14, 501–2
Ls^a antigen, 413–15
Lu3 antigen, 200–1
Lu4 antigen, 202–3
Lu5 antigen, 203–4
Lu6 antigen, 204–6
Lu7 antigen, 206–7
Lu8 antigen, 207–9
Lu9 antigen, 209–10
Lu11 antigen, 210–11
Lu12 antigen, 212–13
Lu13 antigen, 213–14
Lu14 antigen, 214–16
Lu16 antigen, 216
Lu17 antigen, 217–18
Lu20 antigen, 221–2
Lu21 antigen, 223–4
Lu^a antigen, 197–9
Lu^b antigen, 199–200
Lutheran blood group system, 193–224
    amino acid sequence, 194
    antigens, number of, 193
    carrier molecule, 194
    comparison of three types of Lu(a-b-)
        phenotypes, 196
    database accession numbers, 194
    disease association, 196
    expression, 193
    function, 196
        gene, 193
        molecular basis
            of antigens, 195
            of recessive Lu(a-b-) phenotype, 196
        phenotypes, 196
        terminology, 193
    LW^a antigen, 375–7
    LW^{ab} antigen, 377–8
    LW^b antigen, 378–80

M

M antigen, 34–6
M^c antigen, 48–9
McC^a antigen, 446–7
McC^b antigen, 451–2
MAM antigen, 516–18
MAR antigen, 185–6
MARS antigen, 102–4
Matthews antigen *see* Js^b antigen
M^e antigen, 57
Membrane proteins, 10
MER2 antigen, 465–6
2-Mercaptoethanol (2-ME), 13
M^g antigen, 53–5
Mi.I *see* Vw antigen
Mi.II *see* Hut antigen
Mi^a antigen, 46–8
MINY antigen, 89–91
Mit antigen, 73–4
MNS blood group system, 29–104
    amino acid sequence, 30
    antigens, number of, 29
    carrier molecule, 30
    disease association, 32
    expression, 29
    function, 32
    gene, 29
    GeneBank accession numbers, 29–30
    hybrid glycophorin molecules,
        phenotypes and associated
        antigens, 32–3
    molecular basis:
        of antigens involving single nucleotide
            mutations, 31
        of other phenotypes, 33
    phenotypes, 32
    terminology, 29
Mo^a antigen, 316–18
Mol antigen *see* FPTT antigen
Mt^a antigen, 58–9
Mull antigen *see* Lu9 antigen
Mur antigen, 51–3

MUT antigen, 51
M^v antigen, 69–71

N

N antigen, 36–8
'N' antigen, 83–4
NFLD antigen, 323–5
Nob antigen, 78–80
Nou antigen, 177
Nunhart antigen *see* Jn^a antigen
Ny^a antigen, 64–5

O

OK blood group system, 461–5
  amino acid sequence, 462
  antigens, number of, 461
  carrier molecule, 462
  database accession numbers, 461
  disease association, 463
  expression, 461
  function, 463
  gene, 461
  molecular basis of antigens, 462
  phenotypes, 463
  terminology, 461
Ok^a antigen, 463–4
Or antigen, 84–5
Os^a antigen, 96–7

P

P antigen, 477–8
P blood group system, 105–92
  antigens, number of, 105
  carrier molecule, 105
  disease association, 106
  expression, 105
  function, 106
  gene, 105
  phenotypes, 106
  terminology, 105
P1 antigen, 106–8
PEL antigen, 514–15
Pelletier antigen *see* PEL antigen
Peltz antigen *see* Ku antigen
Phenotypes, 4
  molecular basis of, 11
P^k antigen, 495
Plasmodium falciparum, 300
Polyagglutination, 546
  types and reactions with group O RBCs
    and lectins, 545

Proteins, altered on Rh$_{null}$ RBCs, 119
Pseudo-discrepancies between genotype and
  phenotype, causes of, 548

R

Raph antigen *see* MER2 antigen
RAPH blood group system, 465–6
  antigens, 4
    number of, 465
  carrier molecule, 465
  disease association, 465
  expression, 465
  gene, 465
  terminology, 465
RAZ antigen, 263–4
Rb^a antigen, 310–11
RBC membrane:
  components, model of, 4
  proteins, 543
Rd antigen, 351
Rg1 antigen, 389
Rg2 antigen, 390
Rh blood group system, 109–92
  amino acid sequence, 110
  antigens, number of, 109
  carrier molecule, 111
  comparison of Rh$_{null}$ and Rh$_{mod}$ RBCs, 120
  database accession numbers, 110
  disease association, 113
  expression, 109
  function, 113
  gene, 109
  gene map, 110
  molecular basis:
    of phenotypes, 116
    of Rh$_{null}$ and Rh$_{mod}$ phenotypes, 119
  phenotypes, 113
  proteins altered on Rh$_{null}$ RBCs, 119
  rearranged *RHD* and *RHCE*, 118
  RHE variants, 118
  RHAG, 110
  RhAG, 111–12
  RhCE and RhD, 110
  RhD and RhCE, 111
  terminology, 109
Rh complexes, serological
  reactions of, 114
rh' antigen *see* C antigen
rh" antigen *see* E antigen
Rh$_0$ antigen *see* D antigen
Rh7 *see* Ce antigen
Rh11 *see* E^w antigen
Rh26 (c-like) antigen, 157–8

Rh27 *see* cE antigen
Rh28 *see* hr^H antigen
Rh29 antigen, 160–1
Rh30 *see* Go^a antigen
Rh31 *see* hr^B antigen
Rh32 antigen, 164–5
Rh33 antigen, 166–7
Rh34 *see* Hr^B antigen
Rh35 antigen, 168–9
Rh36 *see* Be^a antigen
Rh37 *see* Evans antigen
Rh39 antigen, 172–3
Rh40 *see* Tar antigen
Rh41 antigen, 175
Rh42 antigen, 175–6
Rh43 *see* Crawford antigen
Rh44 *see* Nou antigen
Rh45 *see* Riv antigen
Rh46 *see* Sec antigen
Rh47 *see* Dav antigen
Rh48 *see* JAL antigen
Rh49 *see* STEM antigen
Rh50 *see* FPTT antigen
Rh51 *see* MAR antigen
Rh52 *see* BARC antigen
Rh53 *see* JAHK antigen
Rh54 *see* DAK antigen
Rh55 *see* LOCR antigen
RhCE antigens, 112
RHCE haplotypes and associated
    information, 117
RhD antigens, 112
Rhesus Similarity Index, 128
rh$_i$ antigen *see* Ce antigen
Ri^a antigen, 62–3
Riv antigen, 178

S

s antigen, 40–1
S antigen, 38–9
S. Allen antigen *see* JAL antigen
S1 phenotypes, relationship, 454
SAT antigen, 92–4
Sc1 antigen, 347–8
Sc2 antigen, 348–9
Sc3 antigen, 349–50
Sc4 antigen, 351–2
Scianna blood group system, 344–52
    amino acid sequence, 344
    antigens, number of, 344
    carrier molecule, 345
    database accession number, 344
    disease association, 346

expression, 344
function, 346
gene, 344
molecular basis:
    of antigens, 345
    of Sc$_{null}$ phenotypes, 346
phenotypes, 346
terminology, 344
s^D antigen, 72–3
Sd^a antigen, 511–12
Sec antigen, 179–80
SGRO antigen *see* K13 antigen
Shabalala *see* hr^s antigen
    see also Hr antigen
Sl3 antigen, 453
Sl^a antigen, 448–9
SSEA-4 *see* LKE antigen
St^a antigen, 59–61
STEM antigen, 182–3
SW1 antigen, 330–1
Sw^a antigen, 321–2
Swain–Langley antigen *see* Sl^a antigen

T

Tamm-Horsfall glycoprotein antigen
    see Sd^a antigen
Tar antigen, 173–4
Tc^a antigen, 424–5
Tc^b antigen, 425–6
Tc^c antigen, 427–8
Thornton antigen *see* Rh42 antigen
Total Rh antigen *see* Rh29
TOU antigen, 261–2
Tr^a antigen, 328–9
Transfusion reaction, 13
Trihexosylceramide *see* P^k antigen
TSEN antigen, 87–9

U

U antigen, 41–2
Ul^a antigen, 242–4
UMC antigen, 436–7
Unnamed blood group collection, 14, 499–500
    antigens, number of, 499
    carrier molecule, 499
    terminology, 499
Uromodulin antigen *see* Sd^a antigen

V

V antigen, 144–6
Vel antigen, 503–4

Vg^a antigen, 319–20
Vil antigen, 452–3
VLAN antigen, 260–1
V. M. (Marshall) antigen *see* K18 antigen
Vr antigen, 55–7
VS antigen, 152–3
Vw antigen, 49–51

W

WARR antigen, 311–12
Wb antigen, 412–13
Wd^a antigen, 308–9
WES^a antigen, 433–4
WES^b antigen, 434–5
WH antigen, 388
Wiel antigen *see* D^w antigen
Willis antigen *see* C^w antigen
Wj antigen *see* AnWj antigen
Wk^a antigen *see* K17 antigen
Wr^a antigen, 305–6
Wr^b antigen, 306–8
Wu antigen, 314–15

X

Xg blood group system, 338–43
  antigens, number of, 338
  carrier molecule, 340
  CD99 amino acid sequence, 338
  database accession numbers, 338
  disease association, 340

expression, 338
function, 340
gene, 338
phenotype, 340
phenotypic relationship of Xg^a and CD99
  antigens, 341
terminology, 338
Xg^a amino acid sequence, 339
Xg^a antigen, 341–2

Y

Yk^a antigen, 449–50
Yt blood group system, 332–7
  amino acid sequence, 332
  antigens, number of, 332
  carrier molecule, 333
  database accession numbers, 332
  disease association, 334
  expression, 332
  function, 334
  gene, 332
  molecular basis of antigens, 333
  phenotype, 334
  terminology, 332
Yt^a antigen, 334–6
Yt^b antigen, 336–7

Z

ZZAP, 13

CPSIA information can be obtained at www.ICGtesting.com
Printed in the USA
BVOW042351160212

283134BV00002B/2/P

9 780125 865852